Unit Price Estimating Methods

Third Edition

Editors

Phillip R. Waier, PE, CSI

John H. Chiang, PE

Unit Price Estimating Methods

Third Edition

RSMeans

Copyright 2002
Construction Publishers & Consultants
63 Smiths Lane
Kingston, MA 02364-0800
(781) 422-5000

The editors for this book were John Chiang, Phillip Waier, Mary Green, and Andrea Keenan. The production manager was Michael Kokernak. The production coordinator was Marion Schofield. The electronic publishing specialists was Jonathan Forgit. The proofreader was Robin Richardson. The book and cover were designed by Norman R. Forgit.

Printed in the United States of America

10 9 8 7 6 5 4 3 2

Library of Congress Cataloging in Publication Data

ISBN 0-87629-662-2

 Reed Construction Data

Table of Contents

Foreword

In addition to collecting and publishing construction cost data for over 60 years, Means has also developed a popular series of estimating and other construction reference books. *Means Unit Price Estimating* was one of the first, a response to customers who sought information on how to create unit price estimates, division by division, and how to use published costs most effectively. The book has been used by novice and practiced estimators to build and enhance their skills.

This book explains and demonstrates the methods and procedures of Unit Price Estimating, from the plans and specifications to the estimate summary. All aspects of the estimate, including the site visit, the quantity takeoff, pricing and bidding, are covered in detail. A complete sample estimate is presented to demonstrate the application of these estimating practices.

All construction costs used as examples in this book are from the 2002 edition of *Means Building Construction Cost Data*. Many pages, tables and charts from the annual cost guide are reproduced in order to show the origin and development of the cost data used in the estimating examples. *Means Building Construction Cost Data* is the leading source of thorough and up-to-date costs for the construction industry.

This book follows the logical progression of the estimating process. Initial chapters include analysis of the plans and specifications, the site visit, and evaluation of collected site data. The quantity takeoff, determination of costs, and final pricing are thoroughly covered, reflecting the importance and required attention to detail for each procedure. Also included are some strategies and principles that may be applied upon completion of the estimate. In addition, the format and origin of costs in *Means Building Construction Cost Data* are explained.

The final section of the book is devoted to a detailed discussion of estimating for each CSI MasterFormat division. A sample building project estimate is carried out, by division, to the estimate summary. This process demonstrates the proven estimating techniques described in the book. These are the techniques used by a majority of construction estimators and represent sound estimating practice.

Features of this new third edition include the fully updated sample estimates, as well as a new focus on computer estimating, which provides an overview of the available computer estimating products and their capabilities, from the most basic to quite sophisticated programs. The use of published cost data and its integration into these systems is also covered.

Further enhancements to the original book include an expanded Appendix, with reference tables on a selection of commonly used construction materials.

It is the editors' intention that the new features of the third edition, combined with the proven methods established in the first and second, will provide readers with a comprehensive course in unit price estimating.

Introduction

Estimates are an important part of everyday life. We estimate the amount of food required to feed our family for a week, or travel time to a desired destination. The preparation of a construction estimate requires the same kind of analysis. In general, we have to determine the amount of materials needed to complete a project, and the time required to install those materials, before assigning a cost to both.

Each estimate contains three interdependent variables:

- Quantity
- Quality
- Cost

Construction documentation in the form of plans and specifications dictates the quality and quantities of materials required. Cost is then determined based on these two elements. If a specific cost/budget must be maintained, then either the quantity or quality of the components is adjusted to meet the cost requirement.

Unit Price Estimating

What is the correct or accurate cost of a given construction project? Is it the total price the owner pays the contractor? Might not another reputable contractor perform the same work for a different cost, whether higher or lower? In fact, there isn't one *correct* estimated cost for a given project. There are too many variables in construction. At best, the estimator can determine a close approximation of what the final costs will be. The resulting accuracy of the approximation is affected by the amount of detail provided, the amount of time spent on the estimate, and the skill of the estimator. The estimator's skill is based on his or her working knowledge of how buildings are constructed. The estimator's creativity may lead to significant differences between estimates prepared by equally qualified individuals. Using a new method to accomplish a particular work activity (in the form of a new piece of equipment or new methodology) may significantly impact final costs.

The accuracy is determined by the cost accountant at the end of the project. If the project was completed in the allotted time and the profit met the expectation provided in the estimate, then it was accurate.

It is apparent that building construction estimating is not as simple as blindly applying material and labor prices and arriving at a magic figure. The purpose of this text is to make the estimating process easier and more organized for the experienced estimator, and to provide less experienced estimators with a basis for sound estimating practice.

Who Uses Unit Price Estimates?

A unit price estimate is the most detailed and accurate type of construction estimate. It is used by architects, engineers, contractors, and many owners/facilities managers in their daily activities.

Architects/Engineers

Design professionals perform unit price estimates for the following reasons.

- To evaluate change orders. When a change order is required, the A/E requests a proposal from the contractor. The proposal is usually supported by a detailed estimate. The A/E may perform a unit price estimate to validate or dispute the costs before approving the change.
- To verify a schedule of payments. Periodically, it may be necessary to validate a contractor's schedule of payments (the cost related to various elements of work) for invoicing purposes.

Contractors

Most of the estimates a contractor performs are unit price. Except in the early design stages, owners usually require "hard number" estimates (the price for which the contractor will execute a contract to do the work). Therefore, the contractor must perform the most accurate type of estimate—a unit price estimate. Any binding estimate is usually a unit price estimate.

Facilities Manager

When facilities managers/owners want an accurate estimate on which to base an authorization for expenditure or a loan application, a unit price estimate may be prepared. Owners (or their representatives) also evaluate contractors' proposals for work or change orders to ongoing projects based on unit price estimates.

When accuracy is required, a unit price estimate is the answer.

Chapter 1

Estimate Types

Chapter 1

Estimate Types

Construction estimators use four basic types of estimates. These types may be referred to by different names and may not be recognized by all as definitive, but most estimators will agree that each type has its place in the construction estimating process. The type of estimate performed is related to the amount of design information available. As the project proceeds through the various stages of design (from schematic design to design development to contract documents), the type of estimate changes and the accuracy of the estimate increases. Figure 1. 1 graphically demonstrates the relationship of required time versus resulting accuracy for these four basic estimate types.

1. **Order of Magnitude Estimate:** The order of magnitude estimate could be loosely described as an educated guess. It is also known as a "napkin estimate," because it is often the result of a conversation between a contractor (or developer) and a client/owner over lunch or dinner, in which an estimate is created on the nearest piece of paper, usually a napkin. Order of Magnitude Estimates can be completed in a matter of minutes. Accuracy is plus or minus 20%.

2. **Square Foot and Cubic Foot Estimates:** This type of estimate is most often useful when only the proposed size and use of a planned building is known. Very little information is required. Performing a breakout for this type of estimate enables the designer and estimator to adjust components for the proposed use of the structure (hospital, factory, school, apartments), type of foundation (slab on grade, spread footing, piles), and superstructure (steel, concrete, or a combination) and to focus the cost more closely to the final price. Accuracy of the square foot estimate is plus or minus 15%.

3. **Assemblies (or Systems) Estimate:** A systems estimate is best used as a budgetary tool in the planning stages of a project. Accuracy is expected at plus or minus 10%.

4. **Unit Price Estimate:** Working drawings and full specifications are required to complete a unit price estimate. It is the most accurate of the four types but is also the most time-consuming. Used primarily for bidding purposes, accuracy is plus or minus 5%.

Order of Magnitude Estimates

The Order of Magnitude Estimate can be completed with only a minimum of information. The proposed use and size of the planned structure should be known and may be the only requirement. The "units" to describe the structure can be very general, and need not be well-defined. For example: "An office building for a small service company in a suburban industrial park will cost about $650,000." This type of statement (or estimate) can be made after a few minutes of thought used to draw upon experience and to make comparisons with similar projects from the past. While this rough figure might be appropriate for a project in one region of the country, an adjustment may be required for a change of location and for cost changes over time (price changes, inflation, etc.).

Figure 1.2, from *Means Building Construction Cost Data*, shows examples of a different approach to the Order of Magnitude Estimate. This format is based on unit of use. Please note at the bottom of the categories "Hospitals" and "Housing" that costs are given "per bed or person," "per rental unit," and "per apartment." This data does not require that details of the proposed project be known in order to determine rough costs; the only required information is the intended use of the building and its approximate size. What is lacking in accuracy (plus or minus 20%) is more than compensated by the minimal time required to complete the Order of Magnitude Estimate—a matter of minutes.

Square Foot and Cubic Foot Estimates

The use of Square Foot and Cubic Foot Estimates is most appropriate prior to the preparation of plans or preliminary drawings, when budgetary parameters are being analyzed and established. Please refer again to Figure 1.2 and note that costs for each type of project are presented first as "Total project costs" by square foot and by cubic foot. These costs are then broken down into different construction components, and then into the relationship of each component to the project as a whole, in terms of costs per square foot. This breakdown enables the designer, planner, or estimator to adjust certain components according to the unique requirements of the proposed project.

Historical data for square foot costs of new construction are plentiful (see *Means Building Construction Cost Data*, Division 17). However, the best source of

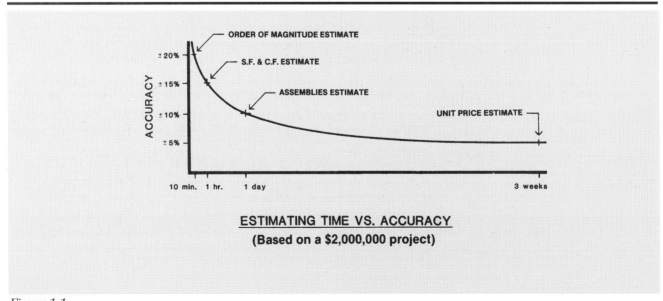

ESTIMATING TIME VS. ACCURACY
(Based on a $2,000,000 project)

Figure 1.1

square foot costs is the estimator's own cost records for similar projects, adjusted to the parameters of the project in question. While helpful for preparing preliminary budgets, Square Foot and Cubic Foot Estimates can also be useful as checks against other, more detailed estimates. While slightly more time is required than with Order of Magnitude Estimates, a greater accuracy (plus or minus 15%) is achieved due to more specific definition of the project. A Square Foot Estimate is consistent with the amount of design information available at the schematic design phase.

Assemblies (Systems) Estimates

Rising design and construction costs in recent years have made budgeting and cost efficiency increasingly important in the early stages of building projects. Never before has the estimating process had such a crucial role in the initial planning. Unit Price Estimating, because of the time and detailed information required, is not suited as a budgetary or planning tool. A faster and more cost-effective method is needed for the planning phase of a building project; this is the "Assemblies," or "Systems Estimate." An Assemblies Estimate is usually prepared when the architect completes the design development plans.

The Systems method is a logical, sequential approach which reflects how a building is constructed. Seven "UNIFORMAT II" major groups organize building construction into major components that can be used in Assemblies Estimates. These UNIFORMAT II divisions are listed below:

Assemblies Major Groups:

A - Substructure

B - Shell

C - Interiors

D - Services

E - Equipment & Furnishings

F - Special Construction & Demolition

G - Building Sitework

Each major group is further broken down into systems. Each of these systems incorporates several different items into an assemblage that is commonly used in building construction. Figure 1.3 is an example of a typical system, in this case "Drywall Partitions/Wood Stud Framing" (see *Means Assemblies Cost Data*).

In the Assemblies format, a construction component may appear in more than one division. For example, concrete is found in A – Substructure, as well as in B and G (see list above). Conversely, each division may incorporate many different areas of construction, and the labor of different trades.

A great advantage of the Assemblies Estimate is that the estimator/designer is able to substitute one system for another during design development and can quickly determine the cost differential. The owner can then anticipate accurate budgetary requirements before final details and dimensions are established.

Final design details of the building project are required for a Unit Price Estimate. The Assemblies method does not require such details, but the estimators who use it must have a solid background knowledge of construction materials and methods, building code requirements, design options, and budgetary restrictions.

17100	S.F. & C.F. Costs			UNIT	UNIT COSTS			% OF TOTAL		
					1/4	MEDIAN	3/4	1/4	MEDIAN	3/4
400	0500	Masonry	R17100-100	S.F.	5.65	11.10	17.20	6%	10%	15.50%	**400**
	2720	Plumbing			2.71	5.20	9.80	3.60%	6.70%	8%
	2730	Heating & ventilating			4.64	6.70	12.95	6.20%	7.40%	13.50%
	2900	Electrical			4.46	6.60	10.10	6.30%	8%	11.70%
	3100	Total: Mechanical & Electrical		↓	11.30	20.50	33.50	15.20%	25.50%	32.70%
410	0010	**GARAGES, PARKING**	R17100-100	S.F.	23	33	59				**410**
	0020	Total project costs		C.F.	2.21	3	4.37			
	2720	Plumbing		S.F.	.61	1.03	1.59	2.60%	3.40%	3.90%
	2900	Electrical			.96	1.46	2.14	4.30%	5.20%	6.30%
	3100	Total: Mechanical & Electrical		↓	1.28	3.77	4.83	6.50%	9.40%	12.80%
	3200									
	9000	Per car, total cost		Car	9,900	12,500	16,100			
430	0010	**GYMNASIUMS**	R17100-100	S.F.	65.50	83.50	106				**430**
	0020	Total project costs		C.F.	3.26	4.17	5.40			
	1800	Equipment		S.F.	1.38	2.81	5.60	2.10%	3.40%	6.70%
	2720	Plumbing			4.14	5.10	6.10	5.40%	7.30%	7.90%
	2770	Heating, ventilating, air conditioning			4.45	6.80	13.60	9%	11.10%	22.60%
	2900	Electrical			4.95	6.15	8.15	6.60%	8.30%	10.70%
	3100	Total: Mechanical & Electrical		↓	15.35	21	28.50	20.60%	26.20%	29.40%
	3500	See also division 11480								
460	0010	**HOSPITALS**	R17100-100	S.F.	135	157	233				**460**
	0020	Total project costs		C.F.	9.50	11.75	16.95			
	1800	Equipment		S.F.	3.19	6.15	10.55	2.50%	3.80%	6%
	2720	Plumbing			11.25	15.10	19.45	7.80%	9.40%	11.80%
	2770	Heating, ventilating, air conditioning			15.85	21	28.50	8.40%	14.60%	17%
	2900	Electrical			13.50	17.55	27.50	9.90%	12%	14.50%
	3100	Total: Mechanical & Electrical		↓	38.50	51	83.50	26.60%	33.10%	39%
	9000	Per bed or person, total cost		Bed	48,400	103,000	162,000			
	9900	See also division 11700 & 11780								
480	0010	**HOUSING** For the Elderly	R17100-100	S.F.	62	78	95.50				**480**
	0020	Total project costs		C.F.	4.38	6.10	7.80			
	0100	Site work		S.F.	4.30	6.80	9.80	5.10%	8.20%	12.10%
	0500	Masonry			1.88	7.05	10.25	2.10%	7.10%	12.20%
	1800	Equipment			1.48	2.02	3.27	1.90%	3.20%	4.40%
	2510	Conveying systems			1.50	2.02	2.74	1.80%	2.30%	2.90%
	2720	Plumbing			4.58	5.90	7.85	8.10%	9.70%	10.90%
	2730	Heating, ventilating, air conditioning			2.35	3.33	4.98	3.30%	5.60%	7.20%
	2900	Electrical			4.60	6.20	7.95	7.30%	8.50%	10.50%
	3100	Total: Mechanical & Electrical		↓	15.85	18.60	24.50	18.10%	22%	29.10%
	9000	Per rental unit, total cost		Unit	57,500	67,000	74,500			
	9500	Total: Mechanical & Electrical		"	12,300	14,700	17,200			
500	0010	**HOUSING** Public (Low Rise)	R17100-100	S.F.	52	72	94				**500**
	0020	Total project costs		C.F.	4.12	5.75	7.15			
	0100	Site work		S.F.	6.60	9.50	15.40	8.40%	11.70%	16.50%
	1800	Equipment			1.41	2.30	3.67	2.30%	3%	5.10%
	2720	Plumbing			3.53	4.95	6.25	6.80%	9%	11.60%
	2730	Heating, ventilating, air conditioning			1.88	3.65	4	4.20%	6%	6.40%
	2900	Electrical			3.14	4.68	6.50	5.10%	6.60%	8.30%
	3100	Total: Mechanical & Electrical		↓	14.90	19.20	21.50	14.50%	17.60%	26.50%
	9000	Per apartment, total cost		Apt.	57,000	65,000	81,500			
	9500	Total: Mechanical & Electrical		"	12,200	15,000	16,600			
510	0010	**ICE SKATING RINKS**	R17100-100	S.F.	46	80	114				**510**
	0020	Total project costs		C.F.	3.26	3.34	3.85			
	2720	Plumbing		S.F.	1.66	3.11	3.18	3.10%	5.60%	6.70%
	2900	Electrical		↓	4.75	7.30	7.70	6.70%	15%	15.80%

(Reprinted from Means Building Construction Cost Data 2002.)

Figure 1.2

The Assemblies Estimate should not be used as a substitute for the Unit Price Estimate. While the Systems approach can be an invaluable tool in the planning stages of a project, it should be supported by Unit Price Estimating when greater accuracy is required.

Unit Price Estimates

The Unit Price Estimate is the most accurate and detailed of the four estimate types and therefore takes the most time to complete. Detailed working drawings and specifications (contract documents) must be available to the unit price estimator. All decisions regarding the building's materials and methods must be made in order to complete this type of estimate. There are fewer variables, and the estimate can therefore be more accurate. The working drawings and specifications are needed to determine the quantities of materials, equipment, and labor. Current and accurate costs for these items (unit prices) are also necessary. These costs can come from different sources. Wherever possible the estimator should use prices based on experience or cost figures from similar projects. If no records are available, prices may be determined instead from an up-to-date industry source book such as *Means Building Construction Cost Data*.

Because of the detail involved and the need for accuracy, Unit Price Estimates require a great deal of time and expense to complete properly. For this reason, Unit Price Estimating is best suited for construction bidding. It can also be effective for determining certain detailed costs in conceptual budgets or during design development.

Most construction specification manuals and cost reference books such as *Means Building Construction Cost Data* divide all unit price information into the sixteen MasterFormat divisions as adopted by the Construction Specifications Institute:

MasterFormat Construction Index Divisions:

Division 1 - General Requirements

Division 2 - Site Construction

Division 3 - Concrete

Division 4 - Masonry

Division 5 - Metals

Division 6 - Wood and Plastics

Division 7 - Thermal and Moisture Protection

Division 8 - Doors and Windows

Division 9 - Finishes

Division 10 - Specialties

Division 11 - Equipment

Division 12 - Furnishings

Division 13 - Special Construction

Division 14 - Conveying Systems

Division 15 - Mechanical

Division 16 - Electrical

C1010 Partitions

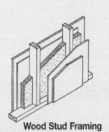

Wood Stud Framing

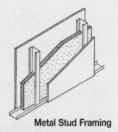

Metal Stud Framing

The Drywall Partitions/Stud Framing Systems are defined by type of drywall and number of layers, type and spacing of stud framing, and treatment on the opposite face. Components include taping and finishing.

Cost differences between regular and fire resistant drywall are negligible, and terminology is interchangeable. In some cases fiberglass insulation is included for additional sound deadening.

System Components	QUANTITY	UNIT	COST PER S.F.		
			MAT.	INST.	TOTAL
SYSTEM C1010 124 1250					
DRYWALL PARTITION,5/8″ F.R.1 SIDE,5/8″ REG.1 SIDE,2″X4″STUDS,16″ O.C.					
Gypsum plasterboard, nailed/screwed to studs, 5/8″F.R. fire resistant	1.000	S.F.	.24	.37	.61
Gypsum plasterboard, nailed/screwed to studs, 5/8″ regular	1.000	S.F.	.24	.37	.61
Taping and finishing joints	2.000	S.F.	.08	.74	.82
Framing, 2 x 4 studs @ 16″ O.C., 10′ high	1.000	S.F.	.41	.75	1.16
TOTAL			.97	2.23	3.20

C1010 124 — Drywall Partitions/Wood Stud Framing

	FACE LAYER	BASE LAYER	FRAMING	OPPOSITE FACE	INSULATION	COST PER S.F.		
						MAT.	INST.	TOTAL
1200	5/8″ FR drywall	none	2 x 4, @ 16″ O.C.	same	0	.97	2.23	3.20
1250				5/8″ reg. drywall	0	.97	2.23	3.20
1300				nothing	0	.69	1.49	2.18
1400		1/4″ SD gypsum	2 x 4 @ 16″ O.C.	same	1-1/2″ fiberglass	1.90	3.44	5.34
1450				5/8″ FR drywall	1-1/2″ fiberglass	1.66	3.02	4.68
1500				nothing	1-1/2″ fiberglass	1.38	2.28	3.66
1600		resil. channels	2 x 4 @ 16″, O.C.	same	1-1/2″ fiberglass	1.67	4.36	6.03
1650				5/8″ FR drywall	1-1/2″ fiberglass	1.54	3.48	5.02
1700				nothing	1-1/2″ fiberglass	1.26	2.74	4
1800		5/8″ FR drywall	2 x 4 @ 24″ O.C.	same	0	1.35	2.82	4.17
1850				5/8″ FR drywall	0	1.11	2.45	3.56
1900				nothing	0	.83	1.71	2.54
1950		5/8″ FR drywall	2 x 4, 16″ O.C.	same	0	1.45	2.97	4.42
1955				5/8″ FR drywall	0	1.21	2.60	3.81
2000				nothing	0	.93	1.86	2.79
2010		5/8″ FR drywall	staggered, 6″ plate	same	0	1.87	3.72	5.59
2015				5/8″ FR drywall	0	1.63	3.35	4.98
2020				nothing	0	1.35	2.61	3.96
2200		5/8″ FR drywall	2 rows-2 x 4	same	2″ fiberglass	2.36	4.08	6.44
2250			16″O.C.	5/8″ FR drywall	2″ fiberglass	2.12	3.71	5.83
2300				nothing	2″ fiberglass	1.84	2.97	4.81
2400	5/8″ WR drywall	none	2 x 4, @ 16″ O.C.	same	0	1.09	2.23	3.32
2450				5/8″ FR drywall	0	1.03	2.23	3.26
2500				nothing	0	.75	1.49	2.24
2600		5/8″ FR drywall	2 x 4, @ 24″ O.C.	same	0	1.47	2.82	4.29
2650				5/8″ FR drywall	0	1.17	2.45	3.62
2700				nothing	0	.89	1.71	2.60

Figure 1.3

This method of organizing the various components of construction provides a standard of uniformity that is widely used by construction industry professionals—architects, engineers, material suppliers, and contractors. An example page from *Means Building Construction Cost Data* is shown in Figure 1.4. This page lists various types of formwork. (Please note that the heading "03100 Concrete Forms & Accessories" denotes the MasterFormat subdivision classification for these items in Division – 3 Concrete.) Each page contains a wealth of information useful in Unit Price Estimating. The type of work to be performed is described in detail: typical crew makeup, daily output, and separate costs for material and installation. Total costs are extended to include the installing contractor's overhead and profit.

Figure 1.5 is a Means' "Condensed Estimate Summary" form. This form can be used after each division has been separately estimated. It serves as a checklist to ensure that estimators have included all divisions, and is a concise means for determining the total costs of the Unit Price Estimate. Please note at the top of the form these items: Total Area, Total Volume, Cost per Square Foot, and Cost per Cubic Foot. This information can prove valuable when the estimator needs Order of Magnitude, or Square Foot and Cubic Foot Estimates for budgeting. This data will also serve as a cross-check for similar projects in the future.

A fifth type of estimate, more accurate than unit price, warrants mention: the Scheduling Estimate. This type involves the application of realistic manpower allocation. A complete Unit Price Estimate is a prerequisite for the preparation of a Scheduling Estimate. A thorough discussion of Scheduling Estimating is beyond the scope of this book, but a brief discussion is included in Chapter Five.

03100 | Concrete Forms & Accessories

03110 | Structural C.I.P. Forms

		CREW	DAILY OUTPUT	LABOR-HOURS	UNIT	MAT.	LABOR	EQUIP.	TOTAL	TOTAL INCL O&P	
420	8850 Maximum R03110 -020				C	157			157	173	420
425	0010 **FORMS IN PLACE, EQUIPMENT FOUNDATIONS** 1 use R03110 -050	C-2	160	.300	SFCA	2.62	8.75		11.37	16.55	425
	0050 2 use		190	.253		1.44	7.40		8.84	13.10	
	0100 3 use R03110 -060		200	.240		1.05	7		8.05	12.05	
	0150 4 use		205	.234		.86	6.85		7.71	11.60	
430	0010 **FORMS IN PLACE, FOOTINGS** Continuous wall, plywood, 1 use R03110 -050	C-1	375	.085	SFCA	2.26	2.42		4.68	6.25	430
	0050 2 use		440	.073		1.25	2.06		3.31	4.58	
	0100 3 use R03110 -060		470	.068		.91	1.93		2.84	4.01	
	0150 4 use		485	.066		.74	1.87		2.61	3.72	
	0500 Dowel supports for footings or beams, 1 use		500	.064	L.F.	.80	1.82		2.62	3.71	
	1000 Integral starter wall, to 4" high, 1 use		400	.080		.85	2.27		3.12	4.46	
	1500 Keyway, 4 use, tapered wood, 2" x 4"	1 Carp	530	.015		.19	.45		.64	.91	
	1550 2" x 6"		500	.016		.28	.48		.76	1.06	
	2000 Tapered plastic, 2" x 3"		530	.015		.52	.45		.97	1.27	
	2050 2" x 4"		500	.016		.67	.48		1.15	1.49	
	2250 For keyway hung from supports, add		150	.053		.80	1.60		2.40	3.37	
	3000 Pile cap, square or rectangular, plywood, 1 use	C-1	290	.110	SFCA	1.85	3.13		4.98	6.90	
	3050 2 use		346	.092		1.02	2.62		3.64	5.20	
	3100 3 use		371	.086		.74	2.45		3.19	4.62	
	3150 4 use		383	.084		.60	2.37		2.97	4.35	
	4000 Triangular or hexagonal caps, plywood, 1 use		225	.142		2.18	4.03		6.21	8.70	
	4050 2 use		280	.114		1.20	3.24		4.44	6.35	
	4100 3 use		305	.105		.87	2.98		3.85	5.60	
	4150 4 use		315	.102		.71	2.88		3.59	5.25	
	5000 Spread footings, plywood, 1 use		305	.105		1.71	2.98		4.69	6.50	
	5050 2 use		371	.086		.95	2.45		3.40	4.85	
	5100 3 use		401	.080		.69	2.26		2.95	4.27	
	5150 4 use		414	.077		.56	2.19		2.75	4.02	
	6000 Supports for dowels, plinths or templates, 2' x 2'		25	1.280	Ea.	3.46	36.50		39.96	60.50	
	6050 4' x 4' footing		22	1.455		6.90	41.50		48.40	71.50	
	6100 8' x 8' footing		20	1.600		13.80	45.50		59.30	85.50	
	6150 12' x 12' footing		17	1.882		24	53.50		77.50	110	
	7000 Plinths, 1 use		250	.128	SFCA	2.50	3.63		6.13	8.40	
	7100 4 use		270	.119	"	.82	3.36		4.18	6.15	
435	0010 **FORMS IN PLACE, GRADE BEAM** Plywood, 1 use R03110 -050	C-2	530	.091	SFCA	1.58	2.65		4.23	5.85	435
	0050 2 use		580	.083		.87	2.42		3.29	4.73	
	0100 3 use R03110 -060		600	.080		.63	2.34		2.97	4.34	
	0150 4 use		605	.079		.51	2.32		2.83	4.17	
440	0010 **FORMS IN PLACE, MAT FOUNDATION** 1 use	C-2	290	.166	SFCA	1.71	4.84		6.55	9.45	440
	0050 2 use		310	.155		.94	4.53		5.47	8.10	
	0100 3 use		330	.145		.67	4.25		4.92	7.35	
	0120 4 use		350	.137		.56	4.01		4.57	6.85	
445	0010 **FORMS IN PLACE, SLAB ON GRADE**										445
	1000 Bulkhead forms with keyway, wood, 1 use, 2 piece	C-1	510	.063	L.F.	.90	1.78		2.68	3.76	
	1050 3 piece		400	.080		1.13	2.27		3.40	4.77	
	1100 4 piece		350	.091		1.13	2.59		3.72	5.30	
	1400 Bulkhead forms w/keyway, 1 piece expanded metal, left in place										
	1410 In lieu of 2 piece form	C-1	1,375	.023	L.F.	1.20	.66		1.86	2.35	
	1420 In lieu of 3 piece form		1,200	.027		1.20	.76		1.96	2.50	
	1430 In lieu of 4 piece form		1,050	.030		1.20	.86		2.06	2.67	
	2000 Curb forms, wood, 6" to 12" high, on grade, 1 use		215	.149	SFCA	1.68	4.22		5.90	8.40	
	2050 2 use		250	.128		.93	3.63		4.56	6.65	
	2100 3 use		265	.121		.67	3.42		4.09	6.10	
	2150 4 use		275	.116		.54	3.30		3.84	5.75	

Figure 1.4

CONDENSED ESTIMATE SUMMARY

SHEET NO.

PROJECT

ESTIMATE NO.

LOCATION   TOTAL AREA/VOLUME   DATE

ARCHITECT   COST PER S.F./C.F.   NO. OF STORIES

PRICES BY: EXTENSIONS BY: CHECKED BY:

DIV.	DESCRIPTION	MATERIAL	LABOR	EQUIPMENT	SUBCONTRACT	TOTAL
1.0	General Requirements					
2.0	Site Construction					
3.0	Concrete					
4.0	Masonry					
5.0	Metals					
6.0	Carpentry					
7.0	Moisture & Thermal Protection					
8.0	Doors, Windows, Glass					
9.0	Finishes					
10.0	Specialties					
11.0	Equipment					
12.0	Furnishings					
13.0	Special Construction					
14.0	Conveying Systems					
15.0	Mechanical					
16.0	Electrical					
	Subtotals					
	Sales Tax %					
	Overhead %					
	Subtotal					
	Profit %					
	Contingency %					
	Adjustments					
	TOTAL BID					

Figure 1.5

Before Starting the Estimate

Chapter 2

Before Starting the Estimate

The "Invitation to Bid"—To the contractor, this can mean the prospect of weeks of hard work with only a chance of bidding success. It can also mean the opportunity to obtain a contract for a successful and lucrative building project. It is not uncommon for the contractor to bid ten or more jobs in order to win just one. The success rate most often depends upon estimating accuracy and thus, the preparation, organization, and care that go into the estimating process.

The first step before starting the estimate is to obtain copies of the plans and specifications in *sufficient quantities*. Most estimators mark up plans with colored pencils and make numerous notes and references. For one trade to estimate from plans that have been used by another is difficult at best and may easily lead to errors. Most often two complete sets are provided by the architect. More sets must be purchased, if needed.

The estimator should be aware of and note any instructions to bidders, which may be included in the specifications or in a separate document. To avoid future confusion, the bid due date, time, and place should be clearly stated and understood upon receipt of the construction documents (plans and specifications). The due date should be marked on a calendar and a schedule. Completion of the estimate should be made as soon as possible, to avoid confusion at the last minute.

If bid security, or a bid bond, is required, then time must be allowed and arrangements made, especially if the bonding capability or capacity of a contractor has not previously been established.

All pre-bid meetings with the owner or architect should be attended, preferably *after* a review of the plans and specifications. Important points are often brought up at such meetings and details clarified. Attendance by all bidders is important, not only to show the owner an interest in the project, but also to ensure equal and competitive bidding. It is to the estimator's advantage to examine and review the plans and specifications before any such meetings and before the initial site visit. It is important to become familiar with the project as soon as possible.

In recent years, specifications, or project manuals, have become massive volumes containing a wealth of information. This author cannot stress enough the importance of reading all contract documents thoroughly. They exist to protect all parties involved in the construction process. The contract documents are written so that the estimators will be bidding equally and competitively, ensuring that all items in a project are included. The contract documents protect the designer (the architect or engineer) by ensuring that all work is supplied and installed as specified. The owner also benefits from thorough and complete construction documents, being guaranteed a measure of quality control and a complete job. Finally, the contractor benefits because the scope of work is well-defined, eliminating the gray areas of what is implied but not stated. "Extras" are more readily avoided. Change orders, if required, are accepted with less argument if the original contract documents are complete, well-stated, and most importantly, read by all concerned parties. Appendix A of this book contains a reproduction of a SPEC-AID, an R.S. Means Co., Inc. publication. This document was created not only to assist designers and planners when developing project specifications, but also as an aid to the estimator. The sections, arranged by MasterFormat division, list thousands of components and variables of the building construction process. The estimator can use this form as a means of documenting a project's requirements, and as a checklist to be sure that all items have been included.

During the first review of the specifications, all items to be priced should be identified and noted. All work to be subcontracted should be examined for "related work" required from other trades. Such work is usually referenced and described in a thorough project specification. "Work by others" or "Not in Contract" should be clearly defined as well as delineated on the drawings. Certain materials are often specified by the designer and purchased by the owner to be installed (labor only) by the contractor. These items should be noted and the responsibilities of each party clearly understood.

The General Conditions, Supplemental Conditions, and Special Conditions sections of the specifications should be examined carefully. These sections describe the items that have a direct bearing on the proposed project, but may not be part of the actual physical construction. An office trailer, temporary utilities, and testing are examples of these kinds of items. Also included in these sections is information regarding completion dates, payment requirements (e.g., retainage), allowances, alternates, and other important project requirements. All of these requirements have a significant bearing on the ultimate costs of the project and must be known and understood prior to performing the estimate. A more detailed discussion of General Conditions is included in Chapter 9.

While analyzing the plans and specifications, the estimator should evaluate the different portions of the project to determine which areas warrant the most attention. For example, if a building is to have a steel framework with a glass and aluminum frame skin, then more time should be spent estimating Division 5 – Metals, and Division 8 – Doors and Windows than Division 6 – Wood and Plastics.

Figures 2.1 and 2.2 are charts showing the relative percentage of different construction components by MasterFormat Divisions and UNIFORMAT II Divisions, respectively. These charts have been developed to represent the average percentages for new construction as a whole. All commonly used building types are included. The estimator should determine, for a given project, the relative proportions of each component, and time should be allocated accordingly. Using the percentages in Figure 2.1, a 10% error in Division 6 would result in approximately a 0.12% error in the project as a whole, while a 10% error in Division 15 would result in an approximate 2.1%

error overall. Thus, more time and care should be given to estimating those areas which contribute more to the cost of the project.

When the overall scope of the work has been identified, the drawings should be examined to confirm the information in the specifications. This is the time to clarify details while reviewing the general content. The estimator should note which sections, elevations, and detail drawings are for which plans. At this point and throughout the whole estimating process, the estimator should note and list any discrepancies between the plans and specifications, as well as any possible omissions. It is often stated in bid documents that bidders are obliged to notify the owner or architect/engineer of any such discrepancies.

Cost Breakdown By MasterFormat Divisions

NO.	DIVISION	%	NO.	DIVISION	%	NO.	DIVISION	%
0159	CONSTRUCTION EQUIPMENT*	7.4%	051	Structural Metal Framing	3.2%	092	Plaster & Gypsum Board	2.0%
023	Earthwork	4.9	052,053	Metal Joists & Decking	8.1	093,094	Tile & Terrazzo	2.7
024,0245	Tunneling & LB Elements	0.1	055	Metal Fabrications	0.8	095,098	Ceilings & Acoustical Treatment	2.9
027	Bases & Pavements	0.3	5	METALS	12.1	096	Flooring	2.4
025,026	Utility Services & Drainage	0.2	061	Rough Carpentry	0.9	097,099	Wall Finishes, Paints & Coatings	1.0
028	Site Improvements	0.1	062	Finish Carpentry	0.3	9	FINISHES	11.0
029	Planting	0.7	6	WOOD & PLASTICS	1.2	10-14	DIVISIONS 10-14	5.7
2	SITE CONSTRUCTION	6.3	071	Dampproofing & Waterproofing	0.1	150-152,154	Plumbing: Piping, Fixtures & Equip.	12.9
031	Conc. Forms & Accessories	3.2	072,078	Thermal, Fire & Smoke	1.3	153,139	Fire Protection/Suppression	3.0
032	Concrete Reinforcement	1.8		Protection		155	Heat Generation Equipment	2.2
033	Cast in Place Concrete	5.4	074,075	Roofing & Siding	1.0	156,158	Air Conditioning & Ventilation	3.0
034	Precast Concrete	2.3	076	Flashing & Sheet Metal	0.3	15	MECHANICAL	21.1
			7	THERMAL & MOISTURE PROT.	2.7	16	ELECTRICAL	11.8
3	CONCRETE	12.7	081-083	Doors & Frames	7.0		TOTAL (Div. 1-16)	100.0%
0405	Basic Masonry Mat. & Methods	0.6						
042	Masonry Units	6.5	088,089	Glazing & Glazed Curtain Walls	1.0	* Percentage for contractor equipment is		
044	Stone	0.3	8	DOORS & WINDOWS	8.0	spread among divisions and included		
4	MASONRY	7.4				above for information only		

Figure 2.1

Cost Breakdown By UNIFORMAT II Divisions

Division No.	Building System	Percentage	Division No.	Building System	Percentage
A	Substructure	6.5%	D10	Services: Conveying	3.7%
B10	Shell: Superstructure	18.0	D20-40	Mechanical	21.1
B20	Exterior Closure	12.2	D50	Electrical	11.8
B30	Roofing	2.7	E10	Equipment	2.0
C	Interiors	17.2	G	Building Sitework	4.8
				Total (Div. A-G)	100.0%

Figure 2.2

When so notified, the designer will most often issue an addendum to the contract documents in order to properly notify all parties concerned and to ensure equal and competitive bidding.

Once familiar with the contract documents, the estimator should notify appropriate subcontractors and vendors to solicit bids. Those subcontractors whose work is affected by the site conditions should accompany the estimator on the job site visit. At least one such visit (and often two or more) is extremely important, especially in cases of renovation and remodeling where existing conditions can have a significant effect on the cost of a project. During the site visit, the estimator should take notes, and possibly photographs, of all information pertinent to the construction and therefore to the project estimate. Pre-printed forms, such as the Job Site Analysis shown in Figures 2.3 and 2.4, can be useful to ensure that all items are included. Please note that most aspects of site work are included, in addition to such items as utilities, taxes, material, and labor availability. This kind of form can be very helpful as a checklist to prevent the omission of items which may be less obvious. If unusual site conditions exist or if questions arise during the takeoff, a second site visit is recommended.

In some areas, questions are likely to arise that cannot be answered clearly by the plans and specifications. Certain items may be omitted, or there may be conflicts or discrepancies within the contract documents. It is crucial that the owner or responsible party be notified quickly (usually in writing) so that these questions may be resolved before causing unnecessary problems. A proper estimate cannot be completed until all such questions are answered.

Perhaps the best way for an estimator to approach a project is to begin with a clear mind and a clear desk. Clutter and confusion in either case can have detrimental effects on the efficiency and accuracy of the estimate.

JOB SITE
ANALYSIS (GENERAL CONTRACTOR)

SHEET NO.

PROJECT		BID DATE
LOCATION		NEAREST TOWN
ARCHITECT	ENGINEER	OWNER

Access, Highway	Surface	Capacity
Railroad Siding	Freight Station	Bus Station
Airport	Motels/Hotels	Hospital
Post Office	Communications	Police
Distance & Travel Time to Site	.	Dock Facilities

Water Source	Amount Available	Quality
Distance from Site	Pipe/Pump Required?	Tanks Required?
Owner	Price (MG)	Treatment Necessary?
Natural Water Availability		Amount

Power Availability	Location	Transformer
Distance	Amount Available	
Voltage	Phase	Cycle	KWH or HP Rate

Temporary Roads	Lengths & Widths
Bridges/Culverts	Number & Size
Drainage Problems	
Clearing Problems	
Grading Problems	
Fill Availability	Distance
Mobilization Time	Cost
Camps or Housing	Size of Work Force
Sewage Treatment	
Material Storage Area	Office & Shed Area

Labor Source	Union Affiliation
Common Labor Supply	Skilled Labor Supply
Local Wage Rates	Fringe Benefits
Travel Time	Per Diem

Taxes, Sales	Facilities	Equipment
Hauling	Transportation	Property
Other		

Material Availability: Aggregates		Cement
Ready Mix Concrete		
Reinforcing Steel	Structural Steel	
Brick & Block	Lumber & Plywood	
Building Supplies	Equipment Repair & Parts	

Demolition: Type	Number	
Size	Equipment Required	
Dump Site	Distance	Dump fees
Permits		

Page 1 of 2

Figure 2.3

19

Clearing: Area | Timber | Diameter | Species

Brush Area	Burn on Site	Disposal Area
Saleable Timber	Useable Timber	Haul
Equipment Required		

Weather: Mean Temperatures

Highs	Lows
Working Season Duration	Bad Weather Allowance
Winter Construction	
Average Rainfall	Wet Season	Dry Season
Stream or Tide Conditions	
Haul Road Problems	
Long-Range Weather	

Soils: Job Borings Adequate?   Test Pits

Additional Borings Needed	Location	Extent
Visible Rock		
U.S. Soil & Agriculture Maps		
Bureau of Mines Geological Data		
County/State Agriculture Agent		
Tests Required		
Ground Water		

Construction Plant Required

Alternate Method

Equipment Available

Rental Equipment   Location

Miscellaneous: Contractor Interest

Subcontractor Interest

Material Fabricator Availability

Possible Job Delays

Political Situation

Construction Money Availability

Unusual Conditions

Summary

Figure 2.4

20

Chapter 3

The Quantity Takeoff

Chapter 3

The Quantity Takeoff

Traditionally, quantities are taken off, or surveyed from the drawings in sequence as the building would be built. Recently, however, the sequence of takeoff and estimating is more often based on the sixteen divisions of the CSI MasterFormat. A thorough and detailed discussion, by MasterFormat division, is included in Chapter 9.

Quantities may be taken off by one person if the project is not too large and time allows. For larger projects, the plans are often split into several disciplines (or divisions) and the work assigned to two or more estimators. In this case a project leader is assigned to coordinate and assemble the estimate.

When working with the plans during the quantity takeoff, consistency is the most important consideration. If each job is approached in the same manner, a pattern will develop, such as moving from the lower floors to the top, clockwise, or counterclockwise. The choice of method is not important, but consistency is. The purpose of being consistent is to avoid duplications as well as omissions and errors. Pre-printed forms provide an excellent means for developing consistent patterns. Figures 3.1 and 3.2 are examples of such forms. The Quantity Sheet (Figure 3.1) is designed purely for quantity takeoff. Note that one set of dimensions can be used for up to four different items. A good example of the use of this form is shown in Chapter 9, Figure 9.37. Figure 3.2, a Consolidated Estimate sheet, is designed to be used for both quantity takeoff and pricing on one form.

Every contractor might benefit from designing custom forms which would be appropriate for specific uses. If employees within the same company all use the same types of forms, then communication, interaction, and coordination of the estimating process will proceed more smoothly. One estimator will be able to understand the work of another. R.S. Means has published two books completely devoted to forms and their use, entitled *Means Forms for Building Construction Professionals* and *Means Forms for Contractors*. Forms, with examples and instructions for use, are included in each book.

If approached logically and systematically, there are a number of short cuts which can help to save time without sacrificing accuracy. Abbreviations simply save the time of writing things out. An abbreviations list, similar to that in the back of *Means Building Construction Cost Data*, might be posted in a conspicuous place, providing a consistent pattern of definitions for use within an office.

QUANTITY SHEET

		SHEET NO.
PROJECT		ESTIMATE NO.
LOCATION	ARCHITECT	DATE
TAKEOFF BY	EXTENSIONS BY	CHECKED BY

DESCRIPTION	NO.	DIMENSIONS		UNIT		UNIT		UNIT

Figure 3.1

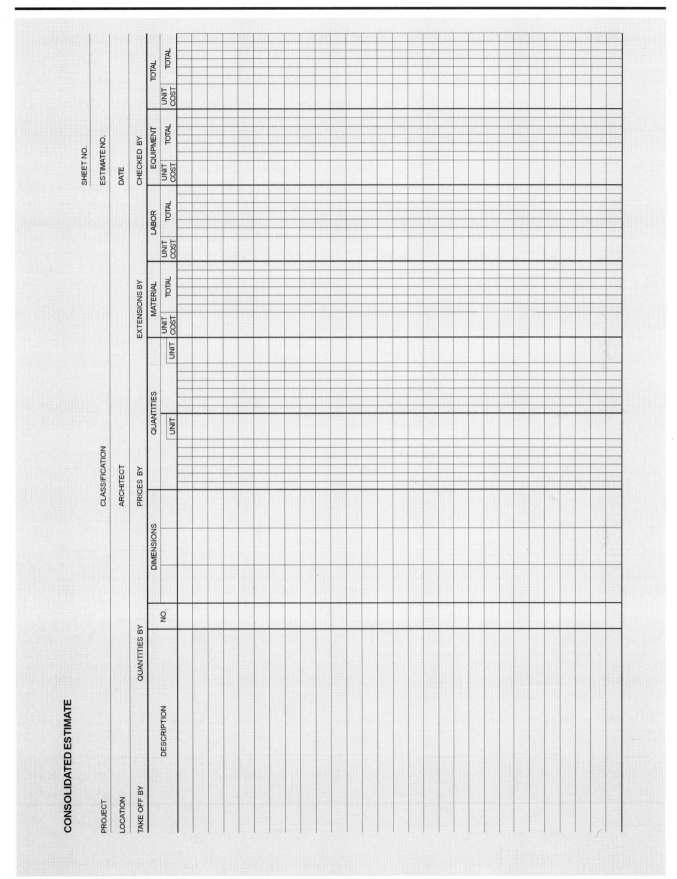

Figure 3.2

25

All dimensions—whether printed, measured, or calculated—that can be used for determining quantities for more than one item should be listed on a separate sheet and posted for easy reference. Posted gross dimensions can also be used to quickly check for order of magnitude errors.

Measurements should be converted to decimal equivalents before calculations are performed to extend the quantities. A chart similar to that shown in Figure 3.3 serves as a quick reference for such conversions. Whether converting from feet and inches to decimals, or from cubic feet to cubic yards, use good judgment and common sense to determine significant digits. Rounding off, or decreasing the number of significant digits, should be done only when it will not statistically affect the resulting product.

The estimator must use good judgment to determine when rounding is appropriate. An overall two or three percent variation in a competitive market can often be the difference between getting or losing a job, or between profit or no profit. The estimator should establish rules for rounding to achieve a consistent level of precision. As a general rule, it is best to not round numbers until the final summary of quantities. The final summary is also the time to convert units (square feet of paving to square yards, linear feet of lumber to board feet, etc.).

Figure 3.4 illustrates an example of the effects of premature rounding. If all items of a project were similarly rounded up for "safety," total project costs may include an unwarranted, or at least unaccountable allowance. An extra 5% may remove a bid from the competition.

Be sure to quantify and include "labor only" items that are not shown on plans. Such items may or may not be indicated in the specifications and might include cleanup, special labor for handling materials, etc.

Conversion of Inches to Decimal Parts per Foot												
	0	1″	2″	3″	4″	5″	6″	7″	8″	9″	10″	11″
0	0	.08	.17	.25	.33	.42	.50	.58	.67	.75	.83	.92
1/8″	.01	.09	.18	.26	.34	.43	.51	.59	.68	.76	.84	.93
1/4″	.02	.10	.19	.27	.35	.44	.52	.60	.69	.77	.85	.94
3/8″	.03	.11	.20	.28	.36	.45	.53	.61	.70	.78	.86	.95
1/2″	.04	.12	.21	.29	.37	.46	.54	.62	.71	.79	.87	.96
5/8″	.05	.14	.22	.30	.39	.47	.55	.64	.72	.80	.89	.97
3/4″	.06	.15	.23	.31	.40	.48	.56	.65	.73	.81	.90	.98
7/8″	.07	.16	.24	.32	.41	.49	.57	.66	.74	.82	.91	.99

Figure 3.3

In general, the quantity takeoff should be organized so that information gathered can be used to future advantage. Scheduling can be made easier if items are taken off and listed by construction phase, or by floor. Material purchasing will similarly benefit.

Units for each item should be consistent throughout the whole project—from takeoff to cost control. In this way, the original estimate can be equitably compared to progress and final cost reports more easily. It will be easier to keep track of a job.

The following list is a summation of the suggestions just mentioned plus a few more rules which will be helpful during the quantity takeoff:

- Use pre-printed forms.
- Transfer carefully and make clear notations of sums carried from one sheet to the next.
- List dimensions consistently.
- Use printed dimensions, otherwise measure dimensions carefully if the scale is known and accurate.
- Add printed dimensions for a single entry.
- Convert feet and inches to decimal feet.
- Do not round off until the final summary of quantities.
- Mark drawings as quantities are determined.
- Be alert for changes in scale, or notes such as "N.T.S." (not to scale).
- Include required items which may not appear in the plans and specs.
- Use building symmetry to avoid repetitive takeoffs.

Effect of Rounding Off Measurements at Quantity Takeoff		Actual	Rounded Off	% Difference
	Length of partition	96'-8"	100'	3.4%
	Height of partition	11'-9"	12'	2.1%
	Total S.F.	1135.87 S.F.	1200 S.F.	5.6%
	Cost ($3.83/S.F.)	$4350.38	$4596.00	5.6%
	+15% overhead	$5002.94	$5285.40	5.6%
	+10% profit	$5503.23	$5813.94	5.6%
	Total cost difference (incl. O&P): $310.71			

Figure 3.4

And perhaps the five most important points:
- Write legibly.
- Be organized.
- Use common sense.
- Be consistent.
- Add explanatory notes to the estimate to provide an audit trail.

Chapter 4

Pricing the Estimate

Chapter 4

Pricing the Estimate

When the quantities have been determined, then prices, or unit costs, must be applied in order to determine the total costs. Depending upon the chosen estimating method (and thus the degree of accuracy required) and the level of detail, these unit costs may be direct or bare costs, or may include overhead, profit, or contingencies. In Unit Price Estimating, the unit costs most commonly used are "bare," or "unburdened." Items such as overhead and profit are usually added to the total direct costs on the bottom line, at the time of the estimate summary.

Sources of Cost Information

One of the most difficult aspects of the estimator's job is determining accurate and reliable bare cost data. Sources for such data are varied, but can be categorized in terms of their relative reliability. The most reliable source of any cost information is the accurate, up-to-date, well-kept records of the estimator's own company. There is no better cost for a particular construction item than the *actual* cost to the contractor of that item from another recent job, modified (if necessary) to meet the requirements of the project being estimated.

Bids from *responsible* subcontractors are the second most reliable source of cost data. Any estimating inaccuracies are essentially absorbed by the subcontractor. A subcontract bid is a known, fixed cost prior to the project. Whether the price is "right" or "wrong" does not matter (as long as it is a responsible bid with no gross errors). The bid is what the appropriate portion of the work will cost. The prime contractor does not have to estimate the work of subcontractors, except for possible verification of the quote.

Quotations by vendors for material costs are, for the same reasons, as reliable as subcontract bids. In this case, however, the estimator must apply estimated labor costs. Thus the "installed" price for a particular item may be more variable. Whenever possible, all price quotations from vendors or subcontractors should be obtained in writing. Qualifications and exclusions should be clearly stated. Inclusions should be checked to be sure that they are complete and as specified. One way to ensure that these requirements are met is to prepare a form on which all subcontractors and vendors must submit quotations. This form can ask all appropriate questions and provide a consistent source of information for the estimator.

The above procedures are ideal, but often in the realistic haste of estimating and bidding, quotations are received orally, in person, by telephone, or by fax. The importance of gathering all pertinent information is heightened because omissions are more likely. A pre-printed form, such as the one shown in Figure 4.1, can be extremely useful to ensure that all required information and qualifications are obtained and understood. How often has the subcontractor stated, "I didn't know that I was supposed to include that?" With the help of such forms, the appropriate questions are asked and answered. An example of the use of this form is shown in Figure 9.34.

If the estimator has no cost records for a particular item and is unable to obtain a quotation, then the next most reliable source of price information is current unit price cost books such as *Means Building Construction Cost Data*. Means presents all such data in the form of national averages; these figures must be adjusted to local conditions. This procedure will be explained in Chapter 7. In addition to being a source of primary costs, unit price books can be useful as a reference or cross-check for verifying costs obtained elsewhere.

Lacking cost information from any of the above-mentioned sources, the estimator may have to rely on data from old books or adjusted records from an old project. While these types of costs may not be very accurate, they may be better than the final alternative—guesswork.

No matter which source of cost information is used, the system and sequence of pricing should be the same as that used for the quantity takeoff. This consistent approach should continue through both accounting and cost control during construction of the project.

Types of Costs

Unit price estimates for building construction may be organized according to the 16 divisions of the CSI MasterFormat. Within each division, the components or individual construction items are identified, listed, and priced. This kind of definition and detail is necessary to complete an accurate estimate. In addition, each "item" can be broken down further into material, labor, and equipment components.

All costs included in a Unit Price Estimate can be divided into two types: direct and indirect. Direct costs are those directly linked to the physical construction of a project—those costs without which the project could not be completed. The material, labor, and equipment costs just mentioned, as well as subcontract costs, are all direct costs. These may also be referred to as "bare," or "unburdened" costs.

Indirect costs are usually added to the estimate at the summary stage and are most often calculated as a percentage of the direct costs. They include such items as sales tax on materials, overhead, profit, and contingencies, etc. It is the indirect costs that generally account for the greatest variation in estimating.

Types of Costs in a Construction Estimate

Direct Costs	Indirect Costs
Material	Taxes
Labor	Overhead
Equipment	Profit
Subcontractors	Contingencies
Project Overhead	

TELEPHONE QUOTATION

	DATE	
PROJECT	TIME	
FIRM QUOTING	PHONE ( )	
ADDRESS	BY	
ITEM QUOTED	RECEIVED BY	

WORK INCLUDED	AMOUNT OF QUOTATION
DELIVERY TIME　　　　　　　　　　　　　　　**TOTAL BID**	

DOES QUOTATION INCLUDE THE FOLLOWING:			If ☐ NO is checked, determine the following:	
STATE & LOCAL SALES TAXES	☐ YES	☐ NO	MATERIAL VALUE	
DELIVERY TO THE JOB SITE	☐ YES	☐ NO	WEIGHT	
COMPLETE INSTALLATION	☐ YES	☐ NO	QUANTITY	
COMPLETE SECTION AS PER PLANS & SPECIFICATIONS	☐ YES	☐ NO	DESCRIBE BELOW	

EXCLUSIONS AND QUALIFICATIONS	
ADDENDA ACKNOWLEDGEMENT　　　　　　**TOTAL ADJUSTMENTS**	
ADJUSTED TOTAL BID	

ALTERNATES	
ALTERNATE NO.	
ALTERNATE NO.	
ALTERNATE NO.	
ALTERNATE NO.	
ALTERNATE NO.	
ALTERNATE NO.	
ALTERNATE NO.	

Figure 4.1

Direct Costs

On the list above, Project Overhead has been listed as a direct cost. Project Overhead represents those costs of a construction project which are usually included in Division 1 – General Requirements. Typical items are the job site office trailer, supervisory labor costs, daily and final cleanup, and temporary heat and power. While these items may not be directly part of the physical structure, the project could not be completed without them. Project Overhead, like all other direct costs, can be separated into material, labor, and equipment components. Figures 4.2 and 4.3 are examples of a form that can help ensure that all appropriate costs are included. Project Overhead will be further discussed below and in Chapter 9.

Material: When quantities have been carefully taken off, estimates of material cost can be very accurate. In order to maintain a high level of accuracy, the unit prices for materials must be reliable and current. The most reliable source of material costs is a quotation from a vendor for the particular job in question. Ideally, the vendor should have access to the plans and specifications for verification of quantities and specified products.

Material pricing appears relatively simple and straightforward. There are, however, certain considerations that the estimator must address when analyzing material quotations. The reputation of the vendor is a significant factor. Can the vendor "deliver," both figuratively and literally? Often estimators may choose not to rely on a "competitive" lower price from an unknown vendor, but will instead use a slightly higher price from a known, reliable vendor. Experience is the best judge for such decisions.

There are many questions that the estimator should ask. How long is the price guaranteed? At the end of that period, is there an escalation clause? Does the price include delivery or sales tax, if required? Note that most of these questions are addressed on the form in Figure 4.1. But more information should be obtained to ensure that a quoted price is accurate and competitive.

The estimator must be sure that the quotation or obtained price is for the materials as per plans and specifications. Architects and engineers may write into the specifications that: a) the particular type or brand of product must be used with no substitution, b) the particular type or brand of product is recommended, but alternate brands may be accepted *upon approval*, or c) no particular type or brand is specified. Depending upon the options, the estimator may be able to find an acceptable, less expensive alternative. In some cases, these substitutions can substantially lower the cost of a project.

When the estimator has received material quotations, there are still other considerations which should have a bearing on the final choice of a vendor. Lead time—the amount of time between order and delivery—must be determined and considered. It does not matter how competitive or low a quote is if the material cannot be delivered to the job site on time. If a delivery date is promised, is there a guarantee or a penalty clause for late delivery?

The estimator should also determine if there are any unusual payment requirements. Cash flow for a company can be severely affected if a large material purchase, thought to be payable in 30 days (90 days 10 years ago!) is delivered C.O.D. Truck drivers may not allow unloading until payment has been received. Such requirements must be determined during the estimating stage so that the cost of borrowing money, if necessary, can be included.

PROJECT OVERHEAD SUMMARY										

PROJECT
OVERHEAD SUMMARY
PROJECT

SHEET NO.

ESTIMATE NO.

LOCATION   ARCHITECT   DATE

QUANTITIES BY: PRICES BY: EXTENSIONS BY: CHECKED BY:

DESCRIPTION	QUANTITY	UNIT	MATERIAL/EQUIPMENT		LABOR		TOTAL COST	
			UNIT	TOTAL	UNIT	TOTAL	UNIT	TOTAL
Job Organization: Superintendent								
Project Manager								
Timekeeper & Material Clerk								
Clerical								
Safety, Watchman & First Aid								
Travel Expense: Superintendent								
Project Manager								
Engineering: Layout								
Inspection/Quantities								
Drawings								
CPM Schedule								
Testing: Soil								
Materials								
Structural								
Equipment: Cranes								
Concrete Pump, Conveyor, Etc.								
Elevators, Hoists								
Freight & Hauling								
Loading, Unloading, Erecting, Etc.								
Maintenance								
Pumping								
Scaffolding								
Small Power Equipment/Tools								
Field Offices: Job Office								
Architect/Owner's Office								
Temporary Telephones								
Utilities								
Temporary Toilets								
Storage Areas & Sheds								
Temporary Utilities: Heat								
Light & Power								
Water								
PAGE TOTALS								

Figure 4.2

DESCRIPTION	QUANTITY	UNIT	MATERIAL/EQUIPMENT		LABOR		TOTAL COST	
			UNIT	TOTAL	UNIT	TOTAL	UNIT	TOTAL
Totals Brought Forward								
Winter Protection: Temp. Heat/Protection								
Snow Plowing								
Thawing Materials								
Temporary Roads								
Signs & Barricades: Site Sign								
Temporary Fences								
Temporary Stairs, Ladders & Floors								
Photographs								
Clean Up								
Dumpster								
Final Clean Up								
Punch List								
Permits: Building								
Misc.								
Insurance: Builders Risk								
Owner's Protective Liability								
Umbrella								
Unemployment Ins. & Social Security								
Taxes								
City Sales Tax								
State Sales Tax								
Bonds								
Performance								
Material & Equipment								
Main Office Expense								
Special Items								
TOTALS:								

Figure 4.3

If unable to obtain the quotation of a vendor from whom the material would be purchased, the estimator has other sources for obtaining material prices. These include, in order of reliability:

1. Current price lists from manufacturers' catalogs. Be sure to check that the list is for "contractor prices."
2. Cost records from previous jobs. Historical costs must be updated for present conditions.
3. Reputable and current annual unit price cost books, such as *Means Building Construction Cost Data*. Such books usually represent national averages and must be factored to local markets.

No matter which price source is used, the estimator must be sure to include any costs over the actual cost of the material. The above-mentioned concerns regarding vendor quotations should also be taken into consideration.

Labor: In order to determine the labor cost for each unit of construction, the estimator must know two pieces of information: first, the labor rate (hourly wage or salary) of the worker, and second, how many units this worker can produce or install in a given time period—in other words, the output or productivity. Wage rates are known going into a project but productivity may be very hard to determine. The best source of labor productivity (and therefore labor costs) is the estimator's well-kept records from previous projects.

To estimators working for contractors, construction labor rates for employers will be known, well-documented, and constantly updated. Estimators for owners, architects, or engineers must determine labor rates from outside sources. Unit price cost books, such as *Means Building Construction Cost Data*, provide national average union labor rates by trade and base the unit costs for labor on these averages. Figure 4.4 shows national average union rates for the construction industry (based on January 1, 2002). Figure 4.5 lists national average open shop (non-union) rates, again based on January 1, 2002.

If more accurate rates are required, the estimator has alternate sources. Union locals can provide rates (as well as negotiated increases) for a particular location. This source requires the estimator to call the union hall for each trade. Employer bargaining groups can usually provide labor cost data, but this data may not be continually updated. R.S. Means Co., Inc. publishes *Means Labor Rates for the Construction Industry* on an annual basis. This book lists the labor rates by trade for 314 U.S. and Canadian cities. Other sources of wage rate information are the published Federal prevailing wage rates, known as *Davis Bacon Wages*, and the state prevailing wage rates. These rates are the minimum rate that must be paid to workers on Federal- or State-funded projects.

The above sources are for union rates. Determination of "open shop" labor rates is much more difficult. There are often organizations in larger cities representing open shop contractors. These organizations may have records of local pay scales, but ultimately the wage rates are determined by each contractor.

Unit labor costs are the least predictable of all costs for building projects. It is important to determine as accurately as possible both the prevailing wages and the productivity. If there are no company records for productivity, cost data books, such as *Means Building Construction Cost Data*, and productivity reference books, such as *Means Productivity Standards for Construction 3rd Edition*, can be invaluable. Included with the listing for each individual construction item is the designation of typical crew make-up, together with the productivity or output—the amount of work that the crew will produce. Figure 4.6, a typical

Abbr.	Trade	A Base Rate Incl. Fringes		B Workers' Comp. Ins.	C Average Fixed Overhead	D Overhead	E Profit	F Total Overhead & Profit	G	H Rate with O & P	I
		Hourly	Daily					%	Amount	Hourly	Daily
Skwk	Skilled Workers Average (35 trades)	$30.95	$247.60	16.8%	16.5%	13.0%	10%	56.3%	$17.40	$48.35	$386.80
	Helpers Average (5 trades)	22.75	182.00	18.5		11.0		56.0	12.75	35.50	284.00
	Foreman Average, Inside ($.50 over trade)	31.45	251.60	16.8		13.0		56.3	17.70	49.15	393.20
	Foreman Average, Outside ($2.00 over trade)	32.95	263.60	16.8		13.0		56.3	18.55	51.50	412.00
Clab	Common Building Laborers	23.45	187.60	18.1		11.0		55.6	13.05	36.50	292.00
Asbe	Asbestos/Insulation Workers/Pipe Coverers	33.45	267.60	16.2		16.0		58.7	19.65	53.10	424.80
Boil	Boilermakers	36.25	290.00	14.7		16.0		57.2	20.75	57.00	456.00
Bric	Bricklayers	30.50	244.00	16.0		11.0		53.5	16.30	46.80	374.40
Brhe	Bricklayer Helpers	23.50	188.00	16.0		11.0		53.5	12.55	36.05	288.40
Carp	Carpenters	30.00	240.00	18.1		11.0		55.6	16.70	46.70	373.60
Cefi	Cement Finishers	28.70	229.60	10.6		11.0		48.1	13.80	42.50	340.00
Elec	Electricians	35.45	283.60	6.7		16.0		49.2	17.45	52.90	423.20
Elev	Elevator Constructors	37.10	296.80	7.7		16.0		50.2	18.60	55.70	445.60
Eqhv	Equipment Operators, Crane or Shovel	32.35	258.80	10.6		14.0		51.1	16.55	48.90	391.20
Eqmd	Equipment Operators, Medium Equipment	31.20	249.60	10.6		14.0		51.1	15.95	47.15	377.20
Eqlt	Equipment Operators, Light Equipment	29.80	238.40	10.6		14.0		51.1	15.25	45.05	360.40
Eqol	Equipment Operators, Oilers	26.65	213.20	10.6		14.0		51.1	13.60	40.25	322.00
Eqmm	Equipment Operators, Master Mechanics	32.80	262.40	10.6		14.0		51.1	16.75	49.55	396.40
Glaz	Glaziers	30.00	240.00	13.8		11.0		51.3	15.40	45.40	363.20
Lath	Lathers	28.75	230.00	11.1		11.0		48.6	13.95	42.70	341.60
Marb	Marble Setters	30.10	240.80	16.0		11.0		53.5	16.10	46.20	369.60
Mill	Millwrights	31.75	254.00	10.6		11.0		48.1	15.25	47.00	376.00
Mstz	Mosaic & Terrazzo Workers	29.25	234.00	9.8		11.0		47.3	13.85	43.10	344.80
Pord	Painters, Ordinary	27.15	217.20	13.8		11.0		51.3	13.95	41.10	328.80
Psst	Painters, Structural Steel	27.90	223.20	48.4		11.0		85.9	23.95	51.85	414.80
Pape	Paper Hangers	27.10	216.80	13.8		11.0		51.3	13.90	41.00	328.00
Pile	Pile Drivers	29.80	238.40	24.9		16.0		67.4	20.10	49.90	399.20
Plas	Plasterers	28.10	224.80	15.8		11.0		53.3	15.00	43.10	344.80
Plah	Plasterer Helpers	23.70	189.60	15.8		11.0		53.3	12.65	36.35	290.80
Plum	Plumbers	35.95	287.60	8.3		16.0		50.8	18.25	54.20	433.60
Rodm	Rodmen (Reinforcing)	34.25	274.00	28.3		14.0		68.8	23.55	57.80	462.40
Rofc	Roofers, Composition	26.60	212.80	32.6		11.0		70.1	18.65	45.25	362.00
Rots	Roofers, Tile & Slate	26.75	214.00	32.6		11.0		70.1	18.75	45.50	364.00
Rohe	Roofers, Helpers (Composition)	19.80	158.40	32.6		11.0		70.1	13.90	33.70	269.60
Shee	Sheet Metal Workers	35.10	280.80	11.7		16.0		54.2	19.00	54.10	432.80
Spri	Sprinkler Installers	36.20	289.60	8.7		16.0		51.2	18.55	54.75	438.00
Stpi	Steamfitters or Pipefitters	36.20	289.60	8.3		16.0		50.8	18.40	54.60	436.80
Ston	Stone Masons	30.65	245.20	16.0		11.0		53.5	16.40	47.05	376.40
Sswk	Structural Steel Workers	34.25	274.00	39.8		14.0		80.3	27.50	61.75	494.00
Tilf	Tile Layers	29.15	233.20	9.8		11.0		47.3	13.80	42.95	343.60
Tilh	Tile Layers Helpers	23.35	186.80	9.8		11.0		47.3	11.05	34.40	275.20
Trlt	Truck Drivers, Light	24.30	194.40	14.9		11.0		52.4	12.75	37.05	296.40
Trhv	Truck Drivers, Heavy	25.00	200.00	14.9		11.0		52.4	13.10	38.10	304.80
Sswl	Welders, Structural Steel	34.25	274.00	39.8		14.0		80.3	27.50	61.75	494.00
Wrck	*Wrecking	23.45	187.60	41.2		11.0		78.7	18.45	41.90	335.20

*Not included in averages

Figure 4.4

		A		B	C	D	E	F	G	H	I
		Base Rate Incl. Fringes		Work-ers' Comp. Ins.	Average Fixed Over-head	Over-head	Profit	Total Overhead & Profit		Rate with O & P	
Abbr.	Trade	Hourly	Daily					%	Amount	Hourly	Daily
Skwk	Skilled Workers Average (35 trades)	$21.25	$170.00	16.8%	16.5%	27.0%	10%	70.3%	$14.95	$36.20	$289.60
	Helpers Average (5 trades)	15.65	125.20	18.5		25.0		70.0	10.95	26.60	212.80
	Foreman Average, inside ($.50 over trade)	21.75	174.00	16.8		27.0		70.3	15.30	37.05	296.40
	Foreman Average, Outside ($2.00 over trade)	23.25	186.00	16.8		27.0		70.3	16.35	39.60	316.80
Clab	Common Building Laborers	15.25	122.00	18.1		25.0		69.6	10.60	25.85	206.80
Asbe	Asbestos/Insulation Workers/Pipe Coverers	22.10	176.80	16.2		30.0		72.7	16.05	38.15	305.20
Boil	Boilermakers	23.95	191.60	14.7		30.0		71.2	17.05	41.00	328.00
Bric	Bricklayers	21.35	170.80	16.0		25.0		67.5	14.40	35.75	286.00
Brhe	Bricklayer Helpers	16.45	131.60	16.0		25.0		67.5	11.10	27.55	220.40
Carp	Carpenters	21.00	168.00	18.1		25.0		69.6	14.60	35.60	284.80
Cefi	Cement Finishers	20.10	160.80	10.6		25.0		62.1	12.50	32.60	260.80
Elec	Electricians	23.75	190.00	6.7		30.0		63.2	15.00	38.75	310.00
Elev	Elevator Constructors	24.85	198.80	7.7		30.0		64.2	15.95	40.80	326.40
Eqhv	Equipment Operators, Crane or Shovel	22.30	178.40	10.6		28.0		65.1	14.50	36.80	294.40
Eqmd	Equipment Operators, Medium Equipment	21.55	172.40	10.6		28.0		65.1	14.05	35.60	284.80
Eqlt	Equipment Operators, Light Equipment	20.55	164.40	10.6		28.0		65.1	13.40	33.95	271.60
Eqol	Equipment Operators, Oilers	18.40	147.20	10.6		28.0		65.1	12.00	30.40	243.20
Eqmm	Equipment Operators, Master Mechanics	22.65	181.20	10.6		28.0		65.1	14.75	37.40	299.20
Glaz	Glaziers	21.30	170.40	13.8		25.0		65.3	13.90	35.20	281.60
Lath	Lathers	20.15	161.20	11.1		25.0		62.6	12.60	32.75	262.00
Marb	Marble Setters	21.05	168.40	16.0		25.0		67.5	14.20	35.25	282.00
Mill	Millwrights	22.25	178.00	10.6		25.0		62.1	13.80	36.05	288.40
Mstz	Mosaic & Terrazzo Workers	20.50	164.00	9.8		25.0		61.3	12.55	33.05	264.40
Pord	Painters, Ordinary	19.30	154.40	13.8		25.0		65.3	12.60	31.90	255.20
Psst	Painters, Structural Steel	19.80	158.40	48.4		25.0		99.9	19.80	39.60	316.80
Pape	Paper Hangers	19.25	154.00	13.8		25.0		65.3	12.55	31.80	254.40
Pile	Pile Drivers	20.85	166.80	24.9		30.0		81.4	16.95	37.80	302.40
Plas	Plasterers	19.65	157.20	15.8		25.0		67.3	13.20	32.85	262.80
Plah	Plasterer Helpers	16.60	132.80	15.8		25.0		67.3	11.15	27.75	222.00
Plum	Plumbers	23.75	190.00	8.3		30.0		64.8	15.40	39.15	313.20
Rodm	Rodmen (Reinforcing)	22.95	183.60	28.3		28.0		82.8	19.00	41.95	335.60
Rofc	Roofers, Composition	18.35	146.80	32.6		25.0		84.1	15.45	33.80	270.40
Rots	Roofers, Tile & Slate	18.45	147.60	32.6		25.0		84.1	15.50	33.95	271.60
Rohe	Roofers, Helpers (Composition)	13.65	109.20	32.6		25.0		84.1	11.50	25.15	201.20
Shee	Sheet Metal Workers	23.15	185.20	11.7		30.0		68.2	15.80	38.95	311.60
Spri	Sprinkler Installers	23.90	191.20	8.7		30.0		65.2	15.60	39.50	316.00
Stpi	Steamfitters or Pipefitters	23.90	191.20	8.3		30.0		64.8	15.50	39.40	315.20
Ston	Stone Masons	20.85	166.80	16.0		25.0		67.5	14.05	34.90	279.20
Sswk	Structural Steel Workers	22.95	183.60	39.8		28.0		94.3	21.65	44.60	356.80
Tilf	Tile Layers	20.40	163.20	9.8		25.0		61.3	12.50	32.90	263.20
Tilh	Tile Layers Helpers	16.35	130.80	9.8		25.0		61.3	10.00	26.35	210.80
Trlt	Truck Drivers, Light	17.25	138.00	14.9		25.0		66.4	11.45	28.70	229.60
Trhv	Truck Drivers, Heavy	17.75	142.00	14.9		25.0		66.4	11.80	29.55	236.40
Sswl	Welders, Structural Steel	22.95	183.60	39.8		28.0		94.3	21.65	44.60	356.80
Wrck	*Wrecking	15.70	125.60	41.2		25.0		92.7	14.55	30.25	242.00

*Not included in averages

Figure 4.5

051 | Structural Metal Framing

051 200 | Structural Steel

			CREW	MAKEUP	DAILY OUTPUT	LABOR HOURS	UNIT	
220	7450	W 14 x 176	E-2	1 Struc. Steel Foreman 4 Struc. Steel Workers 1 Equip. Oper. (crane) 1 Equip. Oper. Oiler 1 Crane, 90 Ton	912	.061	L.F.	220
230	0010	**LIGHTWEIGHT FRAMING**						230
	0400	Angle framing, 4" and larger	E-4	1 Struc. Steel Foreman 3 Struc. Steel Workers 1 Gas Welding Machine	3,000	.011	Lb.	
	0450	Less than 4" angles			1,800	.018	Lb.	
	0600	Channel framing, 8" and larger			3,500	.009	Lb.	
	0650	Less than 8" channels	↓	↓	2,000	.016	Lb.	
	1000	Continuous slotted channel framing system, minimum	2 Sswk	2 Structural Steel Workers	2,400	.007	Lb.	
	1200	Maximum	"	"	1,600	.010	Lb.	
	1300	Cross bracing, rods, 3/4" diameter	E-3	1 Struc. Steel Foreman 1 Struc. Steel Worker 1 Welder 1 Gas Welding Machine 1 Torch, Gas & Air	700	.034	Lb.	
	1310	7/8" diameter			700	.034	Lb.	
	1320	1" diameter			700	.034	Lb.	
	1330	Angle, 5" x 5" x 3/8"			2,800	.009	Lb.	
	1350	Hanging lintels, average	↓	↓	850	.028	Lb.	
	1380	Roof frames, 3'-0" square, 5' span	E-2	1 Struc. Steel Foreman 4 Struc. Steel Workers 1 Equip. Oper. (crane) 1 Equip. Oper. Oiler 1 Crane, 90 Ton	4,200	.013	Lb.	
	1400	Tie rod, not upset, 1-1/2" to 4" diameter, with turnbuckle	2 Sswk	2 Structural Steel Workers	800	.020	Lb.	
	1420	No turnbuckle			700	.023	Lb.	
	1500	Upset, 1-3/4" to 4" diameter, with turnbuckle			800	.020	Lb.	
	1520	No turnbuckle			700	.023	Lb.	
	1600	Tubular aluminum framing for window wall, minimum			600	.027	Lb.	
	1800	Maximum	↓	↓	500	.032	Lb.	
232	0010	**LINTELS** Plain steel angles, under 500 lb.	1 Bric	1 Bricklayer	500	.016	Lb.	232
	0100	500 to 1000 lb.			600	.013	Lb.	
	0200	1,000 to 2,000 lb.			600	.013	Lb.	
	0300	2,000 to 4,000 lb.			600	.013	Lb.	
	2000	Steel angles, 3-1/2" x 3", 1/4" thick, 2'-6" long			50	.160	Ea.	
	2100	4'-6" long			45	.178	Ea.	
	2500	3-1/2" x 3-1/2" x 5/16", 5'-0" long			44	.182	Ea.	
	2600	4" x 3-1/2", 1/4" thick, 5'-0" long			40	.200	Ea.	
	2700	9'-0" long			35	.229	Ea.	
	2800	4" x 3-1/2" x 5/16", 7'-0" long			38	.211	Ea.	
	2900	5" x 3-1/2" x 5/16", 10'-0" long	↓	↓	35	.229	Ea.	
238	0010	**SPACE FRAME** Steel 4' modular, 40' to 70' spans, 5.5 psf, minimum	E-2	1 Struc. Steel Foreman 4 Struc. Steel Workers 1 Equip. Oper. (crane) 1 Equip. Oper. Oiler 1 Crane, 90 Ton	556	.101	S.F.	238
	0200	Maximum			365	.153	S.F.	
	0400	5' modular, 4.5 psf minimum			585	.096	S.F.	
	0500	Maximum	↓	↓	405	.138	S.F.	
240	0010	**STEEL CUTTING** Hand burning, including preparation,						240
	0020	torch cutting, and grinding, no staging						
	0100	Steel to 1/2" thick	E-14	1 Welder Foreman 1 Gas Welding Machine	70	.114	L.F.	
	0150	3/4" thick			50	.160	L.F.	
	0200	1" thick	↓	↓	45	.178	L.F.	

Figure 4.6

page from *Means Productivity Standards for Construction 3rd Edition*, includes this data and indicates the required number of labor-hours for each "unit" of work for the designated task.

The estimator who has neither company records nor the sources described above must put together the appropriate crews and determine the expected output or productivity. This type of estimating should only be attempted based upon strong experience and considerable exposure to construction methods and practices.

Equipment: Over recent years, construction equipment has become much more sophisticated, not only because of the incentive of reduced labor costs, but also as a response to new, highly technological construction methods and materials. As a result, equipment costs represent an increasing percentage of total project costs in building construction. Estimators must carefully address the equipment and all related expenses. Costs for equipment can be divided into two categories.

1. Rental, lease, or ownership costs. These costs may be determined based on hourly, daily, weekly, monthly, or annual increments. These fees or payments only buy the "right" to use the equipment.
2. Operating costs. Once the "right" of use is obtained, costs are incurred for actual use, or operation. These costs include fuel, lubrication, maintenance, and parts.

Equipment costs as described above do not include the operator's wages. However, some cost books and suppliers may include the operator in the quoted price for equipment as an "operated" rental cost. The estimator must be aware of what is and what is not included.

Quotations for equipment rental or lease costs can be obtained from local dealers and suppliers, or even from manufacturers. These costs can fluctuate and should be updated regularly. Ownership costs must be determined within a company. There are many considerations beyond the up-front purchase price; these facts must be taken into account when figuring the cost of owning equipment. Interest rates and amortization schedules should be studied prior to the purchase. Insurance costs, storage fees, maintenance, taxes, and licenses, all added together, can become a significant percentage of the cost of owning equipment. These costs must be anticipated prior to purchase in order to properly manage the ownership.

When purchasing equipment, the owner should be aware of some basic principles of accounting (or hire a good accountant). In particular, various methods of depreciation (a way of quantifying loss of value to the owner over time) can have varied effects at tax time.

Figure 4.7 shows the effects of three "textbook" examples of depreciation methods: "Straight Line," "Sum of Years Digit," and "Declining Balance." Notice that different dollar amounts are depreciated in different years. These are very simplistic examples. A tax planning strategy should be developed in order to determine the appropriate and most advantageous method. An accountant should be consulted due to the current, complicated, ever-changing tax laws. The possibilities of accelerated methods, investment tax credits, and other incentives should also be considered.

Equipment ownership costs apply to both leased and owned equipment. The operating costs of equipment, whether rented, leased, or owned, are available from the following sources (listed in order of reliability):

1. The company's own records.

2. Annual cost books containing equipment operating costs, such as *Means Building Construction Cost Data*.

3. Manufacturers' estimates.

4. Textbooks dealing with equipment operating costs.

These operating costs consist of fuel, lubrication, expendable parts replacement, and minor maintenance. For financial analysis purposes, the equipment ownership and operating costs should be listed separately. In this way, the decision to rent, lease, or purchase can be decided project by project.

There are two commonly used methods for including equipment costs in a construction estimate. The first is to include the equipment as a part of the construction task for which it is used. In this case, costs are included in each line item as a separate unit price. The advantage of this method is that costs are allocated to the division or task that actually incurs the expense. As a result, more accurate records can be kept for each construction component. The disadvantage of this method occurs in the pricing of equipment (e.g., tower crane, personnel hoist, etc.) that may be used by many different trades for different tasks. Duplication of costs can occur in this instance.

The second method for including equipment costs in the estimate is to keep all such costs separate and to include them in Division 1 as a part of Project Overhead. The advantage of this method is that all equipment costs are grouped together, and that items used by all trades are included (without duplication). The disadvantage is that for future estimating purposes, equipment costs will be known only by job and not by unit of construction. Under these circumstances, omissions can easily occur.

Whichever method is used, the estimator must be consistent, and must be sure that all equipment costs are included, but not duplicated. The estimating method should be the same as that chosen for cost monitoring and accounting, so that the data will be available for future projects.

Subcontractors: In essence, subcontractor quotations should be solicited and analyzed in the same way as material quotes. A primary concern is that the bid covers the work as per plans and specifications, and that all appropriate work alternates and allowances are included. Any exclusions should be clearly stated and explained. If the bid is received orally, a form such as that in Figure 4.1 will help to ensure that all is included. Oral bids/quotes should be followed up by fax and hard copy confirmation. Any unique scheduling or payment

Depreciation Strategies for Equipment — Based upon a $15,000 Purchase Price with No Allowance for Salvage	Methods of Depreciation		
	Year	Straight Line	Sum of Years Digit	Declining Balance
	1	$ 3,000	$ 5,000	$ 6,000.00
	2	3,000	4,000	3,600.00
	3	3,000	3,000	2,160.00
	4	3,000	2,000	1,296.00
	5	3,000	1,000	777.60
	Total	$15,000	$15,000	$13,833.60

Figure 4.7

requirements must be noted and evaluated prior to submission of the prime bid. Such requirements could affect or restrict the normal progress of the project, and should therefore be known in advance.

The estimator should note how long the subcontract bid will be honored. This time period usually varies from 30 to 90 days and is often included as a condition in complete bids. The general contractor may have to define the time limits of the prime bid based upon certain subcontractors. The estimator must also note any escalation clauses that may be included in subcontractor bids.

Reliability is another factor to be considered when soliciting and evaluating subcontractor bids. Reliability cannot be measured or priced until the project is actually under construction. Most general contractors stay with the same subcontractors for just this reason. A certain unspoken communication exists in these established relationships and usually has a positive effect on the performance of the work. Such familiarity, however, can often erode the competitive nature of bidding. To be competitive with the prime bid, the estimator should always obtain comparison subcontract prices, whether these prices come from another subcontractor or are prepared by the estimator.

The estimator should question and verify the bonding capability and capacity of unfamiliar subcontractors. Taking such action may be necessary when bidding in a new location. Other than word of mouth, these inquiries may be the only way to confirm subcontractor reliability.

For major subcontract items such as mechanical, electrical, and conveying systems, it may be necessary to make up spreadsheets in order to tabulate inclusions and omissions. This procedure ensures that all cost considerations are included in the "adjusted" quotation. Time permitting, the estimator should make a takeoff and price these major subcontract items to compare with the subcontractor bids. If time does not permit a detailed takeoff, the estimator should at least budget the work. An Assemblies Estimate is ideal for this purpose.

Project Overhead: Project overhead includes those items as specified in Division 1 – General Requirements. It also includes those items required for the actual construction of the project, but not necessarily applicable to another specific MasterFormat division. As seen in Figures 4.2 and 4.3, project overhead covers items from project supervision to cleanup, from temporary utilities to permits. All may not agree that certain items (such as equipment or scaffolding) should be included as Project Overhead, and might prefer to list such items in another division. Ultimately, it is not important *where* each item is incorporated into the estimate, but that *every item is included somewhere*, and not in two locations.

Project overhead often includes time-related items; equipment rental, supervisory labor, and temporary utilities are examples. The cost for these items depends upon the duration of the project. A preliminary schedule should, therefore, be developed *prior* to completion of the estimate so that time-related items can be property counted. This will be further discussed in Chapter 5.

Bond requirements for a project are usually specified in the General Conditions portion of the specification. Costs for bonds are based on total project costs and are determined at the estimate summary stage. A discussion of different types of bonds is included below.

Indirect Costs

The direct costs of a project must be itemized, tabulated and totalled before the indirect costs can be applied to the estimate. The indirect costs are almost always defined as a percentage of direct costs and include:

1. Sales Tax (if required)
2. Office or Operating Overhead (vs. Project Overhead)
3. Profit
4. Contingencies
5. Bonds (often included as Project Overhead)

Sales Tax: Sales tax varies from state to state and often from city to city within a state (see Figure 4.8). Larger cities may have a sales tax in addition to the state sales tax. Some localities also impose separate sales taxes on labor and equipment.

When bidding takes place in unfamiliar locations, the estimator should check with local agencies regarding the amount and the method of payment of sales tax. Local authorities may require owners to withhold payments to out-of-state contractors until payment of all required sales tax has been verified. Sales tax is often taken for granted or even omitted and, as can be seen in Figure 4.8, can be as much as 7% of material costs. Indeed, this can represent a significant portion of the project's total cost. Conversely, some clients and/or their projects may be tax exempt. If this fact is unknown to the estimator, a large dollar amount for sales tax might be needlessly included in a bid.

Office or Operating Overhead: Office overhead, or the cost of doing business, is perhaps one of the main reasons why so many contractors are unable to realize a profit, or even to stay in business. If a contractor does not know the costs of operating the business, then, more than likely, these costs will not be recovered. Many companies survive, and even turn a profit, by simply adding a certain percentage for overhead to each job, without knowing how the percentage is derived or what is included. When annual volume changes significantly, whether by increase or decrease, the previously used percentage for overhead may no longer be valid. Often when such a volume change

R01100-090 Sales Tax by State

State sales tax on materials is tabulated below (5 states have no sales tax). Many states allow local jurisdictions, such as a county or city, to levy additional sales tax.

Some projects may be sales tax exempt, particularly those constructed with public funds.

State	Tax (%)	State	Tax (%)	State	Tax (%)	State	Tax (%)
Alabama	4	Illinois	6.25	Montana	0	Rhode Island	7
Alaska	0	Indiana	5	Nebraska	5	South Carolina	5
Arizona	5	Iowa	5	Nevada	6.5	South Dakota	4
Arkansas	4.625	Kansas	4.9	New Hampshire	0	Tennessee	6
California	7	Kentucky	6	New Jersey	6	Texas	6.25
Colorado	2.9	Louisiana	4	New Mexico	5	Utah	4.75
Connecticut	6	Maine	5.5	New York	4	Vermont	5
Delaware	0	Maryland	5	North Carolina	4	Virginia	4.5
District of Columbia	5.75	Massachusetts	5	North Dakota	5	Washington	6.5
Florida	6	Michigan	6	Ohio	5	West Virginia	6
Georgia	4	Minnesota	6.5	Oklahoma	4.5	Wisconsin	5
Hawaii	4	Mississippi	7	Oregon	0	Wyoming	4
Idaho	5	Missouri	4.225	Pennsylvania	6	Average	4.70 %

Figure 4.8

occurs, the owner finds that the company is not doing as well as before and cannot determine the reasons. Chances are, overhead costs are not being fully recovered. As an example, Figure 4.9 lists office costs and expenses for a "typical" construction company for a year. It is assumed that the anticipated annual volume of the company is $10 million. Each of the items is described briefly below.

Owner: This includes only a reasonable base salary and does not include profits.

Engineer/Estimator: Since the owner is primarily on the road getting business, this is the person who runs the daily operation of the company and is responsible for estimating.

Assistant Estimator: This person is the "number cruncher," performing most quantity takeoffs and some pricing.

Project Manager: This is the person who runs the projects from the office and acts as the liaison between the owner and the field.

General Superintendent: The general super's post is perhaps the most important position in the company. This person is responsible for the day-to-day progress of all projects—the nuts and bolts. All field personnel are handled by this person.

Bookkeeper/Office Manager: This is the overworked, underpaid person who actually runs the company. Every company must have one.

Secretary/Receptionist: This is the assistant to the person who actually runs the company.

Office Worker Insurance & Taxes: These costs are for main office personnel only and, for this example, are calculated as 38% of the total salaries based on the following breakdown:

Workers' Compensation	.46%
FICA	7.65%
Unemployment	7%
Medical & other insurance	13%
Profit sharing, pension, etc.	10%
Rounded to	38%

Physical Plant Expenses: Whether the office, warehouse, and yard are rented or owned, roughly the same costs are incurred. Telephone and utility costs will vary depending on the size of the building and the type of business. Office equipment includes items such as copy machine rental, typewriters, etc.

Professional Services: Accountant fees are primarily for quarterly audits and year-end statements. Legal fees go towards collecting and contract disputes. Advertising includes the Yellow Pages, promotional materials, etc.

Miscellaneous: There are many expenses that could be placed in this category. included in the example are just a few.

Uncollected Receivables: This amount can vary greatly, often depending upon the overall economic climate. Depending upon the timing of "uncollectables," cash flow can be severely restricted and can cause serious financial problems, even for large companies. Sound cash planning and anticipation of such possibilities can help to prevent severe repercussions.

While the office example used here is feasible within the industry, keep in mind that it is hypothetical and that conditions and costs vary from company to company.

ANNUAL MAIN OFFICE EXPENSES

Salaries

Owner	$ 80,000
Engineer/Estimator	50,000
Assistant Estimator	35,000
Project Manager	60,000
General Superintendent	50,000
Bookkeeper/Office Manager	28,000
Secretary/Receptionist	18,000

Office Worker Benefits

Workers' Compensation		
FICA & Unemployment	38% of	
Medical Insurance	Salaries	121,980
Miscellaneous Benefits		

Physical Plant

Office & Warehouse (Rental)	30,000
Utilities	2,400
Telephone	3,000
Office Equipment	3,500
Office Supplies	2,000
Auto & Truck (4 vehicles)	24,000

Professional Services

Accounting	4,000
Legal	2,000
Advertising	4,000

Miscellaneous

Dues	1,500
Seminars & Travel	2,500
Entertainment & Gifts	3,000
Uncollected Receivables (1%)	$100,000
TOTAL ANNUAL EXPENSES	**$624,880**

Figure 4.9

In order for this company to stay in business without losses (profit is not yet a factor), not only must all direct construction costs be paid, but an additional $624,880 must be recovered during the year in order to operate the office. Remember that the anticipated volume is $10 million for the year. Office overhead costs, therefore, will be approximately 6.3% of annual volume for this example. The most common method for recovering these costs is to apply this percentage to each job over the course of the year. The percentage may be applied in two ways:

1. Office overhead applied as a percentage of total project costs. This is probably the most commonly used method and is appropriate where material and labor costs are not separated.

2. Office overhead applied as a percentage of labor costs only. This method requires that labor and material costs be estimated separately. As a result, material handling charges are also more easily applied.

The second method described above allows for more precision in the estimate. This method assumes that office expenses are more closely related to labor costs than to total project costs. For example, assume that two companies have the same total annual volume. Company A builds projects that are material-intensive (90% materials, 10% labor). Company B builds projects that are very labor-intensive (10% materials, 90% labor). In order to manage the large labor force, the office (and overhead) expense of Company B will be much greater than that of Company A. As a result, the applicable overhead percentage of B is greater than that of A based on equal annual volumes. For argument's sake, the overhead percentage of *total costs* for Company A is 3%, for Company B, 10%. If company A then gets projects that are more labor-intensive, an allowance of 3% becomes too low and costs will not be recovered. Likewise, if Company B starts to build material-intensive projects, 10% will be too high an overhead figure and bids may no longer be competitive. Office overhead may be more precisely recovered if it is figured as a percentage of labor costs, rather than total costs. In order to do this, a company must determine the ratio of material to labor costs from its historical records. In the example of Figure 4.9, assume that for this company, the ratio is 50/50. Total annual labor costs would be anticipated to be $4,437,500, as calculated below. As a percentage of labor, office overhead will be:

Annual Volume	$10,000,000
Anticipated Overhead (6.3%)	-625,000
Anticipated Profit (5%)	-500,000
Total Bare Costs	8,875,000
Labor (50% of Bare Costs)	$ 4,437,500

$$\frac{\$624,880}{\$4,437,500} = 14\%$$

By applying this overhead percentage (14%) to labor costs, the company is assured of recovering office expenses even if the ratio of material to labor changes significantly.

The estimator must also remember that if volume changes significantly, then the percentage for office overhead should be recalculated for current conditions. The same is true if there are changes in office staff. Remember that salaries are the major portion of office overhead costs. It should be noted that a percentage is commonly applied to material costs, for handling, in addition to and regardless of the method of recovering office overhead costs. This percentage is more easily calculated if material costs are estimated and listed separately.

Profit: Determining a fair and reasonable percentage to be included for profit is not an easy task. This responsibility is usually left to the owner or chief estimator. Experience is crucial in anticipating what profit the market will bear. The economic climate, competition, knowledge of the project, and familiarity with the architect or owner, all affect the way in which profit is determined. Chapter 6 will show one way to mathematically determine profit margin based on historical bidding information. As with all facets of estimating, experience is the key to success.

Contingencies: Like profit, contingencies can be difficult to quantify. Especially appropriate in preliminary budgets, the addition of a contingency is meant to protect the contractor as well as to give the owner a realistic estimate of project costs.

A contingency percentage should be based on the number of "unknowns" in a project. This percentage should be inversely proportional to the amount of planning detail that has been done for the project. If complete plans and specifications are supplied, and the estimate is thorough and precise, then there is little need for a contingency. Figure 4.10, from *Means Building Construction Cost Data*, lists suggested contingency percentages that may be added to an estimate based on the stage of planning and development.

As an estimate is priced and each individual item is rounded up or "padded," this is, in essence, adding a contingency (see Figure 3.4). This method can cause problems, however, because the estimator can never be quite sure of what is the actual cost and what is the "padding," or safety margin for each item. At the summary, the estimator cannot determine exactly how much has been included as a contingency for the whole project. A much more accurate and controllable approach is the precise pricing of the estimate and the addition of one contingency amount at the bottom line.

Bonds: Bonding requirements for a project will be specified in Division 1 General Requirements, and will be included in the construction contract. Various types of bonds may be required. Listed below are a few common types:

Bid Bond. A form of bid security executed by the bidder or principle and by a surety (bonding company) to guarantee that the bidder will enter into a contract within a specified time and furnish any required Performance or Labor and Material Payment bonds.

Completion Bond. Also known as "Construction" or "Contract" bond. The guarantee by a surety that the construction contract will be completed and that it will be clear of all liens and encumbrances.

Labor and Material Payment Bond. The guarantee by a surety to the owner that the contractor will pay for all labor and materials used in the performance of the contract as per the construction documents. The claimants under the bond are those having direct contracts with the contractor or any subcontractor.

Performance and Payment Bond. (1) A guarantee that a contractor will perform a job according to the terms of the contracts. (2) A bond of the contractor in which a surety guarantees to the owner that the work will be performed in accordance with the contract documents. Except where prohibited by statute, the performance bond is frequently combined with the labor and material payment bond. The payment bond guarantees that

01100 | Summary

01107	Professional Consultant	CREW	DAILY OUTPUT	LABOR-HOURS	UNIT	MAT.	LABOR	EQUIP.	TOTAL	TOTAL INCL O&P	
700	1300	4 person crew	A-8	1	32	Day		1,025		1,025	1,575	700
	1500	Aerial surveying, including ground control, minimum fee, 10 acres				Total					5,500	
	1510	100 acres									9,100	
	1550	From existing photography, deduct				↓					1,340	
	1600	2' contours, 10 acres				Acre					440	
	1650	20 acres									300	
	1800	50 acres									90	
	1850	100 acres									80	
	2000	1000 acres									17.01	
	2050	10,000 acres				↓					11.01	
	2150	For 1' contours and										
	2160	dense urban areas, add to above				Acre					40%	
	3000	Inertial guidance system for										
	3010	locating coordinates, rent per day				Ea.					4,000	

01200 | Price & Payment Procedures

01250	Contract Modification Procedures	CREW	DAILY OUTPUT	LABOR-HOURS	UNIT	MAT.	LABOR	EQUIP.	TOTAL	TOTAL INCL O&P	
200	0010	CONTINGENCIES for estimate at conceptual stage				Project					20%	200
	0050	Schematic stage									15%	
	0100	Preliminary working drawing stage (Design Dev.)									10%	
	0150	Final working drawing stage				↓					3%	
300	0010	CREWS For building construction, see How To Use This Book										300
500	0010	JOB CONDITIONS Modifications to total										500
	0020	project cost summaries										
	0100	Economic conditions, favorable, deduct				Project					2%	
	0200	Unfavorable, add									5%	
	0300	Hoisting conditions, favorable, deduct									2%	
	0400	Unfavorable, add									5%	
	0500	General Contractor management, experienced, deduct									2%	
	0600	Inexperienced, add									10%	
	0700	Labor availability, surplus, deduct									1%	
	0800	Shortage, add									10%	
	0900	Material storage area, available, deduct									1%	
	1000	Not available, add									2%	
	1100	Subcontractor availability, surplus, deduct									5%	
	1200	Shortage, add									12%	
	1300	Work space, available, deduct									2%	
	1400	Not available, add				↓					5%	
600	0010	OVERTIME For early completion of projects or where	R01100 -110									600
	0020	labor shortages exist, add to usual labor, up to				Costs		100%				

01255	Cost Indexes	CREW	DAILY OUTPUT	LABOR-HOURS	UNIT	MAT.	LABOR	EQUIP.	TOTAL	TOTAL INCL O&P	
200	0010	CONSTRUCTION COST INDEX (Reference) over 930 zip code locations in										200
	0020	The U.S. and Canada, total bldg cost, min. (Clarksdale, MS)				%					66.20%	
	0050	Average									100%	
	0100	Maximum (New York, NY)				↓					134.60%	
400	0010	HISTORICAL COST INDEXES (Reference) Back to 1952										400

Figure 4.10a

01200 | Price & Payment Procedures

01255	Cost Indexes		CREW	DAILY OUTPUT	LABOR-HOURS	UNIT	2002 BARE COSTS				TOTAL INCL O&P	
							MAT.	LABOR	EQUIP.	TOTAL		
500	0010	**LABOR INDEX** (Reference) For over 930 zip code locations in										500
	0020	the U.S. and Canada, minimum (Clarksdale, MS)				%		33.80%				
	0050	Average				↓		100%				
	0100	Maximum (New York, NY)						163.30%				
600	0010	**MATERIAL INDEX** (Reference) For over 930 zip code locations in										600
	0020	the U.S. and Canada, minimum (Elizabethtown, KY)				%	92.10%					
	0040	Average				↓	100%					
	0060	Maximum (Ketchikan, AK)					143.90%					

01290	Payment Procedures		CREW	DAILY OUTPUT	LABOR-HOURS	UNIT	MAT.	LABOR	EQUIP.	TOTAL	INCL O&P	
800	0010	**TAXES** Sales tax, State, average	R01100 -090			%	4.70%					800
	0050	Maximum					7%					
	0200	Social Security, on first $80,400 of wages	R01100 -100					7.65%				
	0300	Unemployment, MA, combined Federal and State, minimum						2.10%				
	0350	Average				↓		7%				
	0400	Maximum						8%				

01300 | Administrative Requirements

01310	Project Management/Coordination		CREW	DAILY OUTPUT	LABOR-HOURS	UNIT	2002 BARE COSTS				TOTAL INCL O&P	
							MAT.	LABOR	EQUIP.	TOTAL		
150	0010	**PERMITS** Rule of thumb, most cities, minimum					Job					.50%	150
	0100	Maximum					"					2%	
200	0010	**PERFORMANCE BOND** For buildings, minimum	R01100 -080				Job					.60%	200
	0100	Maximum					"					2.50%	
300	0010	**CONSTRUCTION TIME** Requirements	R01100 -020										300
350	0010	**INSURANCE** Builders risk, standard, minimum	R01100 -040				Job					.22%	350
	0050	Maximum					↓					.59%	
	0200	All-risk type, minimum	R01100 -050									.25%	
	0250	Maximum					↓					.62%	
	0400	Contractor's equipment floater, minimum	R01100 -060				Value					.50%	
	0450	Maximum					"					1.50%	
	0600	Public liability, average					Job					1.55%	
	0800	Workers' compensation & employer's liability, average											
	0850	by trade, carpentry, general					Payroll		18%				
	0900	Clerical							.46%				
	0950	Concrete							16.88%				
	1000	Electrical							6.66%				
	1050	Excavation							10.51%				
	1100	Glazing							13.73%				
	1150	Insulation							16.17%				
	1200	Lathing							11.03%				
	1250	Masonry							15.87%				
	1300	Painting & decorating							13.73%				
	1350	Pile driving							24.78%				
	1400	Plastering							15.69%				
	1450	Plumbing							8.25%				
	1500	Roofing							32.28%				
	1550	Sheet metal work (HVAC)							11.61%				
	1600	Steel erection, structural					↓		39.50%				

Figure 4.10b

50

the contractor will pay all subcontractors and material suppliers. Figure 4.11 shows typical average rates for performance bonds for building and roadway construction projects.

Surety Bond. A legal instrument under which one party agrees to answer to another party for the debt, default, or failure to perform of a third party.

The Paperwork   At the pricing stage of the estimate, there is typically a large amount of paperwork that must be assembled, analyzed, and tabulated. Generally, the information contained in this paperwork is covered by the following major categories:

- Quantity Takeoff sheets for all general contractor items (Figure 3.1)
- Material supplier written quotations
- Material supplier telephone quotations (Figure 4. 1)
- Subcontractor written quotations
- Equipment supplier quotations
- Cost Analysis or Consolidated Cost Analysis sheets (Figures 4.12 and 4.13)
- Estimate Summary sheet (Figures 4.14, 4.15, and 4.16)

A system is needed to efficiently handle this mass of paperwork and to ensure that everything will get transferred (and only once) from the Quantity Takeoff to the Cost Analysis sheets. Some general rules for this procedure are:

- Write on only one side of any document where possible.
- Code each sheet with a large division number in a consistent place, preferably near one of the upper corners.
- Use Telephone Quotation forms for uniformity in recording prices received from any source, not only telephone quotes.
- Document the source of every quantity and price.
- Keep each type of document in its pile (Quantities, Material, Subcontractors, Equipment) filed in order by division number.

Performance Bond

This table shows the cost of a Performance Bond for a construction job scheduled to be completed in 12 months. Add 1% of the premium cost per month for jobs requiring more than 12 months to complete. The rates are "standard" rates offered to contractors that the bonding company considers financially sound and capable of doing the work. Preferred rates are offered by some bonding companies based upon financial strength of the contractor. Actual rates vary from contractor to contractor and from bonding company to bonding company. Contractors should prequalify through a bonding agency before submitting a bid on a contract that requires a bond.

Contract Amount	Building Construction Class B Projects			Highways & Bridges					
				Class A New Construction			Class A-1 Highway Resurfacing		
First $ 100,000 bid	$25.00 per M			$15.00 per M			$9.40 per M		
Next 400,000 bid	$ 2,500	plus	$15.00 per M	$ 1,500	plus	$10.00 per M	$ 940	plus	$7.20 per M
Next 2,000,000 bid	8,500	plus	10.00 per M	5,500	plus	7.00 per M	3,820	plus	5.00 per M
Next 2,500,000 bid	28,500	plus	7.50 per M	19,500	plus	5.50 per M	15,820	plus	4.50 per M
Next 2,500,000 bid	47,250	plus	7.00 per M	33,250	plus	5.00 per M	28,320	plus	4.50 per M
Over 7,500,000 bid	64,750	plus	6.00 per M	45,750	plus	4.50 per M	39,570	plus	4.00 per M

Figure 4.11

- Keep the entire estimate in one or more compartmented folders.
- When an item is transferred to the Cost Analysis sheet, check it off.
- If gross subcontractor quantities are known, pencil in the resultant unit prices to serve as a guide for future projects.

All subcontract costs should be properly noted and listed separately. These costs contain the subcontractor's markups, and will be treated differently from other direct costs when the estimator calculates the general contractor's overhead, profit, and contingency allowance.

After all the unit prices, subcontractor prices, and allowances have been entered on the Cost Analysis sheets, the costs are extended. In making the extensions, ignore the cents column and round all totals to the nearest dollar. In a column of figures, the cents will average out and will not be of consequence. Indeed, for budget-type estimates, the extended figures could be rounded to the nearest $10, or even $100, with the loss of only a small amount of precision. Finally, each subdivision is added and the results checked, preferably by someone other than the person doing the extensions.

It is important to check the larger items for order of magnitude errors. If the total subdivision costs are divided by the building area, the resultant square foot cost figures can be used to quickly pinpoint areas that are out of line with expected square foot costs.

The takeoff and pricing method as discussed has been to utilize a Quantity Sheet for the material takeoff (see Figure 3.1), and to transfer the data to a Cost Analysis form for pricing the material, labor, and subcontractor items (see Figure 4.12).

An alternative to this method is a consolidation of the takeoff task and pricing on a single form. An example, the Consolidated Cost Analysis form, is shown in Figure 4.13. The same sequences and recommendations for completing the Quantity Sheet and Cost Analysis form are to be followed when using the Consolidated Cost Analysis form to price the estimate.

The Estimate Summary

When the pricing of all direct costs is complete, the estimator has two choices: all further price changes and adjustments can be made on the Cost Analysis or Consolidated Estimate sheets, *or* total costs for each subdivision can be transferred to an Estimate Summary sheet so that all further price changes, until bid time, will be done on one sheet.

Unless the estimate has a limited number of items, it is recommended that costs be transferred to an Estimate Summary sheet. This step should be double-checked since an error of transposition may easily occur. Pre-printed forms can be useful. A plain columnar form, however, may suffice.

COST
ANALYSIS

SHEET NO. _____

PROJECT _____ ESTIMATE NO. _____

ARCHITECT _____ DATE _____

TAKE OFF BY: QUANTITIES BY: PRICES BY: EXTENSIONS BY: CHECKED BY:

DESCRIPTION	SOURCE/DIMENSIONS			QUANTITY	UNIT	MATERIAL		LABOR		EQUIPMENT		SUB./TOTAL	
						UNIT COST	TOTAL	UNIT COST	TOTAL	UNIT COST	TOTAL	UNIT COST	TOTAL

Figure 4.12

53

CONSOLIDATED ESTIMATE

SHEET NO.

PROJECT CLASSIFICATION ESTIMATE NO.

LOCATION ARCHITECT DATE

TAKE OFF BY PRICES BY EXTENSIONS BY CHECKED BY

QUANTITIES BY

DESCRIPTION	NO.	DIMENSIONS	QUANTITIES		MATERIAL			LABOR			EQUIPMENT			TOTAL	
			UNIT		UNIT	UNIT COST	TOTAL	UNIT COST	TOTAL		UNIT COST	TOTAL		UNIT COST	TOTAL

Figure 4.13

54

If a company has certain standard listings that are used repeatedly, it would save valuable time to have a custom Estimate Summary sheet printed with the items that need to be listed. The Estimate Summary in Figures 4.14 and 4.15 is an example of a commonly used form. The printed MasterFormat division and subdivision headings act as a checklist to ensure that all required costs are included. Figure 4.16 is a Condensed Estimate Summary form. Appropriate column headings or categories for any Estimate Summary form are:

1. Material
2. Labor
3. Equipment
4. Subcontractor
5. Total

As items are listed in the proper columns, each category is added and appropriate markups applied to the total dollar values. Generally, the sum of each column has different percentages added near the end of the estimate for the indirect costs:

1. Sales Tax
2. Overhead
3. Profit
4. Contingencies

ESTIMATE SUMMARY

PROJECT		SHEET NO.	
LOCATION	TOTAL AREA/VOLUME	ESTIMATE NO.	
ARCHITECT	COST PER S.F./C.F.	DATE	
PRICES BY:	EXTENSIONS BY:	NO. OF STORIES	
		CHECKED BY:	

DIV.	DESCRIPTION	MATERIAL	LABOR	EQUIPMENT	SUBCONTRACT	TOTAL
1.0	**General Requirements**					
	Insurance, Taxes, Bonds					
	Equipment & Tools					
	Design, Engineering, Supervision					
2.0	**Site Construction**					
	Site Preparation, Demolition					
	Earthwork					
	Caissons & Piling					
	Drainage & Utilities					
	Paving & Surfacing					
	Site Improvements, Landscaping					
3.0	**Concrete**					
	Formwork					
	Reinforcing Steel & Mesh					
	Foundations					
	Superstructure					
	Precast Concrete					
4.0	**Masonry**					
	Mortar & Reinforcing					
	Brick, Block, Stonework					
5.0	**Metals**					
	Structural Steel					
	Open-Web Joists					
	Steel Deck					
	Misc. & Ornamental Metals					
	Fasteners, Rough Hardware					
6.0	**Wood & Plastics**					
	Rough					
	Finish					
	Architectural Woodwork					
7.0	**Thermal & Moisture Protection**					
	Water & Dampproofing					
	Insulation & Fireproofing					
	Roofing & Sheet Metal					
	Siding					
	Roof Accessories					
8.0	**Doors & Windows**					
	Doors & Frames					
	Windows					
	Finish Hardware					
	Glass & Glazing					
	Curtain Wall & Entrances					
	PAGE TOTALS					

Figure 4.14

DIV.	DESCRIPTION	MATERIAL	LABOR	EQUIPMENT	SUBCONTRACT	TOTAL
	Totals Brought Forward					
9.0	**Finishes**					
	Studs & Furring					
	Lath, Plaster & Stucco					
	Drywall					
	Tile, Terrazzo, Etc.					
	Acoustical Treatment					
	Floor Covering					
	Painting & Wall Coverings					
10.0	**Specialties**					
	Bathroom Accessories					
	Lockers					
	Partitions					
	Signs & Bulletin Boards					
11.0	**Equipment**					
	Appliances					
	Dock					
	Kitchen					
12.0	**Furnishings**					
	Blinds					
	Seating					
13.0	**Special Construction**					
	Integrated Ceilings					
	Pedestal Floors					
	Pre Fab Rooms & Bldgs.					
14.0	**Conveying Systems**					
	Elevators, Escalators					
	Pneumatic Tube Systems					
15.0	**Mechanical**					
	Pipe & Fittings					
	Plumbing Fixtures & Appliances					
	Fire Protection					
	Heating					
	Air Conditioning & Ventilation					
16.0	**Electrical**					
	Raceways					
	Conductors & Grounding					
	Boxes & Wiring Devices					
	Starters, Boards & Switches					
	Transformers & Bus Duct					
	Lighting					
	Special Systems					
	Subtotals					
	Sales Tax %					
	Overhead %					
	Subtotal					
	Profit %					
	Adjustments/Contingency					
	TOTAL BID					

Figure 4.15

CONDENSED ESTIMATE SUMMARY

		SHEET NO.		
PROJECT		ESTIMATE NO.		
LOCATION	TOTAL AREA/VOLUME	DATE		
ARCHITECT	COST PER S.F./C.F.	NO. OF STORIES		
PRICES BY:	EXTENSIONS BY:	CHECKED BY:		

DIV.	DESCRIPTION	MATERIAL	LABOR	EQUIPMENT	SUBCONTRACT	TOTAL
1.0	General Requirements					
2.0	Site Construction					
3.0	Concrete					
4.0	Masonry					
5.0	Metals					
6.0	Wood & Plastics					
7.0	Thermal & Moisture Protection					
8.0	Doors & Windows					
9.0	Finishes					
10.0	Specialties					
11.0	Equipment					
12.0	Furnishings					
13.0	Special Construction					
14.0	Conveying Systems					
15.0	Mechanical					
16.0	Electrical					
	Subtotals					
	Sales Tax %					
	Overhead %					
	Subtotal					
	Profit %					
	Contingency %					
	Adjustments					
	TOTAL BID					

Figure 4.16

Chapter 5

Pre-Bid Scheduling

Chapter 5

Pre-Bid Scheduling

The need for planning and scheduling is clear once the contract is signed and work commences on the project. However, some scheduling is also important during the bidding stage for the following reasons:

1. To determine if the project can be completed in the allotted or specified time.
2. To determine the time requirements for general conditions items, such as supervision, field office, and watchman.
3. To determine when the building will be enclosed in order to anticipate possible temporary heat and power requirements.

The schedule produced prior to bidding may be a simple bar chart or network diagram that includes overall quantities, probable delivery times, and available manpower. Network scheduling methods, such as the Critical Path Method (CPM) and the Precedence Chart, simplify prebid scheduling because they do not require scaled diagrams.

In the CPM Diagram, the activity is represented by the arrow. The Precedence Diagram, on the other hand, shows the activity in a node with arrows used to denote precedence relationships between the activities. The precedence arrows may be used in different configurations to represent the sequential relationships between activities. Examples of CPM and Precedence Diagrams are shown in Figures 5.1 and 5.2, respectively. In both systems, duration times are indicated along each path. The sequence of activities requiring the most total time represents the shortest possible time in which those activities may be completed.

For example, in Figure 5.1, activities A, B, and C require 20 successive days for completion before activity G can begin. Activity paths for D and E, and for F, are shorter and can be easily fit into the 20-day sequence. Therefore, this 20-day sequence is the shortest possible time for completion of these activities before activity G can begin.

Past experience or a prepared rough schedule may suggest that the allotted time specified in the bidding documents is insufficient to complete the required work. In such cases, a more comprehensive schedule should be produced prior to bidding; this schedule should include the calculations for added overtime or premium time work costs required to meet the completion date.

The following simple example is of a 100' × 200', four-story flat plate building with a basement and 25' square bays. It is based on the data listed below.

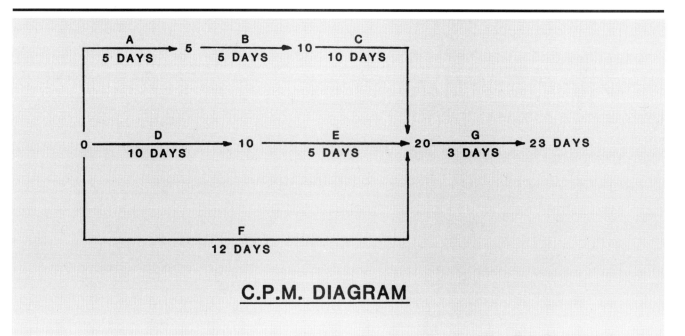

C.P.M. DIAGRAM

Figure 5.1

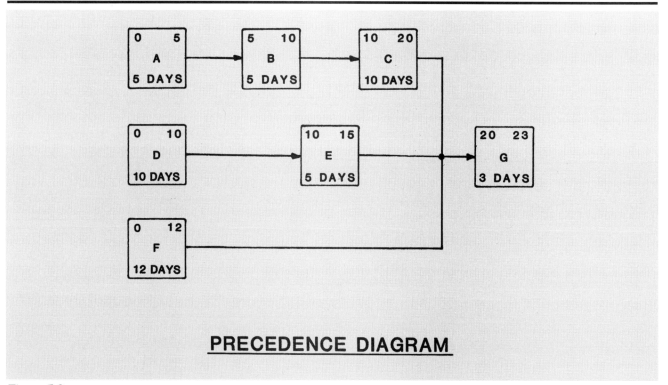

PRECEDENCE DIAGRAM

Figure 5.2

Description	Quantity Output	Duration
Earthwork:		
Bulk Excavation	7,228 C.Y. / 1,000 C.Y./Day	7 Days
Footing Excavation	1,300 C.Y. / 200 C.Y./Day	7 Days
Backfill Bulk	1,300 C.Y. / 240 C.Y./Day	5 Days
Backfill Footings	770 C.Y. / 240 C.Y./Day	3 Days
Forms:		
Spread and Continuous Footings	5,894 S.F. / 400 S.F./Day	15 Days
Walls and Pilasters	10,176 S.F. / 600 S.F./Day	17 Days

Figure 5.3 shows a preliminary schedule for excavation and the foundation. Note that a one day duration time has been added to the estimated time to complete the forming operation for both the footings and walls. The established duration time to excavate, form, reinforce, and place the concrete for the foundation is 32 days, not including a factor for weather.

Our next major category on the schedule is the poured-in-place concrete structure. Because of the large amount of formwork required, a rough

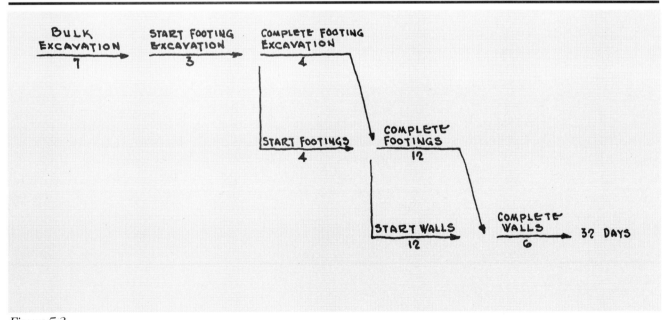

Figure 5.3

schedule may help to produce a more reliable quotation. The most efficient and economical quantity of the forms and number of uses can be determined. Assume the specifications require seven days curing time for the concrete slabs before the forms may be stripped and the slabs reshored. Also assume a twelve man carpenter crew is available with additional laborers to support the carpenters. Durations for each task are developed from derived quantities, productivity rates, and assumed crew sizes. The calculations are shown in Figure 5.4.

A preliminary schedule may be developed to determine the minimum time requirements for forming, reinforcing and placing the concrete. The schedule will also help determine how many sets of forms are required for the most efficient use in the specified curing time. Figure 5.5 is an example of this kind of preliminary schedule.

Repeating the cycle for the roof slab and assuming that one set of floor forms (enough for one floor of the building) is available, the structure would take approximately sixty-eight working days to complete with no allowance for weather. A *quick*, rough estimate of time to erect the same structure may be derived by using the formula below. Exercise care when using the quick method to estimate the duration times of activities that may be overlapped or done concurrently. Crew output is the product of the output per man per day times the number of men.

$$\frac{\text{Form Area}}{\text{Crew Output}} + \frac{\text{Final Reinf.}}{\text{Duration}} + \frac{\text{Final Conc.}}{\text{Duration}} = \text{Total Duration}$$

$$\frac{25,060 \text{ SFCA}}{1,080 \text{ S.F./Day}} + \frac{75,200 \text{ SFCA}}{1,680 \text{ S.F./Day}} + 2 \text{ Days} + 1 \text{ Day} = 71 \text{ Days}$$

$$\text{(first floor)} \qquad \text{(upper floors)}$$

Only the final reinforcing and concrete durations are included because previous durations are not limiting activities. If the same formula is applied to the foundations, approximate total duration time can be determined as follows:

$$\frac{5,894 \text{ SFCA}}{400 \text{ S.F./Day}} + \frac{10,176 \text{ SFCA}}{600 \text{ S.F./Day}} + 1 \text{ Day} + 1 \text{ Day} = 34 \text{ Days}$$

To complete the schedule, probable duration times are derived from the estimated quantities for the remaining divisions of the estimate. The precedence format works well for the preliminary schedules because of the advantage of using arrow placement to show precedence relationships between activities. The relationships between activity nodes (rectangles) can be shown with the arrow configurations as in Figure 5.6.

The computations used in precedence schedules can become complicated. Finish-to-start connectors with lag times are convenient and make calculations easier. The polygons on the diagram are used to show logical delivery times for materials, allowing adequate turn-around time for shop drawings.

Once the preliminary schedule has been developed using a no-scale network diagram depicting activity dependencies, a simple bar chart for the project can be drawn to condense the information.

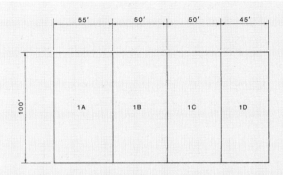

Placement Plan

Duration Derivation

Columns Each

Forms $\dfrac{90 \text{ sfca Ea}}{60 \text{ sfca/day x 12 Carp}} = .125$ days Ea (1st use)

Reinforcing $\dfrac{.13 \text{ tons Ea}}{.375 \text{ tons/day x 4 Rdmn}} = .087$ days Ea

Concrete $\dfrac{1.67 \text{ cy Ea}}{11.5 \text{ cy/day x 8 men}} = .018$ days Ea

Slab Forms 1st Floor

"1A"   Slab Form   5500 sfca
        Edge Form   168 sfca
        _____
                      5668 sfca
        $\dfrac{5668 \text{ sfca}}{90 \text{ sfca/day x 12 Carp}} = 5$ days

"1B & 1C" Slab Form 5000 sfca
           Edge Form   158 sfca
           _____
                       5158 sfca
        $\dfrac{5158 \text{ sfca}}{90 \text{ sfca/day x 12 Carp}} = 4.78$ days

        Columns 10 Ea x .125 days Ea = $\dfrac{1.25 \text{ days}}{6.0 \text{ days}}$

"1D"   Slab Form   4500 sfca
        Edge Form   150 sfca
        _____
                      4650 sfca
        $\dfrac{4650 \text{ sfca}}{90 \text{ sfca/day x 12 Carp}} = 4.3$ days

        Columns 10 Ea x .125 days Ea = $\dfrac{1.25 \text{ days}}{6.0 \text{ days}}$

Reinforcing @ 2.8 psf

"1A" $\dfrac{5500 \text{ sf x 2.8 psf}}{2000}$ $\dfrac{7.7 \text{ tons}}{.7 \text{ tons/day x 5 Rdmn}} = 2$ days

"1B & 1C" $\dfrac{5000 \text{ sf x 2.8 psf}}{2000}$ $\dfrac{7.0 \text{ tons}}{.7 \text{ tons/day = 5 Rdmn}} = 2$ days

"1D" $\dfrac{4500 \text{ sf x 2.8 psf}}{2000}$ $\dfrac{6.3 \text{ tons}}{.7 \text{ tons/day x 5 Rdmn}} = 2$ days

Concrete (Average)

$\dfrac{150 \text{ cy}}{16.5 \text{ cy/day x 9 men}} = 1$ day

2nd & 3rd Floors & Roof

Forms "A" $\dfrac{5668 \text{ sfca}}{140 \text{ sfca/day x 12 Carp}} = 4$ days

"B & C" $\dfrac{5158 \text{ sfca}}{140 \text{ sfca/day x 12 Carp}} = 3$ days

"D" $\dfrac{4650 \text{ sfca}}{140 \text{ sfca/day x 12 Carp}} = 3$ days

Reinforcing and concrete placement durations are the same as the 1st floor.

Figure 5.4

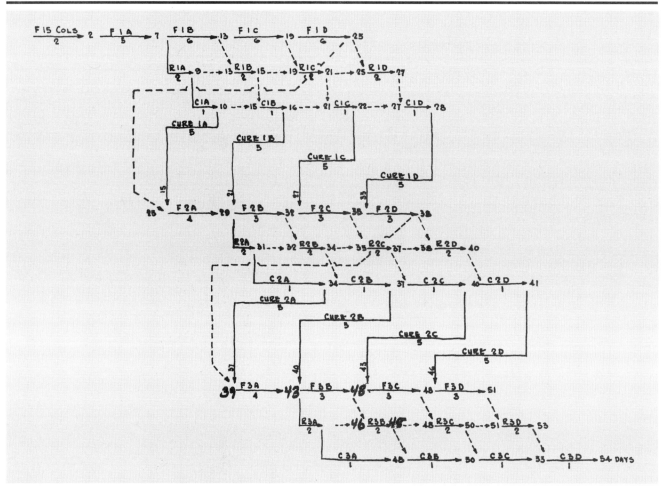

Figure 5.5

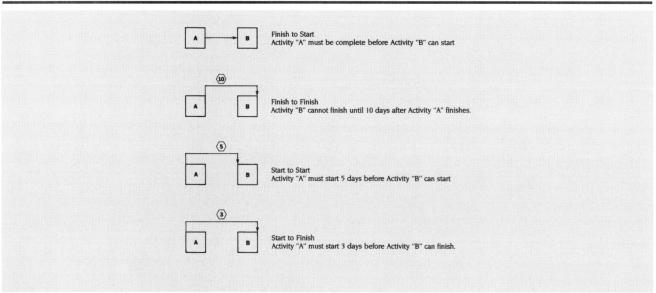

Finish to Start
Activity "A" must be complete before Activity "B" can start

Finish to Finish
Activity "B" cannot finish until 10 days after Activity "A" finishes.

Start to Start
Activity "A" must start 5 days before Activity "B" can start

Start to Finish
Activity "A" must start 3 days before Activity "B" can finish.

Figure 5.6

Chapter 6

Upon Completing the Estimate

Chapter 6

Upon Completing the Estimate

Bidding Strategies

The goal of most contractors is to make as much money as possible on each job, but more importantly, to maximize return on investment on an annual basis. This means making more money by taking fewer jobs at a higher profit.

One measure of successful bidding is how much money is "left on the table," the difference between the low bid and next lowest bid. The contractor who consistently takes jobs by a wide margin below the next bidder is obviously not making as much money as possible. Information on competitive public bidding is accessible. Thus, the amount of money left on the table is easily determined and can be the basis for fine-tuning a future bidding strategy.

Since a contractor cannot physically bid every type of job in a geographic area, a selection process must determine which projects should be bid. This selective process should begin with an analysis of the strengths and weaknesses of the contractor. The following items that must be considered as objectively as possible are:

- Individual strengths of the company's top management.
- Management experience with the type of construction involved, from top management to project superintendents.
- Cost records adequate for the appropriate type of construction.
- Bonding capability and capacity.
- Size of projects with which the company is "comfortable."
- Geographic area that can be managed effectively.
- Unusual corporate assets such as:
  - Specialized equipment availability
  - Reliable and timely cost control systems
  - Strong balance sheet

Market Analysis

Most contractors tend to concentrate on one, or a few, fairly specialized types of construction. From time to time, the company should step back and examine the total picture of the industry they are serving. During this process, the following items should be carefully analyzed.

- Historical trend of the market segment.
- Expected future trend of the market segment.
- Geographic expectations of the market segment.
- Historical and expected competition among other builders.
- Risk involved in the particular market segment.

- Typical size of projects in this market.
- Expected return on investment from the market segment.

If several of these areas are experiencing a downturn, then it might be appropriate to examine an alternate market.

Certain steps should be taken to develop a bid strategy for a particular market. The first is to obtain the bid results of jobs in the prospective geographic area. These results should be set up on a tabular basis. This is fairly easy to do in public jobs since the bid results are normally published (or at least available) from the agency responsible for the project. In private work this step is very difficult, since the bid results are not normally divulged by the owner.

For example, assume a public market where all bid prices and the total number of bidders is known. For each "type" of market sector, create a chart showing the percentage left on the table versus the total number of bidders. When the median figure (percent left on the table) for each number of bidders is connected with a smooth curve, the usual shape of the curve is shown in Figure 6.1.

The exact shape and magnitude of the amounts left on the table will depend on how much risk is involved with that type of work. If the percentages left on the table are high, then the work can be assumed to be very risky; if the percentages are low, the work is probably not too risky.

Bidding Analysis

If a company has been bidding in a particular market, certain information should be collected and recorded as a basis for a bidding analysis. The percentage left on the table should be tabulated, along with the number of bidders for the projects in that market on which the company was the low bidder. By probability, half the bids should be above the median line and half below. (See Figure 6.1.) If more than half are below the line, the company is doing well; if more than half are above, the bidding strategy should be examined. Once the bidding track record for the company has been established, the next step is to reduce the historical percentage left on the table. One method is to create a chart showing, for instance, the last ten jobs on which the company was low bidder and the dollar spread between the low

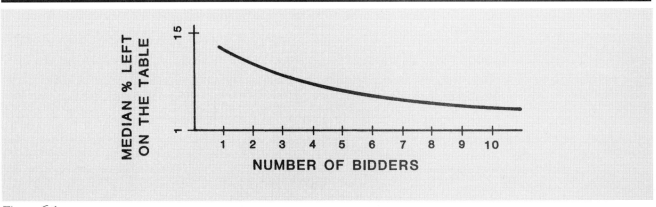

Figure 6.1

and second lowest bid. Next, rank the percentage differences from one to ten (one being the smallest and ten being the largest left on the table). An example is shown in Figure 6.2.

The company's "costs" ($17,170,000) are derived from the company's low bid ($18,887,000) assuming a 10% profit ($1,717,000). The "second bid" is the next lowest bid. The "difference" is the dollar amount between the low bid and the second bid. The differences are then ranked based on the percentage of job "costs" left on the table for each.

$$\text{Median \% Difference} = \frac{5.73 + 6.44}{2} = 6.09\%$$

From Figure 6.2, the median percentage left on the table is 6.09%. To maximize the potential returns on a series of competitive bids, a useful formula for pricing profit is needed. The following formula has proven effective.

$$\text{Normal Profit \%} + \frac{\text{Median \% Difference}}{2} = \text{Adjusted Profit \%}$$

$$10.00 + \frac{6.09}{2} = 13.05\%$$

Now apply this adjusted profit % to the same list of ten jobs as shown in Figure 6.3. Note that the job "costs" remain the same, but the low bids have been revised. Compare the bottom line results of Figure 6.2 to those of Figure 6.3 based on the two profit margins, 10% and 13.05%, respectively.

Job No.	"Cost"	Low Bid	Second Bid	Difference	% Diff.	% Rank	Profit (Assumed at 10%)
1	$ 918,000	$ 1,009,800	$1,095,000	$ 85,200	9.28	10	$ 91,800
2	1,955,000	2,150,500	2,238,000	87,500	4.48	3	195,500
3	2,141,000	2,355,100	2,493,000	137,900	6.44	6	214,100
4	1,005,000	1,105,500	1,118,000	12,500	1.24	1	100,500
5	2,391,000	2,630,100	2,805,000	174,900	7.31	8	239,100
6	2,782,000	3,060,200	3,188,000	127,800	4.59	4	278,200
7	1,093,000	1,202,300	1,282,000	79,700	7.29	7	109,300
8	832,000	915,200	926,000	10,800	1.30	2	83,200
9	2,372,000	2,609,200	2,745,000	135,800	5.73	5	237,200
10	1,681,000	1,849,100	2,005,000	155,900	9.27	9	168,100
	$17,170,000	$18,887,000		$1,008,000			$1,717,000 = 10% of Cost

Figure 6.2

Total volume *drops* from $18,887,000 to $17,333,900.

Net profits *rise* from $1,717,000 to $2,000,900.

Profits rise while volume drops. If the original volume is maintained or even increased, profits would rise even faster. Note how this occurs. By determining a reasonable increase in profit margin, the company has, in effect, raised all bids. By doing so, the company loses two jobs to the second bidder (jobs 4 and 8 in Figure 6.3).

A positive effect of this volume loss is reduced exposure to risk. Since the profit margin is higher, the remaining eight jobs collectively produce more profit than the ten jobs based on the original, lower profit margin. From where did this money come? The money "left on the table" has been reduced from $1,008,000 to $517,100. The whole purpose is to systematically lessen the dollar amount difference between the low bid and the second low bid. This is a hypothetical approach based upon a number of assumptions:

- Bidding must be done within the same market in which data for the analysis was gathered.
- Economic conditions should be stable from the time the data is gathered until the analysis is used in bidding. If conditions change, use of such an analysis should be reviewed.
- Each contractor must make roughly the same number of bidding mistakes. For higher numbers of jobs in the sample, this requirement becomes more probable.
- The company must bid additional jobs if the total annual volume is to be maintained or increased. Likewise, if net total profit margin is to remain constant, fewer jobs need be bid.

Job No.	Company's "Cost"	Revised Low Bid	Second Bid	Adj. Diff.	Profit [10% + 3.05%]		Total
1	$ 918,000	$ 1,037,800	$1,095,000	$ 57,200	$ 91,800 +	$28,000	$ 119,800
2	1,955,000	2,210,100	2,238,000	27,900	195,500 +	59,600	225,100
3	2,141,000	2,420,400	2,493,000	72,600	214,100 +	65,300	279,400
4	(1,005,000)	(1,136,100)	1,118,000 (L)	—	100,500 +	30,600	0
5	2,391,000	2,703,000	2,805,000	102,000	239,100 +	72,900	312,000
6	2,782,000	3,145,100	3,188,000	42,900	278,200 +	84,900	363,100
7	1,093,000	1,235,600	1,282,000	46,400	109,300 +	33,300	142,600
8	(832,000)	(940,600)	926,000 (L)	—	83,200 +	25,400	0
9	2,372,000	2,681,500	2,745,000	63,500	237,200 +	72,300	309,500
10	1,681,000	1,900,400	2,005,000	104,600	168,100 +	51,300	219,400
	$15,333,000	$17,333,900		$517,100			$2,000,900

Figure 6.3

Even though the accuracy of this strategy is based on these criteria, the concept is valid and can be applied, with appropriate and reasonable judgment, to many bidding situations.

Cost Control and Analysis

An internal accounting system should be used by the contractor to logically allocate and track the gathered costs of a construction project. With this information, a cost analysis can be made about each cost center. The cost centers of a project are the major items of construction (e.g., concrete) that can be subdivided by component. This subdivision should coincide with the system and methods of the quantity takeoff. The major purposes of cost control and analysis are:

- To provide management with a system to monitor costs and progress.
- To provide cost feedback to the estimator (s).
- To determine the costs of change orders.
- To be used as a basis for submitting payment requisitions to the owner or his representative.
- To manage cash flow.

A cost control system should be established that is uniform throughout the company and from job to job. The various costs are then consistently allocated. The cost control system might be simplified with a code that provides a different designation for each part of a component cost. The following information should be recorded for each component cost.

- Labor charges in dollars and labor-hours, summarized from weekly time cards, are distributed by code.
- Equipment rental costs are derived from purchase orders or from weekly charges from an equipment company.
- Material charges are determined from purchase orders.
- Appropriate subcontractor charges are allocated.
- Job overhead items may be listed separately or by component.
- Quantities completed to date must also be recorded in order to determine unit costs.

Each component of costs—labor, materials, and equipment—is now calculated on a unit basis by dividing the quantity to date (percentage complete) into the component cost to date. This procedure establishes the actual unit costs to date. The remaining quantities of each component to be completed should be estimated at a unit cost approximating the costs to date. The actual costs to date and the predicted costs are added together to represent the anticipated costs at the end of the project. Typical forms that may be used to develop a cost control system are shown in Figures 6.4 to 6.7.

The analysis of cost centers serves as a useful management tool, providing information on a current basis. Immediate attention is attracted to any center that is operating at a loss. Management can concentrate on this item in an attempt to make it profitable or to minimize the expected loss.

The estimating department can use the unit costs developed in the field as background information for future bidding purposes. Particularly useful are unit labor costs and unit labor-hours (productivity) for the separate components. Current unit labor-hours and labor costs should be integrated into the accumulated historical data.

Frequently, items are added to or deleted from the contract. Accurate cost records are an excellent basis for determining cost changes that result from change orders and requests for additional work.

PERCENTAGE COMPLETE ANALYSIS

PAGE _____

PROJECT _____ DATE _____

ARCHITECT _____ BY _____ FROM _____ TO _____

NO.	DESCRIPTION	ACTUAL OR ESTIMATED	TOTAL PROJECT	THIS PERIOD		PERCENT TOTAL TO DATE										
				QUANTITY	%	QUANTITY	10	20	30	40	50	60	70	80	90	100
		ACTUAL														
		ESTIMATED														
		ACTUAL														
		ESTIMATED														
		ACTUAL														
		ESTIMATED														
		ACTUAL														
		ESTIMATED														
		ACTUAL														
		ESTIMATED														
		ACTUAL														
		ESTIMATED														
		ACTUAL														
		ESTIMATED														
		ACTUAL														
		ESTIMATED														
		ACTUAL														
		ESTIMATED														
		ACTUAL														
		ESTIMATED														
		ACTUAL														
		ESTIMATED														
		ACTUAL														
		ESTIMATED														
		ACTUAL														
		ESTIMATED														
		ACTUAL														
		ESTIMATED														
		ACTUAL														
		ESTIMATED														
		ACTUAL														
		ESTIMATED														
		ACTUAL														
		ESTIMATED														
		ACTUAL														
		ESTIMATED														
		ACTUAL														
		ESTIMATED														
		ACTUAL														
		ESTIMATED														

Figure 6.4

74

JOB
PROGRESS REPORT

SHEET _____ OF _____

PROJECT   JOB NO.

LOCATION   YEARS

WORK ITEM	QUANTITY, $, OR %	BEGINNING BALANCE	MONTHS												ENDING BALANCE
	DATE														
	YEAR														

Figure 6.5

**MATERIAL
COST RECORD**

SHEET NO.

DATE FROM

PROJECT

DATE TO

LOCATION

BY

DATE	NUMBER	VENDOR/DESCRIPTION	QTY.	UNIT PRICE						QTY.	UNIT PRICE						QTY.	UNIT PRICE				

Figure 6.6

LABOR COST RECORD

SHEET NO.

DATE FROM:

PROJECT

DATE TO:

LOCATION

BY:

DATE	CHARGE NO.	DESCRIPTION	HOURS	RATE	AMOUNT	HOURS	RATE	AMOUNT	HOURS	RATE	AMOUNT

Figure 6.7

Cost records require the determination of completed quantities in order to calculate unit costs. These calculations are used to determine the percent completion of each cost center. This percentage is used to calculate the billing of completed items for payment requisitions.

A cost system is only as good as the people responsible for coding and recording the required information. Simplicity is the key word. Do not try to break down the code into very small items unless there is a specific need. Continuous updating of costs is important so that operations which are not in control can be immediately brought to the attention of management.

Productivity and Efficiency

When using a cost control system such as the one described above, the derived unit costs should reflect standard practices. Productivity should be based on a five day, eight hour per day (during daylight hours) work week unless a company's requirements are particularly and normally unique. Installation costs should be derived using normal minimum crew sizes, under normal weather conditions, during the normal construction season. Costs and productivity should also be based on familiar types of construction.

All unusual costs incurred or expected should be separately recorded for each component of work. For example, an overtime situation might occur on every job and in the same proportion. In this case, it would make sense to carry the unit price adjusted for the added cost of premium time. Likewise, unusual weather delays, strike activity, or owner/architect delays should have separate, identifiable cost contributions; these are applied as isolated costs to the activities affected by the delays. This procedure serves two purposes:

- To identify and separate the cost contribution of the delay so that job estimates will not automatically include an allowance for these "non-typical" delays, and
- To serve as a basis for an extra compensation claim and or as justification for reasonable extension of the job.

The use of long-term overtime on almost any construction job is counterproductive; that is, the longer the period of overtime, the lower the actual production rate. There have been numerous studies conducted which come up with slightly different numbers, but all have the same conclusion.

As illustrated in Figure 6.8, there can be a difference between the actual payroll cost per hour and the *effective* cost per hour for overtime work. This is due to the reduced production efficiency with the increase in weekly hours beyond 40. This difference between actual and effective cost is for overtime over a prolonged period. Short-term overtime does not result in as great a reduction in efficiency, and in such cases, cost premiums would approach the payroll costs rather than the effective hourly costs listed in Figure 6.8. As the total hours per week are increased on a regular basis, more time is lost by absenteeism and the accident rate increases.

As an example, assume a project where workers are working 6 days a week, 10 hours per day. From Figure 6.8 (based on productivity studies), the actual productive hours are 51.1 hours. This represents a theoretical production efficiency of 51.1/60 or 85.2%.

Depending upon the locale, overtime work is paid at time and a half or double time. In both cases, the overall *actual* payroll cost (including regular and overtime hours) is determined as follows:

For time and a half:

$$\frac{40 \text{ reg. hrs.} + (20 \text{ overtime hrs.} \times 1.5)}{60 \text{ hrs.}} = 1.167$$

Based on 60 hours, the payroll cost per hour will be 116.7% of the normal rate at 40 hours per week.

For double time:

$$\frac{40 \text{ reg. hrs.} + (20 \text{ overtime hrs.} \times 2)}{60 \text{ hrs.}} = 1.33$$

Payroll cost will be 133% of the normal rate.

However, because the actual productive hours (and thus production efficiency) are reduced to 51.1 hours, the *effective* cost of overtime is:

For time and a half:

$$\frac{40 \text{ reg. hrs.} + (20 \text{ overtime hrs.} \times 1.5)}{51.1 \text{ hrs.}} = 1.37$$

Payroll cost will be 137% of the normal rate.

For double time:

$$\frac{40 \text{ reg. hrs.} + (20 \text{ overtime hrs.} \times 2)}{51.1 \text{ hrs.}} = 1.566$$

Payroll cost will be 156.6% of the normal rate.

Efficiency and Cost Effects of Prolonged Overtime Work								
Days per Week	Hours per Day	Total Hours Worked	Actual Productive Hours	Production Efficiency	Payroll Cost per Hour Overtime After 40 Hrs.		Effective Cost per Hour Overtime After 40 Hrs.	
					@ 1-1/2 times	@ 2 times	@ 1-1/2 times	@ 2 times
	8	40	40.0	100.0%	100.0%	100.0%	100.0%	100.0%
	9	45	43.4	96.5	105.6	111.1	109.4	115.2
5	10	50	46.5	93.0	110.0	120.0	118.3	129.0
	11	55	49.2	89.5	113.6	127.3	127.0	142.3
	12	60	51.6	86.0	116.7	133.3	135.7	155.0
	8	48	46.1	96.0	108.3	116.7	112.8	121.5
	9	54	48.9	90.6	113.0	125.9	124.7	139.1
6	10	60	51.1	85.2	116.7	133.3	137.0	156.6
	11	66	52.7	79.8	119.7	139.4	149.9	174.6
	12	72	53.6	74.4	122.2	144.4	164.2	194.0
	8	56	48.8	87.1	114.3	128.6	131.1	147.5
	9	63	52.2	82.8	118.3	136.5	142.7	164.8
7	10	70	55.0	78.5	121.4	142.9	154.5	181.8
	11	77	57.1	74.2	124.0	148.1	167.3	199.6
	12	84	58.7	69.9	126.2	152.4	180.6	218.1

Figure 6.8

Thus, when figuring overtime, the actual cost per unit of work will be higher than the apparent overtime payroll dollar increase, due to the reduced productivity of the longer work week. These calculations are true only for those pay costs which are determined by hours worked. Costs which are applied weekly or monthly, such as some fringe benefits, will not be similarly affected.

Retainage and Cash Flow

The majority of construction projects have some percentage of retainage held back by the owner until the job is complete. This retainage can range from 5% to as high as 15% or 20% in unusual cases. The most typical retainage is 10%. Since the profit on a given job may be less than the amount of withheld retainage, the contractor must wait longer before a positive cash flow is achieved than if there were no retainage.

Figures 6.9 and 6.10 are graphic and tabular representations of the projected cash flow for a small project. With this kind of projection, the contractor is able to anticipate cash needs throughout the course of the job. Note that at the eleventh of May, before the second payment is received, the contractor has paid out about $25,000 more than has been received. This is the maximum amount of cash (on hand or financed) that is required for the whole project. At this stage of planning, the contractor can determine if there will be adequate cash available or if a loan is needed. In the latter case, the expense of interest could be anticipated and included in the estimate. On larger projects, the projection of cash flow becomes crucial, because unexpected interest expense can quickly erode profit margin.

The General Conditions section of the specifications usually explains the responsibilities of both the owner and the contractor with regard to billing and payments. Even the best planning and projections are contingent upon the owner paying requisitions as anticipated. There is an almost unavoidable adversary relationship between contractor and owner regarding payment during the construction process. However, it is in the best interest of the owner that the contractor be solvent so that delays, complications, and difficulties can be avoided prior to final completion of the project. The interest of both parties is best served if information is shared and communication is open. Both are working toward the same goal: the timely and successful completion of the project.

Life Cycle Costs

Life cycle costing is a valuable method of evaluating the total costs of an economic unit during its entire life. Regardless of whether the unit is a piece of excavating machinery or a manufacturing building, life cycle costing gives the owner an opportunity to took at the economic consequences of a decision. Today, the initial cost of a unit is often not the most important; the operation and maintenance costs of some building types far exceed the initial outlay. Hospitals, for example, may have operating costs within a three-year period that exceed the original construction costs.

Estimators are in the business of initial costs. But what about the other costs that affect any owner: taxes, fuel, inflation, borrowing, estimated salvage at the end of the facility's lifespan, and expected repair or remodeling costs? These costs that may occur at a later date need to be evaluated in advance. The thread that ties all of these costs together is the time value of money. The value of money today is quite different from what it will be tomorrow. $1,000 placed in a savings bank at 5% interest will, in four years, have increased to $1,215.50. Conversely, if $1,000 is needed four years from now, then $822.70 should be

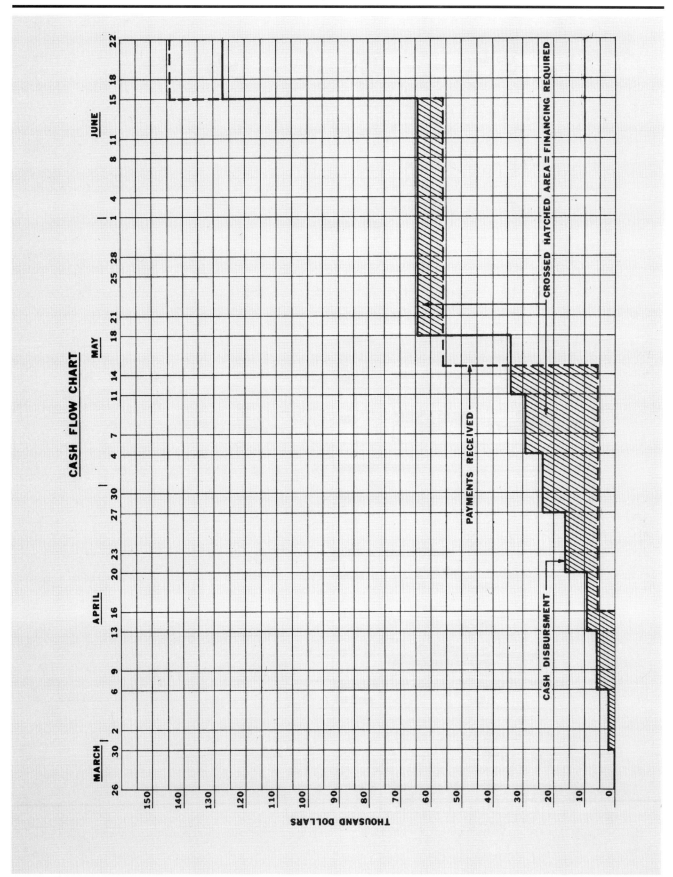

Figure 6.9

Date	Payroll Incl. Taxes (1)	Workers' Comp. (2)	Monthly Billing (3)	Retainage (4)	Subs Billing or Payment (5)	Retainage (6)	Materials Incl. Taxes (7)	Equip. Incl. Taxes (8)	Payments (9)	Accumulated Costs (10)
3-20 Payroll	11926	1148								11926 −
3-30 Monthly Billing			44558	506	708	777				5172 −
4-6 Payroll	3246	255								9369 −
4-13 Payroll	4197	311								40611 −
4-16 Payment									4558	5519 −
4-20 Pay Subs					708					6504 −
4-20 Pay Materials							515			6999 −
4-20 Pay Equipment								965		10826 −
4-20 Payroll	3827	275								15351 −
4-27 Payroll	4525	335								20876 −
4-30 Monthly Billing			51550	5704	5245	582				24407 −
5-4 Payroll	5525	508								26923 +
5-11 Payroll	3531	370								24310 +
5-15 Payment									51350	19065 +
5-18 Payroll	2613	243								3053 −
5-18 Pay Subs					5245					5832 −
5-18 Pay Materials							22118			
5-18 Pay Equipment								779		
6-1 Final Billing			86683	9896	48915	5435				85051 +
6-15 Payment					48915				88803	36136 +
Pay Subs										2545 +
Pay Materials							12685			22715 +
Pay Equipment								736		
	$ 29590	$ 2441	$ 144771	$ 16086	$ 54860	$ 6094	$ 35318	$ 2482	$ 144771	

General Conditions

Permit	150 $3.00	Cash 22715
Supervision – Carpenter Foreman 35 days @ $170		Retainage 16086
Temp. Power + Water		Pay Workers Comp 38799 / 2441
Temp. Office + Storage 6 wks. @ $30		Pay Subs Retainage 36358 / 6094
Clean Up Labor 7 days @ $1.22	7024	30264
		7024
		Gen. Conditions 232240

Figure 6.10

82

placed in an account today. Another way of saying the same thing is that at 5% interest, the value of $1,000 four years from now is only worth $822.70 today.

Using interest and time, future costs are equated to the present by means of a present worth formula. Standard texts in engineering economics have outlined different methods for handling interest and time. A present worth evaluation could be used, or all costs might be converted into an equivalent, uniform annual cost method.

As an example, assume that an excavating machine is purchased for $20,000. Hourly gas and oil charges will be $6.20. A major overhaul of the engine will be needed in three years at a cost of $4,000. After six years, the trade-in value will be $1,500. What is the total *present worth* of the machine over its six-year life if money is borrowed at 12%?

First Cost of the Machine:

Purchase Price =	$20,000.00
Total present worth of Operating and Maintenance @ $6.20/hour = 1,500 hours × $6.20 = $9,300/year × 4.111 (present worth of a uniform series of payment for six years) =	38,232.30
Present worth of a major overhaul three years from now = $4,000 × .7118 (time/value factor) =	2,847.20
Present worth of the salvage at the end of six years = $1,500 × .5066 (time/value factor) =	+ 759.90
The Total Present Worth of the Tractor	$61,839.40

If we consider the uniform annual costs of the machine, then (at 12% interest) the following happens:

The uniform annual cost of the initial investment ($20,000) × .24323 (capital recovery factor) =	$4,864.60
The uniform maintenance and operating costs =	9,300.00
The uniform cost of motor repair in the third year is the present worth times the capital recovery factor, $2,847 × .24323 =	692.52
The equivalent uniform cost of the salvage, $1,500 × .12323 (sinking fund factor) =	+ 184.85
Total Equivalent Uniform Annual Costs	$15,041.97

This is the same as if the total present worth was multiplied by the capital recovery factor: $61,839.40 × .24323 = $15,041.20. If the machine is used 1,500 hours per year, then the hourly base operating cost is approximately $10 per hour.

Another benefit of life cycle costing is having the ability to compare the equivalent uniform costs between two different machines. This comparison can help with the decision to rent or purchase.

Life cycle costs are not always determined easily. The initial investment and the operating costs can be estimated, and maintenance costs can be determined from past experience with similar machines; but anticipating a motor overhaul precisely or knowing the projected salvage price is difficult. Interest rates and time factors are critical in determining life cycle costs. By varying the time intervals and/or the interest rate when comparing the alternatives, a good investment might appear poor, or vice versa. It is therefore essential that realistic, accurate rates be used.

Using Means Building Construction Cost Data

Using Means Building Construction Cost Data

Users of *Means Building Construction Cost Data* are chiefly interested in obtaining quick, reasonable, average prices for building construction items. This is the primary purpose of the annual book—to eliminate guesswork when pricing unknowns. Many people use the cost data, whether for bids or verification of quotations or budgets, without being fully aware of how the prices are obtained and derived. Without this knowledge, this vast resource is not being used to its fullest advantage. In addition to the basic cost data, the book also contains a wealth of information to aid the estimator, the contractor, the designer, and the owner to better plan and manage building construction projects. Productivity data is provided in order to assist with scheduling. National labor rates are analyzed. Tables and charts for location and time adjustments are included and help the estimator to tailor the prices to a specific location. And ultimately, the book is a list of line items for over 20,000 individual components of construction. This information, outlined in the CSI MasterFormat divisions, provides an invaluable checklist to the construction professional to assure that all required items are included in a project.

Format & Data

The major portion of *Means Building Construction Cost Data* is the Unit Price section. This is the primary source of unit cost data and is organized according to the 16 divisions of the MasterFormat. This index was developed by representatives of all parties concerned with the building construction industry and has been adopted by the American Institute of Architects (AIA), the Associated General Contractors of America, Inc. (AGC), and the Construction Specifications Institute, Inc. (CSI).

The MasterFormat Divisions:

Division 1 – General Requirements

Division 2 – Site Construction

Division 3 – Concrete

Division 4 – Masonry

Division 5 – Metals

Division 6 – Wood and Plastics

Division 7 – Thermal and Moisture Protection

Division 8 – Doors and Windows

Division 9 – Finishes

Division 10 – Specialties

Division 11 – Equipment

Division 12 – Furnishings

Division 13 – Special Construction

Division 14 – Conveying Systems

Division 15 – Mechanical

Division 16 – Electrical

In addition to the sixteen MasterFormat divisions of the Unit Price section, Division 17, Square Foot and Cubic Foot Costs, presents consolidated data from over ten thousand actual reported construction projects and provides information based on total project costs as well as costs for major components. Following Division 17 is the Reference Section, which provides supporting cost, design, and reference information for the Unit Price section. Individual materials and construction methods are presented in depth and detail in this section. Also included in this section are Change Orders, Crew Tables, City Cost Index, Abbreviations, and finally, the book index.

The prices as presented in *Means Building Construction Cost Data* are national averages. Material and equipment costs are developed through annual contact with manufacturers, dealers, distributors, and contractors throughout the United States. Means' staff of engineers is constantly updating prices and keeping abreast of changes and fluctuations within the industry. Labor rates are the national average of each trade as determined from union agreements from thirty major U.S. cities. Throughout the calendar year, as new wage agreements are negotiated, labor costs should be factored accordingly.

There are various factors and assumptions on which the costs, as presented in *Means Building Construction Cost Data*, have been based:

Quality – The costs are based on methods, materials, and workmanship in accordance with U.S. Government standards and represent good, sound construction practice.

Overtime – The costs as presented include *no* allowance for overtime. If overtime or premium time is anticipated, labor costs must be factored accordingly.

Productivity – The daily output figures are based on an eight hour workday, during daylight hours. The chart in Figure 6.8 shows that as the number of hours worked per day (over eight) increases, and as the days per week (over five) increase, production efficiency decreases (see Chapter 6).

Size of Project – Costs in *Means Building Construction Cost Data* are based on commercial and industrial buildings which cost $1,000,000 and up. Large residential projects are also included.

Local Factors – Weather conditions, season of the year, local union restrictions, and unusual building code requirements can all have a significant impact on construction costs. The availability of a skilled labor force, sufficient materials, and even adequate energy and utilities will also affect costs. These factors vary in impact and are not dependent upon location. They must be reviewed for each project in every area.

In the presentation of prices in *Means Building Construction Cost Data*, certain rounding rules are employed to make the numbers easy to use without significantly affecting accuracy. The rules are used consistently and are as follows:

Prices From	Rounded to the nearest
$0.01 to $5.00	$.01
$5.01 to $20.00	$.05
$20.01 to $100.00	$.50
$100.01 to $300.00	$1.00
$300.01 to $1,000.00	$5.00
$1,000.01 to $10,000.00	$25.00
$10,000.01 to $50,000.00	$100.00
$50,000.01 and above	$500.00

Unit Price Section

The Unit Price section of *Means Building Construction Cost Data* contains a great deal of information in addition to the unit cost for each construction component. Figure 7.1 is a typical page, showing costs for brick veneer. Note that prices are included for more than ten types of brick and ten bonding patterns and course types. In addition, each type and bond is priced per thousand brick as well as per square foot of wall area. The information and cost data is broken down and itemized in this way to provide not only for the most detailed pricing possible, but also to accommodate different methods of quantity takeoff and estimating.

Within each individual line item, there is a description of the construction component, information regarding typical crews designated to perform the work, and the daily output of each crew. Costs are presented bare, or unburdened, as well as with markups for overhead and profit. Figure 7.2 is a graphic representation of how to use the Unit Price section as presented in *Means Building Construction Cost Data*.

Line Numbers

Every construction item in the Means unit price cost data books has a unique line number. This line number acts as an address so that each item can be quickly located and/or referenced. The numbering system is based on the MasterFormat classification by division. In Figure 7.2, note the bold number in reverse type, 03310. This number represents the MEANS subdivision, in this case, Structural Concrete, of the MasterFormat Division 3 – Concrete. All 16 MasterFormat divisions are organized in this manner. Within each subdivision, the data is broken down into major classifications. These major classifications are listed alphabetically and are designated by bold type for both numbers and descriptions. Each item, or line, is further defined by an individual number. As shown in Figure 7.2, the full line number for each item consists of: MasterFormat Division – MEANS subdivision – a major classification number – an item line number. Each full line number describes a unique construction

			CREW	DAILY OUTPUT	LABOR-HOURS	UNIT	MAT.	LABOR	EQUIP.	TOTAL	TOTAL INCL O&P
04210	**Clay Masonry Units**							**2002 BARE COSTS**			
100 0010	**COMMON BUILDING BRICK** C62, TL lots, material only	R04210 -120									**100**
0020	Standard, minimum					M	270			270	297
0050	Average (select)	▼				"	315			315	345
120 0010	**BRICK VENEER** Scaffolding not included, truck load lots	R04210 -120									**120**
0015	Material costs incl. 3% brick and 25% mortar waste										
0020	Standard, select common, 4" x 2-2/3" x 8" (6.75/S.F.)	R04210 -180	D-8	1.50	26.667	M	360	740		1,100	1,525
0050	Red, 4" x 2-2/3" x 8", running bond			1.50	26.667		400	740		1,140	1,575
0100	Full header every 6th course (7.88/S.F.)	R04210 -500		1.45	27.586		400	765		1,165	1,625
0150	English, full header every 2nd course (10.13/S.F.)			1.40	28.571		395	790		1,185	1,650
0200	Flemish, alternate header every course (9.00/S.F.)			1.40	28.571		395	790		1,185	1,650
0250	Flemish, alt. header every 6th course (7.13/S.F.)			1.45	27.586		400	765		1,165	1,625
0300	Full headers throughout (13.50/S.F.)			1.40	28.571		395	790		1,185	1,650
0350	Rowlock course (13.50/S.F.)			1.35	29.630		395	820		1,215	1,675
0400	Rowlock stretcher (4.50/S.F.)			1.40	28.571		400	790		1,190	1,675
0450	Soldier course (6.75/S.F.)			1.40	28.571		400	790		1,190	1,675
0500	Sailor course (4.50/S.F.)			1.30	30.769		400	850		1,250	1,750
0601	Buff or gray face, running bond, (6.75/S.F.)			1.50	26.667		400	740		1,140	1,575
0700	Glazed face, 4" x 2-2/3" x 8", running bond			1.40	28.571		1,500	790		2,290	2,875
0750	Full header every 6th course (7.88/S.F.)			1.35	29.630		1,475	820		2,295	2,850
1000	Jumbo, 6" x 4" x 12",(3.00/S.F.)			1.30	30.769		1,250	850		2,100	2,675
1051	Norman, 4" x 2-2/3" x 12" (4.50/S.F.)	▼		1.45	27.586	▼	735	765		1,500	1,975
1100	Norwegian, 4" x 3-1/5" x 12" (3.75/S.F.)		D-8	1.40	28.571	M	575	790		1,365	1,850
1150	Economy, 4" x 4" x 8" (4.50 per S.F.)			1.40	28.571		560	790		1,350	1,850
1201	Engineer, 4" x 3-1/5" x 8", (5.63/S.F.)			1.45	27.586		365	765		1,130	1,575
1251	Roman, 4" x 2" x 12", (6.00/S.F.)			1.50	26.667		765	740		1,505	1,975
1300	S.C.R. 6" x 2-2/3" x 12" (4.50/S.F.)			1.40	28.571		900	790		1,690	2,225
1350	Utility, 4" x 4" x 12" (3.00/S.F.)	▼		1.35	29.630		1,075	820		1,895	2,450
1360	For less than truck load lots, add					▼	10			10	11
1400	For battered walls, add							30%			
1450	For corbels, add							75%			
1500	For curved walls, add							30%			
1550	For pits and trenches, deduct							20%			
1999	Alternate method of figuring by square foot										
2000	Standard, sel. common, 4" x 2-2/3" x 8", (6.75/S.F.)		D-8	230	.174	S.F.	2.69	4.82		7.51	10.35
2020	Standard, red, 4" x 2-2/3" x 8", running bond (6.75/SF)			220	.182		2.69	5.05		7.74	10.70
2050	Full header every 6th course (7.88/S.F.)			185	.216		3.13	6		9.13	12.65
2100	English, full header every 2nd course (10.13/S.F.)			140	.286		4.02	7.90		11.92	16.55
2150	Flemish, alternate header every course (9.00/S.F.)			150	.267		3.57	7.40		10.97	15.30
2200	Flemish, alt. header every 6th course (7.13/S.F.)			205	.195		2.84	5.40		8.24	11.40
2250	Full headers throughout (13.50/S.F.)			105	.381		5.35	10.55		15.90	22
2300	Rowlock course (13.50/S.F.)			100	.400		5.35	11.10		16.45	23
2350	Rowlock stretcher (4.50/S.F.)			310	.129		1.80	3.57		5.37	7.50
2400	Soldier course (6.75/S.F.)			200	.200		2.69	5.55		8.24	11.45
2450	Sailor course (4.50/S.F.)			290	.138		1.80	3.82		5.62	7.85
2600	Buff or gray face, running bond, (6.75/S.F.)			220	.182		2.85	5.05		7.90	10.90
2700	Glazed face brick, running bond			210	.190		9.90	5.30		15.20	19
2750	Full header every 6th course (7.88/S.F.)			170	.235		11.55	6.50		18.05	22.50
3000	Jumbo, 6" x 4" x 12" running bond (3.00/S.F.)			435	.092		3.53	2.55		6.08	7.80
3050	Norman, 4" x 2-2/3" x 12" running bond, (4.5/S.F.)			320	.125		3.50	3.46		6.96	9.15
3100	Norwegian, 4" x 3-1/5" x 12" (3.75/S.F.)			375	.107		2.11	2.95		5.06	6.85
3150	Economy, 4" x 4" x 8" (4.50/S.F.)			310	.129		2.50	3.57		6.07	8.25
3200	Engineer, 4" x 3-1/5" x 8" (5.63/S.F.)			260	.154		2.05	4.26		6.31	8.80
3250	Roman, 4" x 2" x 12" (6.00/S.F.)			250	.160		4.52	4.43		8.95	11.75
3300	SCR, 6" x 2-2/3" x 12" (4.50/S.F.)			310	.129		4.14	3.57		7.71	10.05
3350	Utility, 4" x 4" x 12" (3.00/S.F.)	▼		450	.089	▼	3.20	2.46		5.66	7.30
3400	For cavity wall construction, add							15%			
3450	For stacked bond, add							10%			
3500	For interior veneer construction, add							15%			
3550	For curved walls, add							30%			

Figure 7.1

element. For example, in Figure 7. 1, the line number for Norman brick veneer, running bond (per square foot) is 04210-120-3050.

Line Description

Each line has a text description of the item for which costs are listed. The description may be self-contained and all inclusive. If indented, items are delineations (by size, color, material, etc.) or breakdowns of previously described items. An index is provided in the back of *Means Building Construction Cost Data* to aid in locating particular items.

Crew

For each construction element (each line item) a minimum typical crew is designated as appropriate to perform the work. The crew may include one or more trades, foremen, craftsmen and helpers, and any equipment required for proper installation of the described item. If an individual trade installs the item using only hand tools, the smallest efficient number of tradesmen will be indicated (1 Carp, 2 Carp, etc.). Abbreviations for trades are shown in Figure 7.3. If more than one trade is required to install the item and/or if powered equipment is needed, a crew number will be designated (B-5, D-3, etc.) A complete listing of crews is presented following the Reference Section in the back of *Means Building Construction Cost Data* (see Figure 7.4). Each crew breakdown contains the following components:

1. Number and type of workers designated.
2. Number, size, and type of any equipment required.
3. Hourly labor costs listed two ways: bare (base rate including fringe benefits) and including installing contractor's overhead and profit (billing rate). See Figure 7.3.
4. Daily equipment costs, based on the weekly equipment rental cost divided by 5, plus the hourly operating cost, times 8 hours. This cost is listed two ways: as a bare cost and with a 10% markup to cover handling and management costs.
5. Labor and equipment are broken down further: cost per labor-hour for labor, and cost per labor-hour for the equipment.
6. The total daily labor-hours for the crew.
7. The total bare costs per day for the crew, including equipment.
8. The total daily cost of the crew including the installing contractor's overhead and profit.

The total daily cost of the craftsmen involved or the total daily crew indicated is used to calculate the unit installation cost for each item (for both bare costs and cost including overhead and profit).

The crew designation does not mean that this is the only crew that can perform the work. Crew size and content have been developed and chosen based on practical experience and feedback from contractors and represent a labor and equipment makeup commonly found in the industry. The most appropriate crew for a given task is best determined based on particular project requirements. Unit costs may vary if crew sizes or content are significantly changed.

Figure 7.5 is a page from Division 01590 of *Means Building Construction Cost Data*. This type of page lists the equipment costs used in the presentation and calculation of the crew costs and unit price data. Rental costs are shown as daily, weekly, and monthly rates. The Hourly Operating Cost represents the cost of fuel, lubrication, and routine maintenance. The column on the right in

How to Use the Unit Price Pages

The following is a detailed explanation of a sample entry in the Unit Price Section. Next to each bold number below is the item being described with appropriate component of the sample entry following in parenthesis. Some prices are listed as bare costs, others as costs that include overhead and profit of the installing contractor. In most cases, if the work is to be subcontracted, the general contractor will need to add an additional markup (R.S. Means suggests using 10%) to the figures in the column "Total Incl. O&P."

Division Number/Title (03300/Cast-In-Place Concrete)

Use the Unit Price Section Table of Contents to locate specific items. The sections are classified according to the CSI MasterFormat (1995 Edition).

Line Numbers (03310 240 3900)

Each unit price line item has been assigned a unique 12-digit code based on the CSI MasterFormat classification.

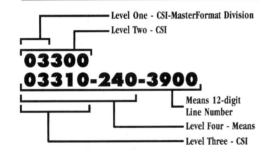

— Level One - CSI-MasterFormat Division
— Level Two - CSI
03300
03310-240-3900
— Means 12-digit Line Number
— Level Four - Means
— Level Three - CSI

Description (Concrete-In-Place, etc.)

Each line item is described in detail. Sub-items and additional sizes are indented beneath the appropriate line items. The first line or two after the main item (in boldface) may contain descriptive information that pertains to all line items beneath this boldface listing.

Items which include the symbol **CN** are updated in the Key Material Price Section of *The Change Notice* quarterly publication.

Reference Number Information

R03310 -010 You'll see reference numbers shown in bold rectangles at the beginning of some sections. These refer to related items in the Reference Section, visually identified by a vertical gray bar on the edge of pages.

The relation may be: (1) an estimating procedure that should be read before estimating, (2) an alternate pricing method, or (3) technical information.

The "R" designates the Reference Section. The numbers refer to the MasterFormat classification system.

It is strongly recommended that you review all reference numbers that appear within the section in which you are working.

Note: Not all reference numbers appear in all Means publications.

	03300	Cast-In-Place Concrete										
	03310	**Structural Concrete**	CREW	DAILY OUTPUT	LABOR-HOURS	UNIT	MAT.	LABOR	EQUIP.	TOTAL	TOTAL INCL O&P	
240	0010	**CONCRETE IN PLACE** Including forms, reinforcing										240
	0050	steel, including finishing unless otherwise indicated										
	3800	Footings, spread under 1 C.Y.	C-14C	38.07	2.942	C.Y.	100	84.50	.94	185.44	244	
	3850	Over 5 C.Y.		81.04	1.382		92	40	.44	132.44	164	
	3900	Footings, strip, 18" x 9", plain		41.04	2.729		91	78.50	.87	170.37	225	
	3950	36" x 12", reinforced		61.55	1.820		92.50	52.50	.58	145.58	185	
	4000	Foundation mat, under 10 C.Y.		38.67	2.896		124	83.50	.93	208.43	268	
	4050	Over 20 C.Y.		56.67	1.986		110	57	.64	167.64	205	
	4200	Grade walls, 8" thick, 8' high	C-14D	45.83	3.164		126	130	15.65	271.65	360	
	4250	14' high		27.26	7.337		149	219	26.50	394.50	540	
	4260	12" thick, 8' high		64.32	3.109		111	93	11.15	215.15	279	
	4270	14' high		40.01	4.999		119	149	17.95	285.95	385	

Figure 7.2

		A		B	C	D	E	F	G	H	I
		Base Rate Incl. Fringes		Workers' Comp. Ins.	Average Fixed Over-head	Over-head	Profit	Total Overhead & Profit		Rate with O & P	
Abbr.	Trade	Hourly	Daily					%	Amount	Hourly	Daily
Skwk	Skilled Workers Average (35 trades)	$30.95	$247.60	16.8%	16.5%	13.0%	10%	56.3%	$17.40	$48.35	$386.80
	Helpers Average (5 trades)	22.75	182.00	18.5		11.0		56.0	12.75	35.50	284.00
	Foreman Average, Inside ($.50 over trade)	31.45	251.60	16.8		13.0		56.3	17.70	49.15	393.20
	Foreman Average, Outside ($2.00 over trade)	32.95	263.60	16.8		13.0		56.3	18.55	51.50	412.00
Clab	Common Building Laborers	23.45	187.60	18.1		11.0		55.6	13.05	36.50	292.00
Asbe	Asbestos/Insulation Workers/Pipe Coverers	33.45	267.60	16.2		16.0		58.7	19.65	53.10	424.80
Boil	Boilermakers	36.25	290.00	14.7		16.0		57.2	20.75	57.00	456.00
Bric	Bricklayers	30.50	244.00	16.0		11.0		53.5	16.30	46.80	374.40
Brhe	Bricklayer Helpers	23.50	188.00	16.0		11.0		53.5	12.55	36.05	288.40
Carp	Carpenters	30.00	240.00	18.1		11.0		55.6	16.70	46.70	373.60
Cefi	Cement Finishers	28.70	229.60	10.6		11.0		48.1	13.80	42.50	340.00
Elec	Electricians	35.45	283.60	6.7		16.0		49.2	17.45	52.90	423.20
Elev	Elevator Constructors	37.10	296.80	7.7		16.0		50.2	18.60	55.70	445.60
Eqhv	Equipment Operators, Crane or Shovel	32.35	258.80	10.6		14.0		51.1	16.55	48.90	391.20
Eqmd	Equipment Operators, Medium Equipment	31.20	249.60	10.6		14.0		51.1	15.95	47.15	377.20
Eqlt	Equipment Operators, Light Equipment	29.80	238.40	10.6		14.0		51.1	15.25	45.05	360.40
Eqol	Equipment Operators, Oilers	26.65	213.20	10.6		14.0		51.1	13.60	40.25	322.00
Eqmm	Equipment Operators, Master Mechanics	32.80	262.40	10.6		14.0		51.1	16.75	49.55	396.40
Glaz	Glaziers	30.00	240.00	13.8		11.0		51.3	15.40	45.40	363.20
Lath	Lathers	28.75	230.00	11.1		11.0		48.6	13.95	42.70	341.60
Marb	Marble Setters	30.10	240.80	16.0		11.0		53.5	16.10	46.20	369.60
Mill	Millwrights	31.75	254.00	10.6		11.0		48.1	15.25	47.00	376.00
Mstz	Mosaic & Terrazzo Workers	29.25	234.00	9.8		11.0		47.3	13.85	43.10	344.80
Pord	Painters, Ordinary	27.15	217.20	13.8		11.0		51.3	13.95	41.10	328.80
Psst	Painters, Structural Steel	27.90	223.20	48.4		11.0		85.9	23.95	51.85	414.80
Pape	Paper Hangers	27.10	216.80	13.8		11.0		51.3	13.90	41.00	328.00
Pile	Pile Drivers	29.80	238.40	24.9		16.0		67.4	20.10	49.90	399.20
Plas	Plasterers	28.10	224.80	15.8		11.0		53.3	15.00	43.10	344.80
Plah	Plasterer Helpers	23.70	189.60	15.8		11.0		53.3	12.65	36.35	290.80
Plum	Plumbers	35.95	287.60	8.3		16.0		50.8	18.25	54.20	433.60
Rodm	Rodmen (Reinforcing)	34.25	274.00	28.3		14.0		68.8	23.55	57.80	462.40
Rofc	Roofers, Composition	26.60	212.80	32.6		11.0		70.1	18.65	45.25	362.00
Rots	Roofers, Tile & Slate	26.75	214.00	32.6		11.0		70.1	18.75	45.50	364.00
Rohe	Roofers, Helpers (Composition)	19.80	158.40	32.6		11.0		70.1	13.90	33.70	269.60
Shee	Sheet Metal Workers	35.10	280.80	11.7		16.0		54.2	19.00	54.10	432.80
Spri	Sprinkler Installers	36.20	289.60	8.7		16.0		51.2	18.55	54.75	438.00
Stpi	Steamfitters or Pipefitters	36.20	289.60	8.3		16.0		50.8	18.40	54.60	436.80
Ston	Stone Masons	30.65	245.20	16.0		11.0		53.5	16.40	47.05	376.40
Sswk	Structural Steel Workers	34.25	274.00	39.8		14.0		80.3	27.50	61.75	494.00
Tilf	Tile Layers	29.15	233.20	9.8		11.0		47.3	13.80	42.95	343.60
Tilh	Tile Layers Helpers	23.35	186.80	9.8		11.0		47.3	11.05	34.40	275.20
Trlt	Truck Drivers, Light	24.30	194.40	14.9		11.0		52.4	12.75	37.05	296.40
Trhv	Truck Drivers, Heavy	25.00	200.00	14.9		11.0		52.4	13.10	38.10	304.80
Sswl	Welders, Structural Steel	34.25	274.00	39.8		14.0		80.3	27.50	61.75	494.00
Wrck	*Wrecking	23.45	187.60	41.2		11.0		78.7	18.45	41.90	335.20

*Not included in averages

Figure 7.3

Crew No.	Bare Costs		Incl. Subs O & P		Cost Per Labor-Hour	
Crew B-23B	Hr.	Daily	Hr.	Daily	Bare Costs	Incl. O&P
1 Labor Foreman (outside)	$25.45	$203.60	$39.60	$316.80	$26.70	$41.08
1 Laborer	23.45	187.60	36.50	292.00		
1 Equip. Operator (medium)	31.20	249.60	47.15	377.20		
1 Drill Rig, Wells		1993.00		2192.30		
1 Pickup Truck, 3/4 Ton		84.00		92.40		
1 Pump, Cntfgl, 6"		198.20		218.00	94.80	104.28
24 L.H., Daily Totals		$2916.00		$3488.70	$121.50	$145.36

Crew No.	Bare Costs		Incl. Subs O & P		Cost Per Labor-Hour	
Crew B-24	Hr.	Daily	Hr.	Daily	Bare Costs	Incl. O&P
1 Cement Finisher	$28.70	$229.60	$42.50	$340.00	$27.38	$41.90
1 Laborer	23.45	187.60	36.50	292.00		
1 Carpenter	30.00	240.00	46.70	373.60		
24 L.H., Daily Totals		$657.20		$1005.60	$27.38	$41.90

Crew No.	Bare Costs		Incl. Subs O & P		Cost Per Labor-Hour	
Crew B-25	Hr.	Daily	Hr.	Daily	Bare Costs	Incl. O&P
1 Labor Foreman	$25.45	$203.60	$39.60	$316.80	$25.75	$39.69
7 Laborers	23.45	1313.20	36.50	2044.00		
3 Equip. Oper. (med.)	31.20	748.80	47.15	1131.60		
1 Asphalt Paver, 130 H.P.		1290.00		1419.00		
1 Tandem Roller, 10 Ton		168.40		185.25		
1 Roller, Pneumatic Wheel		237.80		261.60	19.28	21.20
88 L.H., Daily Totals		$3961.80		$5358.25	$45.03	$60.89

Crew No.	Bare Costs		Incl. Subs O & P		Cost Per Labor-Hour	
Crew B-25B	Hr.	Daily	Hr.	Daily	Bare Costs	Incl. O&P
1 Labor Foreman	$25.45	$203.60	$39.60	$316.80	$26.20	$40.31
7 Laborers	23.45	1313.20	36.50	2044.00		
4 Equip. Oper. (medium)	31.20	998.40	47.15	1508.80		
1 Asphalt Paver, 130 H.P.		1290.00		1419.00		
2 Rollers, Steel Wheel		336.80		370.50		
1 Roller, Pneumatic Wheel		237.80		261.60	19.42	21.37
96 L.H., Daily Totals		$4379.80		$5920.70	$45.62	$61.68

Crew No.	Bare Costs		Incl. Subs O & P		Cost Per Labor-Hour	
Crew B-25C	Hr.	Daily	Hr.	Daily	Bare Costs	Incl. O&P
1 Labor Foreman	$25.45	$203.60	$39.60	$316.80	$26.37	$40.57
3 Laborers	23.45	562.80	36.50	876.00		
2 Equip. Oper. (medium)	31.20	499.20	47.15	754.40		
1 Asphalt Paver, 130 H.P.		1290.00		1419.00		
1 Rollers, Steel Wheel		168.40		185.25	30.38	33.42
48 L.H., Daily Totals		$2724.00		$3551.45	$56.75	$73.99

Crew No.	Bare Costs		Incl. Subs O & P		Cost Per Labor-Hour	
Crew B-26	Hr.	Daily	Hr.	Daily	Bare Costs	Incl. O&P
1 Labor Foreman (outside)	$25.45	$203.60	$39.60	$316.80	$26.50	$41.20
6 Laborers	23.45	1125.60	36.50	1752.00		
2 Equip. Oper. (med.)	31.20	499.20	47.15	754.40		
1 Rodman (reinf.)	34.25	274.00	57.80	462.40		
1 Cement Finisher	28.70	229.60	42.50	340.00		
1 Grader, 30,000 Lbs.		431.60		474.75		
1 Paving Mach. & Equip.		1451.00		1596.10	21.39	23.53
88 L.H., Daily Totals		$4214.60		$5696.45	$47.89	$64.73

Crew No.	Bare Costs		Incl. Subs O & P		Cost Per Labor-Hour	
Crew B-27	Hr.	Daily	Hr.	Daily	Bare Costs	Incl. O&P
1 Labor Foreman (outside)	$25.45	$203.60	$39.60	$316.80	$23.95	$37.28
3 Laborers	23.45	562.80	36.50	876.00		
1 Berm Machine		174.40		191.85	5.45	6.00
32 L.H., Daily Totals		$940.80		$1384.65	$29.40	$43.28

Crew No.	Bare Costs		Incl. Subs O & P		Cost Per Labor-Hour	
Crew B-28	Hr.	Daily	Hr.	Daily	Bare Costs	Incl. O&P
2 Carpenters	$30.00	$480.00	$46.70	$747.20	$27.82	$43.30
1 Laborer	23.45	187.60	36.50	292.00		
24 L.H., Daily Totals		$667.60		$1039.20	$27.82	$43.30

Crew No.	Bare Costs		Incl. Subs O & P		Cost Per Labor-Hour	
Crew B-29	Hr.	Daily	Hr.	Daily	Bare Costs	Incl. O&P
1 Labor Foreman (outside)	$25.45	$203.60	$39.60	$316.80	$25.46	$39.25
4 Laborers	23.45	750.40	36.50	1168.00		
1 Equip. Oper. (crane)	32.35	258.80	48.90	391.20		
1 Equip. Oper. Oiler	26.65	213.20	40.25	322.00		
1 Gradall, 3 Ton, 1/2 C.Y.		736.40		810.05	13.15	14.47
56 L.H., Daily Totals		$2162.40		$3008.05	$38.61	$53.72

Crew No.	Bare Costs		Incl. Subs O & P		Cost Per Labor-Hour	
Crew B-30	Hr.	Daily	Hr.	Daily	Bare Costs	Incl. O&P
1 Equip. Oper. (med.)	$31.20	$249.60	$47.15	$377.20	$27.07	$41.12
2 Truck Drivers (heavy)	25.00	400.00	38.10	609.60		
1 Hyd. Excavator, 1.5 C.Y.		705.00		775.50		
2 Dump Trucks, 16 Ton		941.20		1035.30	68.59	75.45
24 L.H., Daily Totals		$2295.80		$2797.60	$95.66	$116.57

Crew No.	Bare Costs		Incl. Subs O & P		Cost Per Labor-Hour	
Crew B-31	Hr.	Daily	Hr.	Daily	Bare Costs	Incl. O&P
1 Labor Foreman (outside)	$25.45	$203.60	$39.60	$316.80	$25.16	$39.16
3 Laborers	23.45	562.80	36.50	876.00		
1 Carpenter	30.00	240.00	46.70	373.60		
1 Air Compr., 250 C.F.M.		135.00		148.50		
1 Sheeting Driver		10.20		11.20		
2-50 Ft. Air Hoses, 1.5" Dia.		9.40		10.35	3.87	4.25
40 L.H., Daily Totals		$1161.00		$1736.45	$29.03	$43.41

Crew No.	Bare Costs		Incl. Subs O & P		Cost Per Labor-Hour	
Crew B-32	Hr.	Daily	Hr.	Daily	Bare Costs	Incl. O&P
1 Laborer	$23.45	$187.60	$36.50	$292.00	$29.26	$44.49
3 Equip. Oper. (med.)	31.20	748.80	47.15	1131.60		
1 Grader, 30,000 Lbs.		431.60		474.75		
1 Tandem Roller, 10 Ton		168.40		185.25		
1 Dozer, 200 H.P.		823.40		905.75	44.48	48.93
32 L.H., Daily Totals		$2359.80		$2989.35	$73.74	$93.42

Crew No.	Bare Costs		Incl. Subs O & P		Cost Per Labor-Hour	
Crew B-32A	Hr.	Daily	Hr.	Daily	Bare Costs	Incl. O&P
1 Laborer	$23.45	$187.60	$36.50	$292.00	$28.62	$43.60
2 Equip. Oper. (medium)	31.20	499.20	47.15	754.40		
1 Grader, 30,000 Lbs.		431.60		474.75		
1 Roller, Vibratory, 29,000 Lbs.		434.60		478.05	36.09	39.70
24 L.H., Daily Totals		$1553.00		$1999.20	$64.71	$83.30

Crew No.	Bare Costs		Incl. Subs O & P		Cost Per Labor-Hour	
Crew B-32B	Hr.	Daily	Hr.	Daily	Bare Costs	Incl. O&P
1 Laborer	$23.45	$187.60	$36.50	$292.00	$28.62	$43.60
2 Equip. Oper. (medium)	31.20	499.20	47.15	754.40		
1 Dozer, 200 H.P.		823.40		905.75		
1 Roller, Vibratory, 29,000 Lbs.		434.60		478.05	52.42	57.66
24 L.H., Daily Totals		$1944.80		$2430.20	$81.04	$101.26

Crew No.	Bare Costs		Incl. Subs O & P		Cost Per Labor-Hour	
Crew B-32C	Hr.	Daily	Hr.	Daily	Bare Costs	Incl. O&P
1 Labor Foreman	$25.45	$203.60	$39.60	$316.80	$27.66	$42.34
2 Laborers	23.45	375.20	36.50	584.00		
3 Equip. Oper. (medium)	31.20	748.80	47.15	1131.60		
1 Grader, 30,000 Lbs.		431.60		474.75		
1 Roller, Steel Wheel		168.40		185.25		
1 Dozer, 200 H.P.		823.40		905.75	29.65	32.62
48 L.H., Daily Totals		$2751.00		$3598.15	$57.31	$74.96

Crew No.	Bare Costs		Incl. Subs O & P		Cost Per Labor-Hour	
Crew B-33	Hr.	Daily	Hr.	Daily	Bare Costs	Incl. O&P
1 Equip. Oper. (med.)	$31.20	$249.60	$47.15	$377.20	$28.99	$44.11
.5 Laborer	23.45	93.80	36.50	146.00		
.25 Equip. Oper. (med.)	31.20	62.40	47.15	94.30		
14 L.H., Daily Totals		$405.80		$617.50	$28.99	$44.11

Figure 7.4

Figure 7.5 is the Crew's Equipment Cost. These figures represent the costs as included in the bare daily crew costs and are calculated as follows:

Line number:	01590-200-1910
Equipment:	Grader, self-propelled, 30,000 lb.
Rent per Week:	$1,470
Hourly Operating Cost:	$17.20

$$\frac{\text{Weekly rental}}{5 \text{ days per week}} + (\text{Hourly Oper. Cost} \times 8 \text{ hrs./day}) = \frac{\text{Crew's Equipment}}{\text{Cost per Day}}$$

$$\frac{\$1,470}{5} + (\$17.20 \times 8) = \$431.60$$

The Crew's Equipment Cost is basically the daily cost of equipment. This figure is based on rental by the week. Note how the equipment costs in the example above are used in developing the crew cost for Crew B-26 in Figure 7.4. The daily equipment costs, including overhead and profit, contain a 10% fee added to the bare costs to cover the handling and overhead associated with use of the equipment.

The daily output represents the number of units that the designated minimum crew will install in one 8 hour day. (See "Units" section.) These figures have been determined from construction experience under actual working conditions. With a designation of the appropriate crew and determination of the daily output, the unit costs for installation are easily calculated:

$$\frac{\text{Daily Crew Cost (\$/day)}}{\text{Daily Output (units/day)}} = \text{Unit Installation Cost (\$/unit)}$$

The labor-hours per unit can also be easily determined using the data provided. A labor-hour is the equivalent of one worker working for one hour, and is a relative indicator of productivity. To calculate labor-hours per unit:

$$\frac{\text{Total Daily Crew Labor-hours}}{\text{Daily Output}} = \frac{\text{Labor-hours/Day}}{\text{Units/Day}} = \text{Labor-hours per Unit}$$

The resulting units of productivity can be very useful as an aid for scheduling. Manpower allocations are more readily determined and time durations can be calculated based on crew size.

Unit

The unit column (see Figures 7.1 and 7.2) defines the component for which the costs have been calculated. It is this "unit" on which Unit Price Estimating is based. The units, as used, represent standard estimating and quantity takeoff procedures. However, the estimator should always check to be sure that the units taken off are the same as those priced. Note in Figure 7.1 that the same items are priced based on two very different units—per thousand brick and per square foot. A list of standard abbreviations is included at the back of *Means Building Construction Cost Data*.

Bare Costs

The four columns listed under "Bare Costs"—"Material," "Labor," "Equip.," and "Total"—represent the actual cost of construction items to the contractor. In other words, bare costs are those which *do not* include the overhead and profit of the installing contractor, whether a subcontractor or a general contracting company using its own crews.

01500 | Temporary Facilities & Controls

01590 | Equipment Rental

			UNIT	HOURLY OPER. COST	RENT PER DAY	RENT PER WEEK	RENT PER MONTH	CREW EQUIPMENT COST/DAY	
0342	Bucket thumbs	R01590 -100	Ea.	1.25	205	615	1,850	133	200
0345	Grapples			1	202	605	1,825	129	
0350	Gradall type, truck mounted, 3 ton @ 15' radius, 5/8 C.Y.	R02315 -300		24.80	895	2,690	8,075	736.40	
0370	1 C.Y. capacity			28.95	1,050	3,185	9,550	868.60	
0400	Backhoe-loader, 40 to 45 H.P., 5/8 C.Y. capacity	R02315 -400		5.35	188	565	1,700	155.80	
0450	45 H.P. to 60 H.P., 3/4 C.Y. capacity			6.95	205	615	1,850	178.60	
0460	80 H.P., 1-1/4 C.Y. capacity	R02315 -450		9.20	237	710	2,125	215.60	
0470	112 H.P., 1-1/2 C.Y. capacity			12.80	415	1,240	3,725	350.40	
0480	Attachments	R02455 -900							
0482	Compactor, 20,000 lb			1.95	173	520	1,550	119.60	
0485	Hydraulic hammer, 750 ft-lbs			1.40	202	605	1,825	132.20	
0486	Hydraulic hammer, 1200 ft-lbs			1.90	245	735	2,200	162.20	
0500	Brush chipper, gas engine, 6" cutter head, 35 H.P.			3.80	140	420	1,250	114.40	
0550	12" cutter head, 130 H.P.			7.20	208	625	1,875	182.60	
0600	15" cutter head, 165 H.P.			11.70	228	685	2,050	230.60	
0750	Bucket, clamshell, general purpose, 3/8 C.Y.			.70	53.50	160	480	37.60	
0800	1/2 C.Y.			.80	60	180	540	42.40	
0850	3/4 C.Y.			.90	66.50	200	600	47.20	
0900	1 C.Y.			.95	90	270	810	61.60	
0950	1-1/2 C.Y.			1.45	115	345	1,025	80.60	
1000	2 C.Y.			1.60	140	420	1,250	96.80	
1010	Bucket, dragline, medium duty, 1/2 C.Y.			.40	23	69	207	17	
1020	3/4 C.Y.			.45	24.50	73	219	18.20	
1030	1 C.Y.			.45	26	78	234	19.20	
1040	1-1/2 C.Y.			.70	40	120	360	29.60	
1050	2 C.Y.			.75	55	165	495	39	
1070	3 C.Y.			1.10	93.50	280	840	64.80	
1200	Compactor, roller, 2 drum, 2000 lb., operator walking			4.65	143	430	1,300	123.20	
1250	Rammer compactor, gas, 1000 lb. blow			.95	40	120	360	31.60	
1300	Vibratory plate, gas, 13" plate, 1000 lb. blow			.75	38.50	115	345	29	
1350	24" plate, 5000 lb. blow			1.95	66.50	200	600	55.60	
1370	Curb builder/extruder, 14 H.P., gas, single screw			6.05	66.50	200	600	88.40	
1390	Double screw		▼	13.55	110	330	990	174.40	
1750	Extractor, piling, see lines 2500 to 2750								
1860	Grader, self-propelled, 25,000 lb.		Ea.	15.30	420	1,265	3,800	375.40	
1910	30,000 lb.			17.20	490	1,470	4,400	431.60	
1920	40,000 lb.			25.95	740	2,225	6,675	652.60	
1930	55,000 lb.			34.65	1,050	3,130	9,400	903.20	
1950	Hammer, pavement demo., hyd., gas, self-prop., 1000 to 1250 lb.			12.55	415	1,245	3,725	349.40	
2000	Diesel 1300 to 1500 lb.			18.75	555	1,670	5,000	484	
2050	Pile driving hammer, steam or air, 4150 ft.-lb. @ 225 BPM			1.55	282	845	2,525	181.40	
2100	8750 ft.-lb. @ 145 BPM			2.10	470	1,405	4,225	297.80	
2150	15,000 ft.-lb. @ 60 BPM			2.20	505	1,515	4,550	320.60	
2200	24,450 ft.-lb. @ 111 BPM		▼	2.90	550	1,645	4,925	352.20	
2250	Leads, 15,000 ft.-lb. hammers		L.F.	.45	6.65	20	60	7.60	
2300	24,450 ft.-lb. hammers and heavier		"	.65	10.65	32	96	11.60	
2350	Diesel type hammer, 22,400 ft.-lb.		Ea.	13.25	585	1,760	5,275	458	
2400	41,300 ft.-lb.			20.10	690	2,070	6,200	574.80	
2450	141,000 ft.-lb.			33.10	1,550	4,625	13,900	1,190	
2500	Vib. elec. hammer/extractor, 200 KW diesel generator, 34 H.P.			21.10	700	2,095	6,275	587.80	
2550	80 H.P.			39.05	1,025	3,110	9,325	934.40	
2600	150 H.P.			56.40	1,575	4,700	14,100	1,391	
2700	Extractor, steam or air, 700 ft.-lb.			1.70	217	650	1,950	143.60	
2750	1000 ft.-lb.			1.95	315	945	2,825	204.60	
3000	Roller, tandem, gas, 3 to 5 ton			5.15	127	380	1,150	117.20	
3050	Diesel, 8 to 12 ton			4.30	223	670	2,000	168.40	
3100	Towed type, vibratory, gas 12.5 H.P., 2 ton			2.70	255	765	2,300	174.60	
3150	Sheepsfoot, double 60" x 60"		▼	.85	110	330	990	72.80	

Figure 7.5

Material: Material costs are based on the national average contractor purchase price delivered to the job site. Delivered costs are assumed to be within a 20 mile radius of metropolitan areas. No sales tax is included in the material prices because of variations from state to state.

The prices are based on large quantities that would normally be purchased for projects costing $1,000,000 and up. Prices for small quantities must be adjusted accordingly. If more current costs for materials are available for the appropriate location, it is recommended that adjustments be made to the unit costs to reflect any cost difference.

Labor: The unit cost for labor of an item is calculated as shown previously (see "Daily Output"). The daily labor cost of the designated crew is divided by the number of units that the crew will install in one day.

Equipment: The unit cost for equipment is calculated by dividing the daily bare equipment cost by the daily output.

The labor rates used to determine the bare installation costs are shown for 35 standard trades in Figure 7.3, under the column "Base Rate Including Fringes." This rate includes a worker's actual hourly wage plus employer-paid benefits (health insurance, vacation, pension, etc.). The labor rates used are *national average* union rates based on trade union agreements (as of January 1 of the current year) from 30 major cities in the United States. As new wage agreements are negotiated within a calendar year, labor costs should be adjusted.

Total Bare Costs: This column simply represents the arithmetic sum of the bare material, labor, and equipment costs. This total is the average cost to the contractor for the particular item of construction, supplied and installed, or "in place." No overhead and/or profit is included.

Total Including Overhead and Profit

The prices in the "Total Including Overhead and Profit" column might also be called the "billing rate." These prices are, on the average, what the installing contractor would charge for the particular item of work. The term "installing contractor" can refer to either a subcontractor or a general contractor. In effect, this rate reflects the amount the subcontractor would charge to the general contractor or what the general contractor would charge the owner for work performed by the general contractor's own employees. The general contractor normally adds a percentage to subcontractor prices (commonly 10%) for supervision and management.

The "Total Including Overhead and Profit" costs are determined by adding the following two calculations:

1. Bare materials and equipment—increased by 10% for handling. See Figure 7.2.
2. Bare daily crew labor costs—increased to include overhead and profit, then divided by the daily output. See Figures 7.2 and 7.4.

In order to increase crew cost to include overhead and profit, labor and equipment costs are treated separately. 10% is added to the bare equipment cost for handling, management, etc. Labor costs are increased, depending upon trade by percentages for overhead and profit (as shown in Figure 7.3). The resulting rates are listed in the right hand columns of Figure 7.3. Note that the percentage increase for overhead and profit for the average skilled worker is 56.3% of the base rate. The following items are included in the increase for overhead and profit, as shown in Figure 7.3.

Workers' Compensation and Employer's Liability:

Workers' Compensation and Employer's Liability Insurance rates vary from state to state and are tied into the construction trade safety records in that particular state. Rates also vary by trade according to the hazard involved. (See Figures 7.6 and 7.7.) The proper authorities will most likely keep the contractor well informed of the rates and obligations.

State and Federal Unemployment Insurance:

The employer's tax rate is adjusted by a merit-rating system according to the number of former employees applying for benefits. Contractors who find it possible to offer a maximum of steady employment can enjoy a reduction in the unemployment tax rate.

General Requirements	R011	Overhead & Miscellaneous Data

R01100-060 Workers' Compensation Insurance Rates by Trade

The table below tabulates the national averages for Workers' Compensation insurance rates by trade and type of building. The average "Insurance Rate" is multiplied by the "% of Building Cost" for each trade. This produces the "Workers' Compensation Cost" by % of total labor cost, to be added for each trade by building type to determine the weighted average Workers' Compensation rate for the building types analyzed.

Trade	Insurance Rate (% Labor Cost) Range		Average	% of Building Cost Office Bldgs.	Schools & Apts.	Mfg.	Workers' Compensation Office Bldgs.	Schools & Apts.	Mfg.
Excavation, Grading, etc.	3.8% to	24.7%	10.5%	4.8%	4.9%	4.5%	.50%	.51%	.47%
Piles & Foundations	5.9 to	62.5	24.8	7.1	5.2	8.7	1.76	1.29	2.16
Concrete	5.9 to	36.8	16.9	5.0	14.8	3.7	.85	2.50	.63
Masonry	5.3 to	31.1	15.9	6.9	7.5	1.9	1.10	1.19	.30
Structural Steel	5.9 to	112.0	39.5	10.7	3.9	17.6	4.23	1.54	6.95
Miscellaneous & Ornamental Metals	4.9 to	25.3	12.8	2.8	4.0	3.6	.36	.51	.46
Carpentry & Millwork	5.9 to	43.6	18.0	3.7	4.0	0.5	.67	.72	.09
Metal or Composition Siding	5.9 to	32.2	16.5	2.3	0.3	4.3	.38	.05	.71
Roofing	5.9 to	85.9	32.3	2.3	2.6	3.1	.74	.84	1.00
Doors & Hardware	4.1 to	25.5	10.9	0.9	1.4	0.4	.10	.15	.04
Sash & Glazing	5.7 to	38.8	13.7	3.5	4.0	1.0	.48	.55	.14
Lath & Plaster	5.2 to	53.5	15.7	3.3	6.9	0.8	.52	1.08	.13
Tile, Marble & Floors	3.7 to	22.6	9.8	2.6	3.0	0.5	.25	.29	.05
Acoustical Ceilings	3.5 to	26.7	11.0	2.4	0.2	0.3	.26	.02	.03
Painting	5.9 to	27.6	13.7	1.5	1.6	1.6	.21	.22	.22
Interior Partitions	5.9 to	43.6	18.0	3.9	4.3	4.4	.70	.77	.79
Miscellaneous Items	1.1 to	128.0	17.6	5.2	3.7	9.7	.91	.65	1.70
Elevators	2.1 to	17.0	7.7	2.1	1.1	2.2	.16	.08	.17
Sprinklers	2.5 to	19.5	8.7	0.5	—	2.0	.04	—	.17
Plumbing	3.0 to	16.9	8.2	4.9	7.2	5.2	.40	.59	.43
Heat., Vent., Air Conditioning	3.3 to	32.5	11.6	13.5	11.0	12.9	1.57	1.28	1.50
Electrical	2.4 to	10.5	6.7	10.1	8.4	11.1	.68	.56	.74
Total	2.0% to	132.9%	—	100.0%	100.0%	100.0%	16.87%	15.39%	18.88%
		Overall Weighted Average 17.05%							

Figure 7.6

Employer-Paid Social Security (FICA):
The tax rate is adjusted annually by the federal government. It is a percentage of an employee's salary up to a maximum annual contribution.

Builder's Risk and Public Liability:
These insurance rates vary according to the trades involved and the state in which the work is done.

Overhead: The column listed as "Sub's Overhead" provides percentages to be added for office or operating overhead. This is the cost of doing business. The percentages are presented as national averages by trade as shown in Figure 7.3. Note that the operating overhead costs are applied to *labor only* in *Means Building Construction Cost Data*. A detailed discussion of operating, or office, overhead and its application is included in Chapter 4.

Profit: This percentage is the fee added by the contractor to offer both a return on investment and an allowance to cover the risk involved in the type of

Workers' Compensation Insurance Rates by States

The table below lists the weighted average Workers' Compensation base rate for each state with a factor comparing this with the national average of 16.8%.

State	Weighted Average	Factor	State	Weighted Average	Factor	State	Weighted Average	Factor
Alabama	31.9%	190	Kentucky	19.7%	117	North Dakota	14.8%	88
Alaska	11.0	65	Louisiana	27.2	162	Ohio	19.1	114
Arizona	11.0	65	Maine	21.7	129	Oklahoma	21.8	130
Arkansas	14.2	85	Maryland	11.8	70	Oregon	13.6	81
California	18.5	110	Massachusetts	20.4	121	Pennsylvania	22.8	136
Colorado	26.4	157	Michigan	18.5	110	Rhode Island	20.9	124
Connecticut	20.0	119	Minnesota	27.7	165	South Carolina	13.7	82
Delaware	11.5	68	Mississippi	20.4	121	South Dakota	14.5	86
District of Columbia	21.9	130	Missouri	17.6	105	Tennessee	15.8	94
Florida	27.8	165	Montana	17.6	105	Texas	15.8	94
Georgia	24.8	148	Nebraska	15.7	93	Utah	12.9	77
Hawaii	16.7	99	Nevada	17.3	103	Vermont	16.4	98
Idaho	10.7	64	New Hampshire	24.9	148	Virginia	10.9	65
Illinois	23.2	138	New Jersey	9.0	54	Washington	9.2	55
Indiana	7.2	43	New Mexico	16.7	99	West Virginia	13.0	77
Iowa	13.4	80	New York	15.3	91	Wisconsin	13.0	77
Kansas	10.1	60	North Carolina	13.0	77	Wyoming	6.5	39
			Weighted Average for U.S. is	17.0% of payroll = 100%				

Rates in the following table are the base or manual costs per $100 of payroll for Workers' Compensation in each state. Rates are usually applied to straight time wages only and not to premium time wages and bonuses.

The weighted average skilled worker rate for 35 trades is 16.8%. For bidding purposes, apply the full value of Workers' Compensation directly to total labor costs, or if labor is 38%, materials 42% and overhead and profit 20% of total cost, carry 38/80 x 16.8% = 8.0% of cost (before overhead and profit) into overhead. Rates vary not only from state to state but also with the experience rating of the contractor.

Rates are the most current available at the time of publication.

Figure 7.7

construction being bid. The profit percentage may vary from 4% on large, straightforward projects to as much as 25% on smaller, high-risk jobs. Profit percentages are directly affected by economic conditions, the expected number of bidders, and the estimated risk involved in the project. For estimating purposes, *Means Building Construction Cost Data* assumes 10% as a reasonable average profit factor.

Square Foot and Cubic Foot Costs

Division 17 in *Building Construction Cost Data* has been developed to facilitate the preparation of rapid preliminary budget estimates. The cost figures in this division are derived from more than 10,400 actual building projects contained in the R.S. Means data bank of construction costs and include the contractor's overhead and profit. The prices shown *do not* include architectural fees or land costs. Costs for new projects are added to the files each year, while projects over ten years old are discarded. For this reason, certain costs may not show a uniform, annual progression. In no case are all subdivisions of a project listed.

These projects were located throughout the United States and reflect differences in square foot and cubic foot costs due to differences in both labor and material costs, plus differences in the owners' requirements. For instance, a bank in a large city would have different features and costs than one in a rural area. This is true of all the different types of buildings analyzed. All individual cost items were computed and tabulated separately. Thus, the sum of the median figures for Plumbing, H.V.A.C., and Electrical will not normally add up to the total Mechanical and Electrical costs arrived at by separate analysis and tabulation of the projects.

The data and prices presented (as shown in Figure 7.8) are listed as square foot or cubic foot costs and as a percentage of total costs. Each category tabulates the data in a similar manner. The median, or middle figure, is listed. This means that 50% of all projects had lower costs, and 50% had higher costs than the median figure. Figures in the "1/4" column indicate that 25% of the projects had lower costs and 75% had higher costs.

Similarly, figures in the "3/4" column indicate that 75% had lower costs and 25% of the projects had higher costs.

The costs and figures represent all projects and do not take into account project size. As a rule, larger buildings (of the same type and relative location) will cost less to build per square foot than similar buildings of a smaller size. This cost difference is due to economies of scale as well as a lower exterior envelope-to-floor area ratio. A conversion is necessary to adjust project costs based on size relative to the norm. Figure 7.9 lists typical sizes (by square foot) and typical ranges of sizes for different building types.

Figure 7.10 is an Area Conversion Scale that can be used to adjust costs based on project size. For example, a proposed mid-rise office building is to be 34,000 square feet. From Figure 7.9, the typical size is 52,000 square feet. A size factor is calculated:

$$\frac{\text{Proposed Building Area}}{\text{Typical Building Area}} = \frac{34,000 \text{ S.F.}}{52,000 \text{ S.F.}} = 0.65$$

The size factor, applied to the Area Conversion Scale, indicates a cost multiplier of 1.06.

$$\text{Cost Multiplier} \times \text{Median Cost} = \text{Converted Cost}$$
$$1.06 \times \$81 = \$85.86$$

17100 | S.F., C.F. and % of Total Costs

17100 | S.F. & C.F. Costs

				UNIT	UNIT COSTS			% OF TOTAL			
					1/4	MEDIAN	3/4	1/4	MEDIAN	3/4	
600	3100	Total: Mechanical & Electrical	R17100 -100	S.F.	17.65	24.50	36	26%	29.90%	30.50%	600
	3200										
	9000	Per bed or person, total cost		Bed	31,000	38,200	51,000				
610	0010	**OFFICES** Low Rise (1 to 4 story)	R17100 -100	S.F.	60.50	77	102				610
	0020	Total project costs		C.F.	4.37	6.10	8.30				
	0100	Site work		S.F.	4.56	7.75	12	5.30%	9.70%	14%	
	0500	Masonry			2.10	4.91	9.25	2.90%	5.80%	8.70%	
	1800	Equipment			.75	1.38	3.76	1.20%	1.50%	4%	
	2720	Plumbing			2.30	3.48	4.92	3.70%	4.50%	6.10%	
	2770	Heating, ventilating, air conditioning			4.97	6.85	10.15	7.20%	10.50%	11.90%	
	2900	Electrical			5.10	7.10	9.90	7.50%	9.60%	11.10%	
	3100	Total: Mechanical & Electrical		↓	12.05	16.70	24.50	18%	21.80%	26.50%	
620	0010	**OFFICES** Mid Rise (5 to 10 story)	R17100 -100	S.F.	66.50	81	110				620
	0020	Total project costs		C.F.	4.66	5.90	8.55				
	2720	Plumbing		S.F.	2.02	3.12	4.49	2.80%	3.70%	4.50%	
	2770	Heating, ventilating, air conditioning			5.05	7.25	11.55	7.60%	9.40%	11%	
	2900	Electrical			4.95	6.35	9.60	6.50%	8.20%	10%	
	3100	Total: Mechanical & Electrical		↓	12.65	16.15	32	17.90%	22.30%	29.90%	
630	0010	**OFFICES** High Rise (11 to 20 story)	R17100 -100	S.F.	80.50	103	127				630
	0020	Total project costs		C.F.	5.15	7.15	10.25				
	2900	Electrical		S.F.	4.12	5.95	9	5.80%	7%	10.50%	
	3100	Total: Mechanical & Electrical		"	15.05	19.25	35.50	15.40%	17%	34.10%	
640	0010	**POLICE STATIONS**	R17100 -100	S.F.	96.50	129	163				640
	0020	Total project costs		C.F.	7.85	9.45	13.15				
	0500	Masonry		S.F.	11.45	16.95	20.50	7.80%	10.60%	11.30%	
	1800	Equipment			1.59	6.85	11.05	2.10%	5.20%	9.80%	
	2720	Plumbing			5.50	8.90	12.05	5.60%	6.90%	11.30%	
	2770	Heating, ventilating, air conditioning			8.60	11.40	15.50	7%	10.70%	12%	
	2900	Electrical			10.70	16.05	20.50	9.70%	11.90%	14.90%	
	3100	Total: Mechanical & Electrical		↓	34	41	57.50	25.10%	31.30%	36.40%	
650	0010	**POST OFFICES**	R17100 -100	S.F.	75.50	95.50	119				650
	0020	Total project costs		C.F.	4.60	5.90	7				
	2720	Plumbing		S.F.	3.49	4.48	5.45	4.40%	5.60%	5.60%	
	2770	Heating, ventilating, air conditioning			5.45	6.75	7.50	7%	8.40%	10.20%	
	2900	Electrical			6.35	8.85	10.30	7.20%	9.60%	11%	
	3100	Total: Mechanical & Electrical		↓	11.25	21.50	27.50	16.40%	20.40%	22.30%	
660	0010	**POWER PLANTS**	R17100 -100	S.F.	560	715	1,300				660
	0020	Total project costs		C.F.	14.80	30	69				
	2900	Electrical		S.F.	38	80.50	120	9.50%	12.80%	18.40%	
	8100	Total: Mechanical & Electrical		"	94.50	310	690	32.50%	32.60%	52.60%	
670	0010	**RELIGIOUS EDUCATION**	R17100 -100	S.F.	61	79	89.50				670
	0020	Total project costs		C.F.	3.47	4.97	6.45				
	2720	Plumbing		S.F.	2.60	3.69	5.20	4.30%	4.90%	8.40%	
	2770	Heating, ventilating, air conditioning			6.10	6.90	7.60	9.80%	10%	11.80%	
	2900	Electrical			4.96	6.55	8.65	7.90%	9.10%	10.30%	
	3100	Total: Mechanical & Electrical		↓	12.80	22.50	26.50	17.60%	20.50%	22%	
690	0010	**RESEARCH** Laboratories and facilities	R17100 -100	S.F.	93.50	134	195				690
	0020	Total project costs		C.F.	6.95	11.70	15.75				
	1800	Equipment		S.F.	4.16	8.20	19.95	1.60%	5.20%	9.30%	
	2720	Plumbing			9.40	12	19.30	7.30%	9.20%	13.30%	
	2770	Heating, ventilating, air conditioning			8.40	28.50	33.50	7.20%	16.50%	17.50%	
	2900	Electrical			10.75	18.05	31	9.60%	12%	15.80%	
	3100	Total: Mechanical & Electrical		↓	25	47	90	23.90%	35.60%	41.70%	
700	0010	**RESTAURANTS**	R17100 -100	S.F.	90	116	151				700
	0020	Total project costs		C.F.	7.55	9.95	13.10				

Figure 7.8

101

A rough budget cost for the proposed building based on national averages is $85.86 per square foot.

There are two stages of project development when square foot cost estimates are most useful. The first is during the conceptual stage when few, if any, details are available. At this time, square foot costs are useful for ballpark budget purposes. As soon as details become available in the project design, the square foot approach should be discontinued and the project priced more accurately. The second is after the estimate is completed. Square foot costs can be used for verification and as a check against gross errors.

When using the figures in Division 17, it is recommended that the median cost column be consulted for preliminary figures if no additional information is available. When costs have been modified to account for project size and have been converted for location (see City Cost Indexes), the median numbers (as shown in Figure 7.8) should provide a fairly accurate base figure. This figure

Square Foot Base Size							
Building Type	Median Cost per S.F.	Typical Size Gross S.F.	Typical Range Gross S.F.	Building Type	Median Cost per S.F.	Typical Size Gross S.F.	Typical Range Gross S.F.
Apartments, Low Rise	$ 57.00	21,000	9,700 - 37,200	Jails	$174.00	40,000	5,500 - 145,000
Apartments, Mid Rise	72.00	50,000	32,000 - 100,000	Libraries	103.00	12,000	7,000 - 31,000
Apartments, High Rise	82.55	145,000	95,000 - 600,000	Medical Clinics	98.20	7,200	4,200 - 15,700
Auditoriums	95.30	25,000	7,600 - 39,000	Medical Offices	92.25	6,000	4,000 - 15,000
Auto Sales	58.95	20,000	10,800 - 28,600	Motels	70.65	40,000	15,800 - 120,000
Banks	128.00	4,200	2,500 - 7,500	Nursing Homes	92.10	23,000	15,000 - 37,000
Churches	86.05	17,000	2,000 - 42,000	Offices, Low Rise	76.95	20,000	5,000 - 80,000
Clubs, Country	85.80	6,500	4,500 - 15,000	Offices, Mid Rise	80.80	120,000	20,000 - 300,000
Clubs, Social	83.45	10,000	6,000 - 13,500	Offices, High Rise	103.00	260,000	120,000 - 800,000
Clubs, YMCA	83.70	28,300	12,800 - 39,400	Police Stations	129.00	10,500	4,000 - 19,000
Colleges (Class)	112.00	50,000	15,000 - 150,000	Post Offices	95.30	12,400	6,800 - 30,000
Colleges (Science Lab)	164.00	45,600	16,600 - 80,000	Power Plants	716.00	7,500	1,000 - 20,000
College (Student Union)	125.00	33,400	16,000 - 85,000	Religious Education	78.95	9,000	6,000 - 12,000
Community Center	89.70	9,400	5,300 - 16,700	Research	134.00	19,000	6,300 - 45,000
Court Houses	122.00	32,400	17,800 - 106,000	Restaurants	116.00	4,400	2,800 - 6,000
Dept. Stores	53.25	90,000	44,000 - 122,000	Retail Stores	56.60	7,200	4,000 - 17,600
Dormitories, Low Rise	91.95	25,000	10,000 - 95,000	Schools, Elementary	82.45	41,000	24,500 - 55,000
Dormitories, Mid Rise	120.00	85,000	20,000 - 200,000	Schools, Jr. High	84.00	92,000	52,000 - 119,000
Factories	51.60	26,400	12,900 - 50,000	Schools, Sr. High	83.95	101,000	50,500 - 175,000
Fire Stations	90.15	5,800	4,000 - 8,700	Schools, Vocational	83.65	37,000	20,500 - 82,000
Fraternity Houses	88.65	12,500	8,200 - 14,800	Sports Arenas	70.10	15,000	5,000 - 40,000
Funeral Homes	99.10	10,000	4,000 - 20,000	Supermarkets	56.80	44,000	12,000 - 60,000
Garages, Commercial	62.95	9,300	5,000 - 13,600	Swimming Pools	131.00	20,000	10,000 - 32,000
Garages, Municipal	80.55	8,300	4,500 - 12,600	Telephone Exchange	153.00	4,500	1,200 - 10,600
Garages, Parking	33.00	163,000	76,400 - 225,300	Theaters	84.05	10,500	8,800 - 17,500
Gymnasiums	83.30	19,200	11,600 - 41,000	Town Halls	92.45	10,800	4,800 - 23,400
Hospitals	157.00	55,000	27,200 - 125,000	Warehouses	38.10	25,000	8,000 - 72,000
House (Elderly)	77.95	37,000	21,000 - 66,000	Warehouse & Office	44.00	25,000	8,000 - 72,000
Housing (Public)	72.15	36,000	14,400 - 74,400				
Ice Rinks	80.20	29,000	27,200 - 33,600				

Figure 7.9

should then be adjusted according to the estimator's experience, local economic conditions, code requirements and the owner's particular requirements. There is no need to factor the percentage figures, as these should remain relatively constant from city to city.

Building Unit Prices and Assemblies for Your Own Work

(See "Assemblies (Systems) Estimates" in Chapter One.) Using the preceding information about bare costs, overhead, and profit, it should be possible to start to build your own database of unit costs, based on actual takeoff quantities and subcontractor prices. Your own experience as a general contractor or subcontractor should be recorded job by job, taking note of labor-hours, crew size, and material costs in the most accurate manner possible. Noting the jobs' logistics, building height, and quality of finishes, build up your own unit costs particular to your region and type of contracting. This data should be organized so that it is easily manipulated for material, labor, and units (such as LF, SF, CY, SY, EA), and can be changed or updated.

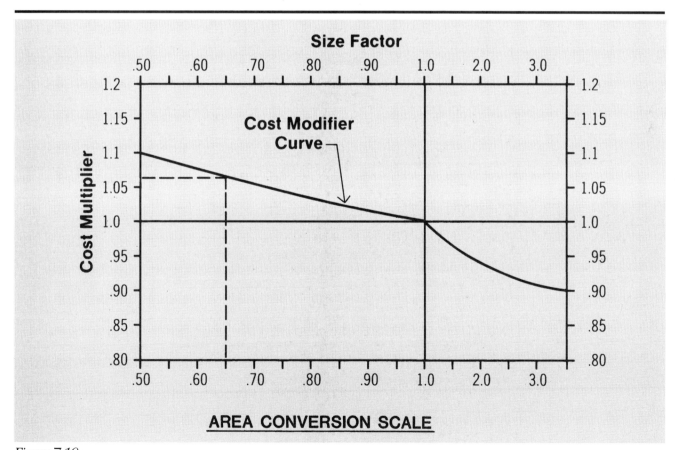

Figure 7.10

Repair & Remodeling

Cost figures in *Means Building Construction Cost Data* are based on new construction utilizing the most cost-effective combination of labor, equipment and material. The work is scheduled in the proper sequence to allow the various trades to accomplish their tasks in an efficient manner.

There are many factors unique to repair and remodeling that can affect project costs. The economy of scale associated with new construction often has no influence on the cost of repair and remodeling. Small quantities of components may have to be custom fabricated at great expense. Work schedule coordination between trades frequently becomes difficult; work area restrictions can lead to subcontractor quotations with start-up and shutdown costs in excess of the cost of the actual work involved. Some of the more prominent factors affecting repair and remodeling projects are listed below.

1. A large amount of cutting and patching may be required to match the existing construction. It is often more economical to remove entire walls rather than create many new door and window openings. This sort of tradeoff has to be carefully analyzed. Matching "existing conditions" may be impossible because certain materials may no longer be manufactured, and substitutions can be expensive. Piping and ductwork runs may not be as simple as they are in the case of new construction. Wiring may have to be snaked through existing walls and floors.

2. Dust and noise protection of adjoining non-construction areas can involve a substantial number of special precautions and may alter normal construction methods.

3. The use of certain equipment may be curtailed as a result of the physical limitations of the project; workmen may be forced to use small equipment or hand tools.

4. Material handling becomes more costly due to the confines of an enclosed building. For multi-story construction, low capacity elevators and stairwells may be the only access to the upper floors.

5. Both existing and completed finish work will need to be protected in order to prevent damage during ensuing work.

6. Work may have to be done on other than normal shifts—around an existing production facility that has to stay in operation throughout the repair and remodeling. Costs for overtime work may be incurred as a result.

7. There may be an increased requirement for shoring and bracing to support the building while structural changes are being made, or to allow for the temporary storage of construction materials on above-grade floors.

These factors and their consequences, can significantly increase the costs for repair and remodeling work as compared with new construction. R.S. Means has developed a method to quantify these factors by adding percentages to the costs of work that is affected. These suggested percentages are shown in Figure 7.11 as minimums and maximums. The estimator must use sound judgment and experience when applying these factors. The effects of each of these factors should be considered in the planning, bidding, and construction phases in order to minimize the potential increased costs associated with repair and remodeling projects.

There are other considerations to be anticipated in estimating for repair and remodeling. Weather protection may be required for existing structures, and during the installation of new windows or heating systems. Pedestrian protection is often required in urban areas. On small projects and because of local conditions, it may be necessary to pay a tradesman for a minimum of four hours for a task that actually requires less time. Unit prices should be used with caution in situations when these kinds of minimum charges may be incurred.

All of the above factors can be anticipated and the basic costs developed before a repair and remodeling project begins. It is the hidden problems, the unknowns, that pose the greatest challenge to the estimator and cause the most anxiety. These problems are often discovered during demolition and may be impossible to anticipate. Projects may be delayed due to these unexpected conditions, and these delays ultimately increase construction costs. Other parts of the project, and thus their cost, are also affected. Only experience, good judgment and a thorough knowledge of the existing structure can help to reduce the number of unknowns and their potential effects.

City Cost Indexes

The unit prices in *Means Building Construction Cost Data* are national averages. When they are to be applied to a particular location, these prices must be adjusted to local conditions. R.S. Means has developed the City Cost Indexes for just that purpose. Following the crews table in the back of the Means *Building Construction Cost Data* are tables of indexes for 719 U.S. and Canadian cities based on a 30 major city average of 100. The figures are broken down into material and installation for the 16 MasterFormat divisions, as shown in Figure 7.12. Please note that for each city there is a weighted average for the material, installation and total indexes. This average is based on the relative contribution of each division to the construction process as a whole. The information in Figure 7.13 does not represent any one building type but instead, all building types as a whole. The figures may be used as a general guide to determine how much time should be spent on each portion of an estimate. When doing an estimate, more time should be spent on the divisions that have a higher percent contribution to the project. Caution should be exercised when using set percentages for projects that have unusually high or low division contributions.

In addition to adjusting costs in *Means Building Construction Cost Data* for particular locations, the City Cost Index can also be used to adjust costs from one city to another. For example, the price of a particular building type is known for City A. In order to budget the costs of the same building type in City B, the following calculation can be made:

$$\frac{\text{City B Index} \times \text{City A Cost}}{\text{City A Index}} = \text{City B Cost}$$

While City Cost Indexes provide a means to adjust prices for location, the Historical Cost Index, as shown in Figure 7.14, provides a means to adjust for time. Using the same principle as above, a time adjustment factor can be calculated:

$$\frac{\text{Index for Year X}}{\text{Index for Year Y}} = \text{Time Adjustment Factor}$$

This time adjustment factor can be used to determine what the budget costs would be for a particular building type in Year X, based on costs for a similar building type known from Year Y. Used in conjunction, the two indexes allow for cost adjustments from one city during a given year to another city in another year (the present or otherwise). For example, an office building built in

San Francisco in 1983 originally cost $1,000,000. How much will a similar building cost in Phoenix in 2002? Adjustment factors are developed as shown previously using data from Figures 7.12 and 7.14:

$$\frac{\text{Phoenix index}}{\text{San Francisco index}} = \frac{89.4}{124.4} = 0.72$$

$$\frac{\text{2002 index}}{\text{1983 index}} = \frac{126.1}{80.2} = 1.57$$

Original cost × location adjustment × time adjustment = Proposed new cost
$1,000,000 × 0.72 × 1.57 = $1,130,400

Reference Tables

Throughout the Unit Price pages of *Means Building Construction Cost Data*, certain line items contain reference numbers (see Figure 7.15) which refer the reader to expanded data and information in the back of the book. The Reference Table section contains over ninety pages of tables, charts, definitions and costs, all of which corroborate the unit price data. The development of many unit costs is explained and detailed in this section. This information can be very helpful to the estimator, particularly when more information is needed about materials and considerations that have gone into a unit price line item. Figure 7.16 is an example of the development of cost (see Figure 7.15 for the line item that references R03310-100).

01200 | Price & Payment Procedures

		01250	Contract Modification Procedures	CREW	DAILY OUTPUT	LABOR-HOURS	UNIT	2002 BARE COSTS				TOTAL INCL O&P	
								MAT.	LABOR	EQUIP.	TOTAL		
400	0010	**FACTORS** Cost adjustments											400
	0100	Add to construction costs for particular job requirements											
	0500	Cut & patch to match existing construction, add, minimum					Costs	2%	3%				
	0550	Maximum						5%	9%				
	0800	Dust protection, add, minimum						1%	2%				
	0850	Maximum						4%	11%				
	1100	Equipment usage curtailment, add, minimum						1%	1%				
	1150	Maximum						3%	10%				
	1400	Material handling & storage limitation, add, minimum						1%	1%				
	1450	Maximum						6%	7%				
	1700	Protection of existing work, add, minimum						2%	2%				
	1750	Maximum						5%	7%				
	2000	Shift work requirements, add, minimum							5%				
	2050	Maximum							30%				
	2300	Temporary shoring and bracing, add, minimum						2%	5%				
	2350	Maximum						5%	12%				
	2400	Work inside prisons and high security areas, add, minimum							30%				
	2450	Maximum							50%				

Figure 7.11

City Cost Indexes

		ARIZONA														ARKANSAS				
DIVISION		PHOENIX			PRESCOTT			SHOW LOW			TUCSON			BATESVILLE			CAMDEN		
		850,853			863			859			856 - 857			725			717		
		MAT.	INST.	TOTAL	MAT.	INST.	TOTAL	MAT.	INST.	TOTAL	MAT.	INST.	TOTAL	MAT.	INST.	TOTAL	MAT.	INST.	TOTAL
01590	EQUIPMENT RENTAL	.0	94.6	94.6	.0	93.5	93.5	.0	94.0	94.0	.0	94.0	94.0	.0	85.1	85.1	.0	85.1	85.1
02	SITE CONSTRUCTION	83.5	99.8	95.9	73.6	99.0	92.9	94.2	99.1	97.9	80.3	99.6	94.9	73.7	83.7	81.3	74.6	83.3	81.2
03100	CONCRETE FORMS & ACCESSORIES	104.3	76.7	80.4	100.9	65.3	70.0	108.4	65.2	70.8	103.7	76.2	79.8	81.7	50.9	54.9	81.3	35.8	41.8
03200	CONCRETE REINFORCEMENT	102.1	79.2	88.7	103.7	75.2	87.1	103.9	60.2	78.5	101.0	78.6	87.9	94.6	52.0	69.8	95.8	28.6	56.7
03300	CAST-IN-PLACE CONCRETE	99.3	84.7	93.4	97.2	71.9	87.0	98.4	71.6	87.6	99.3	84.5	93.3	79.7	50.9	68.1	81.7	37.6	63.9
03	CONCRETE	99.7	79.6	89.7	107.0	69.3	88.2	109.9	66.4	88.2	99.5	79.2	89.4	79.5	51.9	65.8	81.3	36.3	58.9
04	MASONRY	96.2	72.6	81.8	101.9	68.7	81.6	109.1	61.0	79.7	96.7	63.0	76.1	98.9	48.7	68.2	108.7	42.1	68.0
05	METALS	99.6	73.5	90.1	98.1	68.2	87.2	97.8	62.8	85.1	98.9	71.6	88.9	94.8	63.6	83.4	94.8	54.1	79.9
06	WOOD & PLASTICS	102.2	77.7	89.6	102.2	64.2	82.7	106.2	64.3	84.7	102.7	77.7	89.9	85.7	52.5	68.6	85.4	37.1	60.6
07	THERMAL & MOISTURE PROTECTION	108.5	75.9	93.1	109.4	69.8	90.7	109.9	69.1	90.6	109.2	70.1	90.7	97.6	51.1	75.6	97.4	38.6	69.6
08	DOORS & WINDOWS	99.2	75.9	93.6	101.1	64.6	92.3	97.0	60.9	88.3	96.0	75.9	91.1	95.8	48.0	84.3	92.0	31.7	77.5
09200	PLASTER & GYPSUM BOARD	92.0	76.9	82.3	88.9	63.1	72.3	93.2	63.1	73.8	92.3	76.9	82.4	91.2	51.9	65.8	91.2	36.1	55.6
095,098	CEILINGS & ACOUSTICAL TREATMENT	105.6	76.9	86.3	102.9	63.1	76.2	104.2	63.1	76.6	107.0	76.9	86.8	96.1	51.9	66.4	96.1	36.1	55.7
09600	FLOORING	102.4	76.4	96.0	97.2	58.5	87.7	104.1	66.6	94.8	101.5	58.7	91.0	108.8	57.0	96.0	108.6	18.9	86.5
097,099	WALL FINISHES, PAINTS & COATINGS	109.9	69.0	85.9	98.7	55.7	73.5	109.9	55.7	78.2	107.4	55.7	77.2	95.8	50.0	69.0	95.8	39.0	62.6
09	FINISHES	98.7	75.6	86.7	95.1	62.6	78.1	100.2	64.1	81.4	98.4	70.7	84.0	96.8	52.2	73.6	96.9	33.7	64.0
10 - 14	TOTAL DIV. 10000 - 14000	100.0	83.2	96.5	100.0	80.8	96.0	100.0	81.1	96.1	100.0	83.2	96.5	100.0	54.8	90.6	100.0	50.1	89.6
15	MECHANICAL	100.1	83.2	92.6	100.0	79.8	91.1	96.4	76.9	87.7	100.0	74.5	88.7	96.4	48.0	74.8	96.4	36.9	69.9
16	ELECTRICAL	107.2	66.9	79.2	100.0	53.7	67.8	98.7	53.7	67.4	101.6	64.9	76.1	100.1	54.9	68.7	94.3	45.8	60.6
01 - 16	WEIGHTED AVERAGE	99.8	78.2	89.4	99.9	70.7	85.8	100.4	68.5	85.0	98.8	74.3	87.0	94.0	55.4	75.4	94.0	44.7	70.1

		CALIFORNIA																		
DIVISION		SACRAMENTO			SALINAS			SAN BERNARDINO			SAN DIEGO			SAN FRANCISCO			SAN JOSE		
		942,956 - 958			939			923 - 924			919 - 921			940 - 941			951		
		MAT.	INST.	TOTAL	MAT.	INST.	TOTAL	MAT.	INST.	TOTAL	MAT.	INST.	TOTAL	MAT.	INST.	TOTAL	MAT.	INST.	TOTAL
01590	EQUIPMENT RENTAL	.0	101.2	101.2	.0	99.7	99.7	.0	100.9	100.9	.0	97.4	97.4	.0	106.8	106.8	.0	100.0	100.0
02	SITE CONSTRUCTION	97.7	110.6	107.5	117.6	107.3	109.8	72.1	107.8	99.2	102.5	102.1	102.2	132.2	110.7	115.9	137.9	100.6	109.6
03100	CONCRETE FORMS & ACCESSORIES	108.6	123.1	121.2	106.5	126.9	124.2	109.5	114.3	113.7	103.8	112.3	111.2	110.5	139.5	135.7	104.4	138.8	134.3
03200	CONCRETE REINFORCEMENT	100.0	114.5	108.4	107.0	115.0	111.6	105.5	114.2	110.6	110.5	114.1	112.6	112.6	115.6	114.4	103.5	115.3	110.4
03300	CAST-IN-PLACE CONCRETE	110.9	114.5	112.4	104.1	114.7	108.4	71.0	115.5	89.0	107.8	104.2	106.4	129.5	119.8	125.6	123.6	118.5	121.5
03	CONCRETE	116.8	117.3	117.1	125.1	119.2	122.2	81.1	113.9	97.4	116.2	109.1	112.7	127.8	126.8	127.3	121.4	126.1	123.8
04	MASONRY	120.7	115.3	117.4	107.1	124.0	117.4	92.7	113.3	105.3	102.4	111.5	108.0	150.6	132.4	139.5	149.1	128.9	136.7
05	METALS	97.7	100.2	98.6	109.5	102.1	106.8	110.4	100.5	106.8	110.5	99.7	106.6	106.8	107.8	107.2	111.1	108.0	110.0
06	WOOD & PLASTICS	101.7	122.1	112.2	98.9	126.8	113.2	98.6	111.6	105.3	97.5	108.8	103.3	107.6	140.8	124.7	102.0	140.5	121.8
07	THERMAL & MOISTURE PROTECTION	122.1	114.0	118.3	107.8	120.4	113.8	118.3	114.0	116.2	113.9	106.9	110.6	111.4	133.7	122.0	107.3	133.1	119.5
08	DOORS & WINDOWS	117.8	117.2	117.7	102.2	122.6	107.1	100.1	111.1	102.8	104.2	108.9	105.3	110.7	130.1	115.3	93.1	129.9	102.0
09200	PLASTER & GYPSUM BOARD	96.6	122.7	113.5	94.7	127.6	116.0	95.0	112.1	106.1	89.1	109.0	102.0	100.1	141.6	126.9	94.1	141.6	124.8
095,098	CEILINGS & ACOUSTICAL TREATMENT	122.5	122.7	122.6	120.6	127.6	125.3	114.9	112.1	113.0	115.1	109.0	111.0	120.7	141.6	134.8	109.6	141.6	131.1
09600	FLOORING	115.2	112.8	114.6	120.7	121.9	121.0	126.4	105.4	121.2	118.5	116.2	118.0	112.4	121.9	114.8	116.6	121.9	117.9
097,099	WALL FINISHES, PAINTS & COATINGS	115.3	126.1	121.6	113.2	129.7	122.9	110.7	112.8	111.9	111.9	112.8	112.4	116.9	140.6	130.8	114.9	129.7	123.6
09	FINISHES	115.6	122.4	119.2	116.9	127.1	122.2	112.2	112.1	112.2	111.4	113.0	112.3	116.7	137.0	127.3	113.7	135.6	125.1
10 - 14	TOTAL DIV. 10000 - 14000	100.0	132.5	106.8	100.0	133.1	106.9	100.0	114.1	102.9	100.0	114.4	103.0	100.0	136.9	107.7	100.0	136.1	107.5
15	MECHANICAL	100.0	117.4	107.8	96.6	112.1	103.5	96.6	111.3	103.1	100.1	111.8	105.3	100.1	163.1	128.1	100.2	146.6	120.8
16	ELECTRICAL	103.9	100.9	101.8	94.0	121.3	113.0	101.0	108.7	106.4	92.8	94.0	93.6	109.3	151.8	138.8	109.9	145.2	134.4
01 - 16	WEIGHTED AVERAGE	107.5	113.1	110.2	106.4	117.5	111.8	99.1	110.3	104.5	105.3	106.3	105.8	112.3	137.3	124.4	109.7	131.4	120.2

Figure 7.12

Cost Breakdown By MasterFormat Divisions

NO.	DIVISION	%	NO.	DIVISION	%	NO.	DIVISION	%
0159	CONSTRUCTION EQUIPMENT*	7.4%	051	Structural Metal Framing	3.2%	092	Plaster & Gypsum Board	2.0%
023	Earthwork	4.9	052,053	Metal Joists & Decking	8.1	093,094	Tile & Terrazzo	2.7
024,0245	Tunneling & LB Elements	0.1	055	Metal Fabrications	0.8	095,098	Ceilings & Acoustical Treatment	2.9
027	Bases & Pavements	0.3	5	METALS	12.1	096	Flooring	2.4
025,026	Utility Services & Drainage	0.2	061	Rough Carpentry	0.9	097,099	Wall Finishes, Paints & Coatings	1.0
028	Site Improvements	0.1	062	Finish Carpentry	0.3	9	FINISHES	11.0
029	Planting	0.7	6	WOOD & PLASTICS	1.2	10-14	DIVISIONS 10-14	5.7
2	SITE CONSTRUCTION	6.3	071	Dampproofing & Waterproofing	0.1	150-152,154	Plumbing: Piping, Fixtures & Equip.	12.9
031	Conc. Forms & Accessories	3.2	072,078	Thermal, Fire & Smoke Protection	1.3	153,139	Fire Protection/Suppression	3.0
032	Concrete Reinforcement	1.8				155	Heat Generation Equipment	2.2
033	Cast-in Place Concrete	5.4	074,075	Roofing & Siding	1.0	156,158	Air Conditioning & Ventilation	3.0
034	Precast Concrete	2.3	076	Flashing & Sheet Metal	0.3	15	MECHANICAL	21.1
			7	THERMAL & MOISTURE PROT.	2.7	16	ELECTRICAL	11.8
3	CONCRETE	12.7	081-083	Doors & Frames	7.0		TOTAL (Div. 1-16)	100.0%
0405	Basic Masonry Mat. & Methods	0.6						
042	Masonry Units	6.5	088,089	Glazing & Glazed Curtain Walls	1.0	* Percentage for contractor equipment is		
044	Stone	0.3	8	DOORS & WINDOWS	8.0	spread among divisions and included		
4	MASONRY	7.4				above for information only		

Figure 7.13

Historical Cost Indexes

The table below lists both the Means Historical Cost Index based on Jan. 1, 1993 = 100 as well as the computed value of an index based on Jan. 1, 2002 costs. Since the Jan. 1, 2002 figure is estimated, space is left to write in the actual index figures as they become available through either the quarterly "Means Construction Cost Indexes" or as printed in the "Engineering News-Record." To compute the actual index based on Jan. 1, 2002 = 100, divide the Historical Cost Index for a particular year by the actual Jan. 1, 2002 Construction Cost Index. Space has been left to advance the index figures as the year progresses.

Year	Historical Cost Index Jan. 1, 1993 = 100 (Est.)	(Actual)	Current Index Based on Jan. 1, 2002 = 100 (Est.)	(Actual)	Year	Historical Cost Index Jan. 1, 1993 = 100 (Actual)	Current Index Based on Jan. 1, 2002 = 100 (Est.)	(Actual)	Year	Historical Cost Index Jan. 1, 1993 = 100 (Actual)	Current Index Based on Jan. 1, 2002 = 100 (Est.)	(Actual)
Oct 2002					July 1987	87.7	69.5		July 1969	26.9	21.3	
July 2002					1986	84.2	66.8		1968	24.9	19.7	
April 2002					1985	82.6	65.5		1967	23.5	18.6	
Jan 2002	126.1		100.0	100.0	1984	82.0	65.0		1966	22.7	18.0	
July 2001		125.1	99.2		1983	80.2	63.6		1965	21.7	17.2	
2000		120.9	95.9		1982	76.1	60.4		1964	21.2	16.8	
1999		117.6	93.3		1981	70.0	55.5		1963	20.7	16.4	
1998		115.1	91.3		1980	62.9	49.9		1962	20.2	16.0	
1997		112.8	89.5		1979	57.8	45.8		1961	19.8	15.7	
1996		110.2	87.4		1978	53.5	42.4		1960	19.7	15.6	
1995		107.6	85.3		1977	49.5	39.3		1959	19.3	15.3	
1994		104.4	82.8		1976	46.9	37.2		1958	18.8	14.9	
1993		101.7	80.7		1975	44.8	35.5		1957	18.4	14.6	
1992		99.4	78.9		1974	41.4	32.8		1956	17.6	14.0	
1991		96.8	76.8		1973	37.7	29.9		1955	16.6	13.2	
1990		94.3	74.8		1972	34.8	27.6		1954	16.0	12.7	
1989		92.1	73.1		1971	32.1	25.5		1953	15.8	12.5	
1988		89.9	71.3		1970	28.7	22.8		1952	15.4	12.2	

Figure 7.14

108

03200 | Concrete Reinforcement

		03240	Fibrous Reinforcing	CREW	DAILY OUTPUT	LABOR-HOURS	UNIT	2002 BARE COSTS MAT.	LABOR	EQUIP.	TOTAL	TOTAL INCL O&P	
300	0010		**FIBROUS REINFORCING**										300
	0100		Synthetic fibers, add to concrete				Lb.	3.79			3.79	4.17	
	0110		1-1/2 lb. per C.Y.				C.Y.	5.85			5.85	6.45	
	0150		Steel fibers, add to concrete				Lb.	.44			.44	.48	
	0155		25 lb. per C.Y.				C.Y.	11			11	12.10	
	0160		50 lb. per C.Y.					22			22	24	
	0170		75 lb. per C.Y.					34			34	37.50	
	0180		100 lb. per C.Y.					44			44	48.50	

03300 | Cast-In-Place Concrete

		03310	Structural Concrete		CREW	DAILY OUTPUT	LABOR-HOURS	UNIT	2002 BARE COSTS MAT.	LABOR	EQUIP.	TOTAL	TOTAL INCL O&P	
200	0010		**CONCRETE, FIELD MIX** FOB forms 2250 psi	R03310 -080				C.Y.	68			68	75	200
	0020		3000 psi					"	71			71	78	
220	0010		**CONCRETE, READY MIX** Regular weight	R03310 -040										220
	0020		2000 psi					C.Y.	65			65	71.50	
	0100		2500 psi						58			58	64	
	0150		3000 psi CN	R03310 -050					69			69	76	
	0200		3500 psi						71			71	78	
	0300		4000 psi						74			74	81.50	
	0350		4500 psi						75.50			75.50	83	
	0400		5000 psi CN						77			77	84.50	
	0411		6000 psi						88			88	96.50	
	0412		8000 psi						143			143	158	
	0413		10,000 psi						203			203	224	
	0414		12,000 psi						246			246	270	
	1000		For high early strength cement, add						10%					
	1010		For structural lightweight with regular sand, add						25%					
	2000		For all lightweight aggregate, add						45%					
	3000		For integral colors, 2500 psi, 5 bag mix											
	3100		Red, yellow or brown, 1.8 lb. per bag, add					C.Y.	14.55			14.55	16	
	3200		9.4 lb. per bag, add						78.50			78.50	86.50	
	3400		Black, 1.8 lb. per bag, add						17.25			17.25	19	
	3500		7.5 lb. per bag, add						72.50			72.50	79.50	
	3700		Green, 1.8 lb. per bag, add						34.50			34.50	38	
	3800		7.5 lb. per bag, add						165			165	181	
240	0010		**CONCRETE IN PLACE** Including forms (4 uses), reinforcing	R03310 -010										240
	0050		steel, including finishing unless otherwise indicated											
	0300		Beams, 5 kip per L.F., 10' span	R03310 -100	C-14A	15.62	12.804	C.Y.	223	385	46	654	905	
	0350		25' span		"	18.55	10.782		209	325	38.50	572.50	790	
	0500		Chimney foundations, industrial, minimum	R04210 -055	C-14C	32.22	3.476		140	100	1.11	241.11	310	
	0510		Maximum		"	23.71	4.724		163	136	1.51	300.51	395	
	0700		Columns, square, 12" x 12", minimum reinforcing		C-14A	11.96	16.722		242	505	60	807	1,125	
	0720		Average reinforcing			10.13	19.743		340	595	71	1,006	1,400	
	0740		Maximum reinforcing			9.03	22.148		420	670	79.50	1,169.50	1,600	
	0800		16" x 16", minimum reinforcing			16.22	12.330		193	375	44.50	612.50	845	
	0820		Average reinforcing			12.57	15.911		305	480	57	842	1,175	
	0840		Maximum reinforcing			10.25	19.512		420	590	70	1,080	1,475	
	0900		24" x 24", minimum reinforcing			23.66	8.453		163	256	30.50	449.50	620	
	0920		Average reinforcing			17.71	11.293		255	340	40.50	635.50	865	

Figure 7.15

R03310-100 Average C.Y. of Concrete (cont.)

Columns, Square Tied	16″ Square				24″ Square			
	Material		Installation		Material		Installation	
4000 psi concrete	1 C.Y.	$ 74.00	1 C.Y.	$ 39.56	1 C.Y.	$ 74.00	1 C.Y.	$ 25.80
Formwork, 4 uses	81 S.F. @ .80	64.80	81 S.F. @ 3.74	302.94	54 S.F. @ .80	43.20	54 S.F. @ 3.74	201.96
Reinforcing steel, avg.	634 # @ .27	171.18	634 # @ .25	158.50	524 # @ .27	141.48	524 # @ .25	131.00
Total per C.Y.		$309.98		$501.00		$258.68		$358.76

Note: Reinforcing of 16″ and 24″ square columns can vary from 144 lb. to 972 lb. per C.Y.

Columns, Round Tied Reinforced	16″ Diameter				24″ Diameter			
	Material		Installation		Material		Installation	
4000 psi concrete	1 C.Y.	$ 74.00	1 C.Y.	$ 39.56	1 C.Y.	$ 74.00	1 C.Y.	$ 25.80
Formwork, fiber forms	19 L.F. @ 5.95	113.05	19 L.F. @ 6.30	119.70	9 L.F. @ 10.75	96.75	9 L.F. @ 6.80	61.20
Reinforcing steel, avg.	606 # @ .27	163.62	606 # @ .25	151.50	600 # @ .27	162.00	600 # @ .25	150.00
Ties	70 # @ .54	37.80	70 # @ .37	25.90	18 # @ .54	9.72	18 # @ .37	6.66
Total per C.Y.		$388.47		$336.66		$342.47		$243.66

Note: Reinforcing of 16″ and 24″ diameter columns vary from 160 lb. to 1150 lb. per C.Y.

Flat Plate	125 psf, 15′ Span				125 psf, 25′ Span			
	Material		Installation		Material		Installation	
4000 psi concrete	1 C.Y.	$ 74.00	1 C.Y.	$ 21.58	1 C.Y.	$ 74.00	1 C.Y.	$ 18.26
Formwork, 4 uses	61 S.F. @ .76	46.36	61 S.F. @ 2.50	152.50	33 S.F. @ .76	25.08	33 S.F. @ 2.50	82.50
Reinforcing steel	126 # @ .27	34.02	126 # @ .19	23.94	113 # @ .27	30.51	113 # @ .19	21.47
Total per C.Y.		$154.38		$198.02		$129.59		$122.23

Flat Slab with drops	125 psf, 20′ Span				125 psf, 30′ Span			
	Material		Installation		Material		Installation	
4000 psi concrete	1 C.Y.	$ 74.00	1 C.Y.	$ 21.58	1 C.Y.	$ 74.00	1 C.Y.	$ 18.26
Formwork, 4 uses	43 S.F. @ .94	40.42	43 S.F. @ 2.58	110.94	29 S.F. @ .94	27.26	29 S.F. @ 2.58	74.82
Reinforcing steel	119 # @ .27	32.13	119 # @ .19	22.61	131 # @ .27	35.37	131 # @ .19	24.89
Total per C.Y.		$146.55		$155.13		$136.63		$117.97

Grade Wall, 8′ High	8″ Thick				15″ Thick			
	Material		Installation		Material		Installation	
3000 psi concrete	1 C.Y.	$ 69.00	1 C.Y.	$ 13.94	1 C.Y.	$ 69.00	1 C.Y.	$ 11.95
Formwork, 4 uses	81 S.F. @ .80	64.80	81 S.F. @ 2.92	236.52	41 S.F. @ .80	32.80	41 S.F. @ 2.92	119.72
Reinforcing steel	44 # @ .27	11.88	44 # @ .18	7.92	43 # @ .27	11.61	43 # @ .18	7.74
Total per C.Y.		$145.68		$258.38		$113.41		$139.41

Fiberglass Pan Joists, 24″ One Way	125 psf, 15′ Span				125 psf, 25′ Span			
	Material		Installation		Material		Installation	
4000 psi concrete	1 C.Y.	$ 74.00	1 C.Y.	$ 18.26	1 C.Y.	$ 74.00	1 C.Y.	$ 15.82
Formwork, 4 uses	49 S.F. @ .78	38.22	49 S.F. @ 2.92	143.08	40 S.F. @ .78	31.20	40 S.F. @ 2.92	116.80
Reinforcing steel	47 # @ .27	12.69	47 # @ .18	8.46	81 # @ .27	21.87	81 # @ .18	14.58
19″ fiberglass pans	44 S.F. @ 5.54	243.76	44 S.F. @ .41	18.04	37 S.F. @ 5.54	204.98	37 # @ .41	15.17
Total per C.Y.		$368.67		$187.84		$332.05		$162.37

Pile Caps	Under 5 C.Y.				Over 10 C.Y.			
	Material		Installation		Material		Installation	
3000 psi concrete	1.05 C.Y.	$69.00	1 C.Y.	$13.94	1.05 C.Y.	$69.00	1 C.Y.	$ 5.84
Formwork, 4 uses	20 S.F. @ .66	13.20	20 S.F. @ 2.38	47.60	10 S.F. @ .66	6.60	10 S.F. @ 2.38	23.80
Reinforcing steel	52 # @ .27	14.04	52 # @ .15	7.80	48 # @ .27	12.96	48 # @ .15	7.20
Total per C.Y.		$96.24		$69.34		$88.56		$36.84

Slab on Grade	4″ Thick				6″ Thick			
	Material		Installation		Material		Installation	
3500 psi concrete	1.05 C.Y.	$71.00	1 C.Y.	$11.41	1.05 C.Y.	$71.00	1 C.Y.	$ 7.60
Formwork, 4 uses	9.2 L.F. @ .28	2.58	9.2 L.F. @ 1.51	13.89	6 L.F. @ .28	1.68	6 L.F. @ 1.51	9.06
W.W.Fabric	81 S.F. @ .07	5.67	81 S.F. @ .13	10.53	54 S.F. @ .07	3.78	54 S.F. @ .13	7.02
Total per C.Y.		$79.25		$35.83		$76.46		$23.68

Figure 7.16

Computerized Estimating Methods

Computerized Estimating Methods

Computers have clearly added speed, power, and accuracy to construction estimating. They make it possible to produce more estimates in less time, break a job down to a more detailed level for better cost control, manage change orders more easily, test "what if" scenarios with ease, and integrate estimating with other commonly used construction applications. The objective of this chapter is to provide an overview of unit price estimating through the use of computerized estimating methods. Figure 8.1 is a chart showing an overview of the various levels of estimating programs and their features/applications.

Automating the estimating function is an evolutionary process. There are many levels of computerized estimating, which vary in functionality, sophistication, time required to learn, and, most of all, price. Many estimators make the mistake of immediately transitioning their manual estimating into a fully integrated estimating software system without learning the basics of what a computer can do for the estimating process. Successful implementation of an estimating software system will not happen overnight. It usually takes many months of training and user interaction to get a system working to its full capability.

It is recommended that computerized estimating be introduced through a multi-step process. First, learn the basics of computers and estimating through the use of industry-standard spreadsheet software programs, such as *Lotus®* 123, or *Microsoft Excel*. Next, introduce spreadsheet add-on programs which will provide more versatility than a standard spreadsheet program. Finally, consider upgrading to a complete estimating software application which will utilize the computer's full power. This last step usually allows estimating software to be integrated with other related software applications, such as job cost accounting, scheduling, project management, and CAD.

The basis of any computerized estimating system is its unit price cost database. No matter how sophisticated or user-friendly construction estimating software is, its overall success depends upon the completeness, functionality, and accuracy of the cost data, and the methods by which it is used. Without a fully functional database, construction estimating software is nothing more than a very expensive calculator.

One effective way to move through the different levels of computerized estimating is using industry standard data as the foundation of the system. Utilizing *Means CostWorks*, for example, it is possible to seamlessly transition an

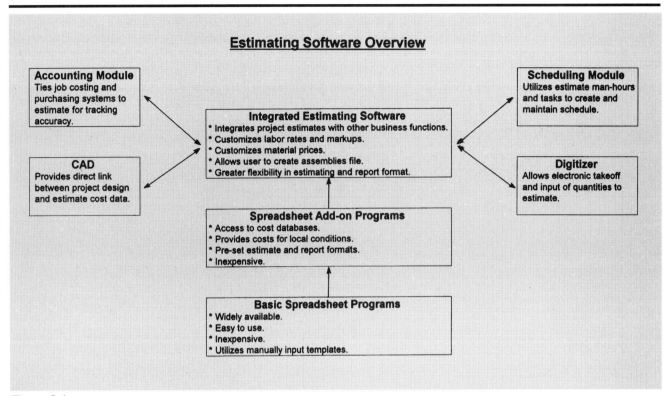

Estimating Software Overview

Accounting Module
Ties job costing and purchasing systems to estimate for tracking accuracy.

CAD
Provides direct link between project design and estimate cost data.

Integrated Estimating Software
* Integrates project estimates with other business functions.
* Customizes labor rates and markups.
* Customizes material prices.
* Allows user to create assemblies file.
* Greater flexibility in estimating and report format.

Scheduling Module
Utilizes estimate man-hours and tasks to create and maintain schedule.

Digitizer
Allows electronic takeoff and input of quantities to estimate.

Spreadsheet Add-on Programs
* Access to cost databases.
* Provides costs for local conditions.
* Pre-set estimate and report formats.
* Inexpensive.

Basic Spreadsheet Programs
* Widely available.
* Easy to use.
* Inexpensive.
* Utilizes manually input templates.

Figure 8.1

114

estimating system from one step to the next, easily updating the cost data and integrating it into thousands of pre-built assemblies. While it is always possible to create a customized cost database from historical job costs, *Means CostWorks* will provide a foundation of costs, as well as the necessary framework to successfully estimate using a computer.

Basic Spreadsheet Programs

A computerized estimating system should perform three basic functions:

1. Calculate costs.
2. Store and manage data.
3. Generate hard copy reports.

Most industry-standard spreadsheet software programs, such as *Lotus*® 123, and *Excel*, meet these requirements. The premise of every spreadsheet program is to automate a calculation sheet. Spreadsheet programs calculate costs in vertical columns and horizontal rows using basic mathematical functions. Figure 8.2 shows a sample screen with a spreadsheet that might be used by a contractor for estimating purposes.

Everyone who has ever done construction estimating has been in the situation of rushing to finish an estimate in order to submit a bid by a certain deadline. Subcontractor quotes usually don't come in until the last minute, and it is a huge task to recalculate costs to complete the bid. Spreadsheet programs can recalculate costs in fractions of a second, which makes it fairly simple to rework the estimate prior to a bid, and allow "what if" analysis for fine tuning.

Another major capability of spreadsheet programs is managing and sorting information. A typical line item of cost data consists of some sort of a line number, description of task, material costs, labor costs, and equipment costs. There is information on crews and productivities which needs to be maintained. Typical construction databases are in the thousands and tens of thousands of line items. Most spreadsheet programs allow information to be entered, managed, and sorted numerically or alphabetically.

When dealing with a client, presentation is very important. Spreadsheets make it possible to print out professional looking estimates, using different fonts, styles and formats. Most versions also have a graphics capability. Charts and graphs can be produced based on information from an estimate spreadsheet, and a company logo can be incorporated into reports.

There are some limitations in using spreadsheets. Information within a spreadsheet must be organized using the same field layout. A significant problem is likely to be difficulty locating certain line items within the database and bringing that information back into the estimate. Scrolling through a database of thousands of line items is a time consuming and cumbersome process. Keeping such a database up-to-date also requires a major investment of time. Furthermore, it is usually not possible to apply a standardized cost database to standard spreadsheets without an add-on program, discussed in the next section. Industry standard databases are important because they provide the ability to update information through database companies such as R.S. Means.

These limitations can be overcome by features offered in the next category of computerized estimating: spreadsheet add-on programs. These are systems that provide the basic features of stand-alone (integrated) estimating systems, but cost only a fraction of a full system. Spreadsheet add-on programs are inexpensive, and usually offer the opportunity to upgrade to a complete stand-alone estimating system in the future.

R.S. Means Incorporated
63 Smiths Lane
Kingston, MA 02364-0800
Assembly Costlist

Quantity	Line Number	Description	Ext. Material Incl O&P	Ext. Installation Incl O&P	Ext. Sub Incl O&P	Ext. Total Incl O&P
A Substructure						
120.000	A10101102500	Strip footing, load 5.1KLF, soil cap 3 KSF, 24"wide x 12"deep, reinf	$1,125	$2,025		$3,150
800.000	A10301202280	Slab on grade, 4" thick, light industrial, reinforced	$1,400	$2,125		$3,525
120.000	A20201101520	Fdn walls, CIP,4' wall height,dir chute,.099 CY/LF,4.8 PLF,8" thick	$1,725	$4,150		$5,875
Totals for A Substructure			**$4,250**	**$8,300**		**$12,550**
B Shell						
800.000	B10201161100	St jsts & dk on brg walls,20'bay,20PSF supimp,13.5"d,40PSF total	$1,050	$560		$1,600
1,100.000	B2010109144O	Conc blk wall - regular weight,hollow,8x8x16",2000 PSI,no core	$1,875	$5,550		$7,425
800.000	B30101202000	Sgl ply memb, EPDM, 45mils, fully adhered	$715	$555		$1,275
Totals for B Shell			**$3,640**	**$6,665**		**$10,300**
Job Totals			**$7,890**	**$14,965**		**$22,850**

Note:The estimate in this figure is organized using UNIFORMAT II.

Figure 8.2

116

Spreadsheet Add-on Programs

Spreadsheet add-on programs allow users to develop more power within their spreadsheet estimating systems. Estimators who require a larger database or coverage of more locations may decide to purchase an add-on program. Specialty contractors who expand into general contracting also find these systems helpful.

CostWorks is an interactive software program designed to organize and simplify construction cost estimating. The CostWorks CD-ROM contains the data from 19 of Means annual cost books. Users can purchase data by book title, or in discounted packages. The Design Professional's Package includes *Square Foot Model Costs, Assemblies Cost Data, Building Construction Cost Data,* and *Interior Cost Data*. All of the cost data and reference information in Means printed cost books is contained in CostWorks, in addition to several other features, such as on-line cost data updates and links to product information on the Internet. The Square Foot and Assemblies CostWorks files provide for summary estimates, and enable users to customize contractor mark-ups, architectural fees, and user fees. See Figure 8.3 for a sample spreadsheet using Means Cost Works.

The advantage of CostWorks is the ability to complete an estimate quickly and print it in a customized format, suitable for presentation to clients. (Users should first be familiar with basic "Windows" navigation techniques and with the organization of Means cost data.)

The CostWorks format is not an estimating program, but is a database of cost information that allows users to select data, apply a quantity to it, apply a localized cost factor, and put it on a *CostList*. The CostList is a listing of data, adjusted for location, with quantities applied to each item. The CostList totals columns so that the sum of the items provides a total project cost. This CostList is the beginning of an estimate. If no modifications are necessary, the CostList *is* the estimate. If modifications to material price, labor burden, or mark-ups are needed, the CostList can be exported to a spreadsheet program (such as *Excel* or *Lotus* 1-2-3) for adjustments.

Many estimators find that spreadsheet add-on programs satisfy all of their estimating needs. They can update their cost data, and create, edit, and print accurate estimates. Others may want even more functionality with a program.

Spreadsheet add-on systems such as CostWorks may fall short in some areas. For example, while unit price cost data is the foundation of all estimating systems, assemblies or building systems developed from unit price cost data take greater advantage of what a computerized system can do. Assemblies make it possible to perform a substantially accurate conceptual level estimate in a fraction of the time it takes to create unit price cost estimates. It is also helpful in taking an estimate from concept to bid, or assemblies to unit price. Spreadsheet programs do not provide this relational database capability. A stand-alone system could take a prebuilt assembly, and break that system down into the unit prices that make it up, whereas spreadsheets with add-ons lack that capability.

Some features that may not be available with spreadsheets/add-ons programs include the ability to automate the quantity takeoff process by using a digitizer or wheel pen, or extracting quantities from a computer-aided design (CAD) program. These types of technologies are only available through complete stand-alone estimating systems, such as offered by Timberline® Software Corp. or Estimating Systems, Inc.

Finally, some estimators are interested in integrating their estimating function into other related software applications, such as job cost accounting, scheduling,

Estimate: Repair Garage
Description: Unit Price Estimating
Location: Michigan Ave
Square Footage: 4000 SF

Date: 2002
City Index: Chicago, IL

Line Number	Description	Quantity	Unit	Labor Hours	Ext. Material	Ext. Labor	Ext. Equipment	Ext. Total	Ext. Total Incl O&P
Division 2 Site Construction									
0206015000100	Borrow, bank run gravel, load at pit, haul 2 mi RT&sprd w/200 HP dozer	75.000	C.Y.	0.047	$970.00	$85.00	$207.00	$1,250.00	$1,425.00
0222087510710	Site dml, pavement removal, bituminous	3,200.000	S.Y.	0.058		$4,500.00	$3,775.00	$8,300.00	$11,100.00
0222576000010	Saw cutting, asphalt, up to 3" deep	720.000	L.F.	0.015	$158.00	$266.00	$194.00	$620.00	$800.00
0231510000010	Backfill, by hand, no compaction, light soil	125.000	C.Y.	0.571		$1,525.00		$1,525.00	$2,375.00
0231510000600	Backfill, by hand, compaction in 6" layers, vibrating plate, add	125.000	C.Y.	0.133		$355.00	$119.00	$475.00	$680.00
0231540001200	Excavating, bulk bank measure, FE loader, track mtd, 1.5 CY cap = 70 CY/hr	140.000	C.Y.	0.021		$77.00	$61.50	$139.00	$186.00
0231590000050	Excavate trench, cont ftg, no sht/dewtrg, 1-4'D,3/8 CY tractor ldr/backhoe	125.000	C.Y.	0.107		$330.00	$140.00	$470.00	$655.00
0231594000700	Excav, util trench, W/chain trencher, 12hp, oper walking,8" W trench,24" D	365.000	L.F.	0.023		$223.00	$84.00	$305.00	$430.00
Totals for Division 2 Site Construction					$1,128.00	$7,361.00	$4,580.50	$13,084.00	$17,651.00

Figure 8.3

118

project management and CAD software. To develop these types of integrations, a complete, stand-alone estimating system is generally required.

Complete Stand-alone Estimating Solutions

A complete stand-alone estimating system does not require any other software in order to function. it incorporates all of the capabilities previously covered, links to electronic quantity takeoff devices, and the ability to "communicate" with other construction software applications. By integrating with the scheduling module, estimate man-hours can be used to create and maintain the project schedule. A link to the accounting module allows job cost accounting. CAD and digitizer programs can also be connected to provide electronic takeoff to estimate. Other features include the ability to customize labor rates and markups, input the company's own material prices, and build assemblies, as well as great flexibility in report formats.

Some estimating systems that fall into this category are developed by companies such as Timberline® Software Corp., MC², Estimating Systems, Inc., G2 Estimator, or Building Systems Design (see figure for 8.4 leading software providers). There are hundreds of these systems, and selecting the appropriate one can be a tedious process that involves collecting printed information, seeing and using demos, and evaluating products and their costs.

Pulsar Estimating Software

Originally designed by R.S. Means Company, Inc., PULSAR makes full use of R.S. Means' Cost Data. Estimating Systems, Inc.'s construction cost estimating software, PULSAR, is used by commercial and government construction offices to create estimates for facilities construction and maintenance. PULSAR was specifically developed for use with Job Order, Delivery Order, and SABER contracts. The software allows estimators to compare and contrast construction bids, and provides a standard basis for contract negotiations. PULSAR is also GSA approved for government purchase and available for use with Windows operating systems.

Summary

There are many steps in the process of automating the estimating process. A standard spreadsheet program may be adequate, an entry level estimating system may be a more appropriate level, or you may find it necessary to obtain the full capabilities of a complete stand-alone estimating system. Estimating software at the lower levels (basic spreadsheet and add-on programs) can be extremely useful, while the cost is moderate. Stand-alone systems do require a significant investment in terms of cost and learning curve, but the potential benefits in terms of accuracy, increased productivity, and control are tremendous.

Means Data for Computer Estimating

Construction costs for software applications.

30 unit price and assemblies cost databases are available through a number of leading estimating software providers, including:

- 3D International
- 4Clicks-Solutions, LLC
- AEPCO, Inc.
- Applied Computer Technologies
- ArenaSoft Estimating
- Ares Corporation
- AssetWork CSI-Maximus
- Benchmark, Inc.
- Building Systems Design, Inc.
- cManagement
- Construction Data Controls, Inc.
- Computerized Micro Solutions
- Conac Group
- CProjects, Inc
- Eagle Point Software
- Estimating Systems, Inc.
- G2 Estimator
- G/C EMUNI, Inc.
- Geac Commercial Systems, Inc.
- Hard Dollar
- IQ Beneco
- Luqs International
- MC^2 Management Computer Controls
- Prism Computer Corporation
- Quest Solutions, Inc.
- Sanders Software, Inc.
- Sinisoft, Inc.
- Timberline Software Corp.
- TMA Systems, Inc.
- US Cost, Inc.
- Vertigraph, Inc.
- Wendlware
- WinEstimator, Inc.

Figure 8.4

Estimating by MasterFormat Division

Chapter 9

Estimating by MasterFormat Division

There are two basic approaches to performing a Unit Price estimate. One is to proceed with the estimate, the quantity takeoff, and pricing in a sequence similar to the order in which a building is constructed. Using this method, concrete for footings is priced very early in the estimate, after excavation but before backfill. The concrete used for upper level floor slabs is estimated much later, after structural steel. This is the approach taken for a systems or assemblies estimate. The advantage to this method is that experienced estimators can visualize the construction process while proceeding with the estimate. As a result, omissions are less likely. The basic disadvantage of this system is that it is not easy to determine and track costs for each division (or subcontract). Costs for a particular trade may be spread throughout the estimate.

The second approach to unit price estimating is by MasterFormat division. Most architectural specifications today are written according to this format, using the sixteen MasterFormat divisions. Trades and subcontracts are generally limited to work within one division (e.g., Mechanical, or Masonry). It makes sense, therefore, that the estimate should also be organized in the MasterFormat system. Using this method, the estimator may have to be more careful to include all required work. Nevertheless, the advantages outweigh potential drawbacks. Most specifications contain references to related work for all divisions. These references serve as an aid to assure completeness, but the estimator for the general contractor still has the responsibility of making sure that subcontractors include all that is specified and required for the appropriate portions of the project. The estimator must also decide what work will be subcontracted and what will be performed by the work force of the general contractor. These decisions can have a significant effect on the final cost of a project. Traditionally, work has been done at a lower cost using the general contractor's own labor force. With the specialization of trades, however, subcontractors can often perform work faster, and thus at a lower cost.

Cost accounting and control is another area affected by the choice of estimating approach. Whichever method is chosen should be used consistently, from estimating to field reporting to final analysis. The first approach, estimating in the sequence of construction, brings the same challenge to cost accounting as it does to estimating—the difficulty of keeping similar items grouped together as the work of single trades.

The second method of estimating and cost accounting, by MasterFormat division, may take longer when it comes to compiling all costs for each division (e.g., until all concrete or site work is complete). However, each trade will be separated and the records will be in accordance with the specifications. Since material purchases and scheduling of manpower and subcontractors are based on the construction sequence, these items should be derived from the project schedule. This schedule is established after and formulated from the unit price estimate, and is the basis of project coordination.

Most project specifications contain a list of alternates which must be included with a submitted bid. These alternates become a series of mini-estimates within the total project estimate. Each alternate may include deductions of some items from the project, originally specified, as well as the addition of other items. Often mistakenly regarded as incidentals, alternates are often left until the project estimate is complete. in order to efficiently determine alternate costs, without performing a completely new estimate for each, the project estimate must be organized with the alternates in mind. If items are to be deducted as part of an alternate, they must be separated in the project estimate. Similarly, when measuring for quantities, pertinent dimensions and quantities for the "adds" of an alternate should be listed separately. If forethought is used, alternates can be estimated quickly and accurately.

The following discussions and the sample estimate are presented by MasterFormat division for the reasons mentioned above. Ultimately, the choice of method is not crucial as long as all items are included and all costs accounted.

Sample Estimate: Project Description

The project for the sample estimate is a three-story office building with a basement parking garage. The 2.6 acre site is located in a suburban area with good access and sufficient material storage area. The proposed building is composite steel frame with an aluminum panel and glass curtain wall system. Plans for the building are shown in Figures 9.1 to 9.15. These plans are provided for illustrative purposes only and represent the major features of the building. In actuality, such a building would require many more sheets of drawings with more plans, details, sections and elevations. Also included would be a full set of specifications. For the purposes of the example, however, these figures provide sufficient information. The quantities as given in the sample estimate represent realistic conditions. Assumptions have also been made for items not shown in Figures 9.1 to 9.15 that would normally be included in the plans and specifications of a project of this type.

Division 1: General Requirements

When estimating by MasterFormat division, the estimator must be careful to include all items which, while not attributable to one trade or to the physical construction of the building, are nevertheless required to successfully complete the project. These items are included in the General Requirements section. Often referred to as the "General Conditions" or "Project Overhead," they are usually set forth in the first part of the specifications. Some requirements may not be directly specified even though they are required to perform the work. Standardized sets of General Conditions have been developed by various segments of the construction industry, such as those by the American Institute of Architects, the Consulting Engineers Council/U.S., National Society of

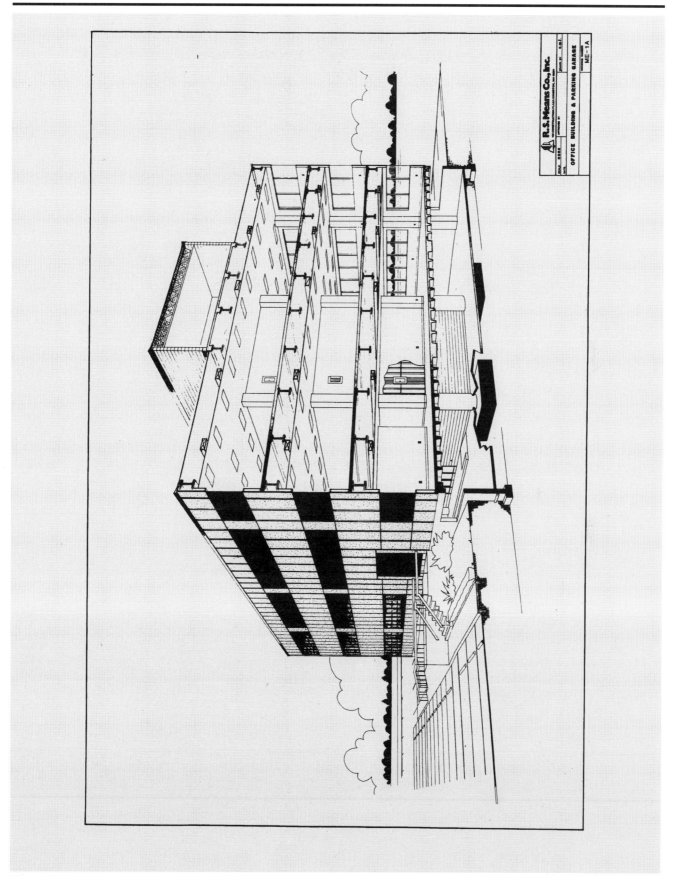

Figure 9.1

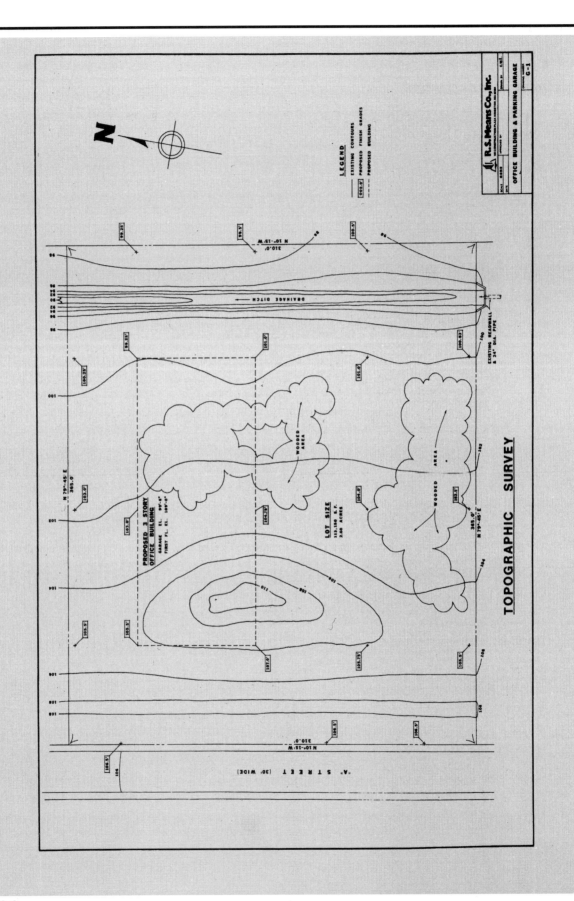

TOPOGRAPHIC SURVEY

Figure 9.2

126

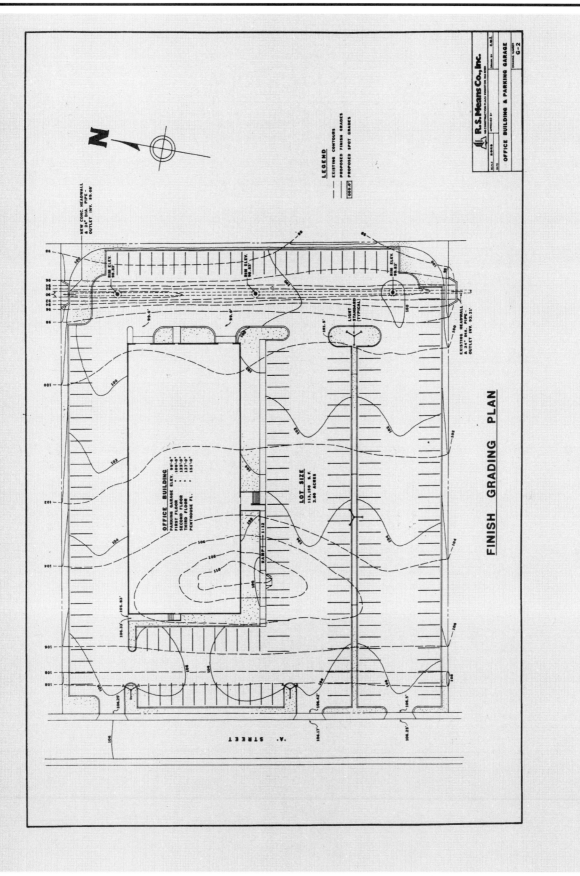

Figure 9.3

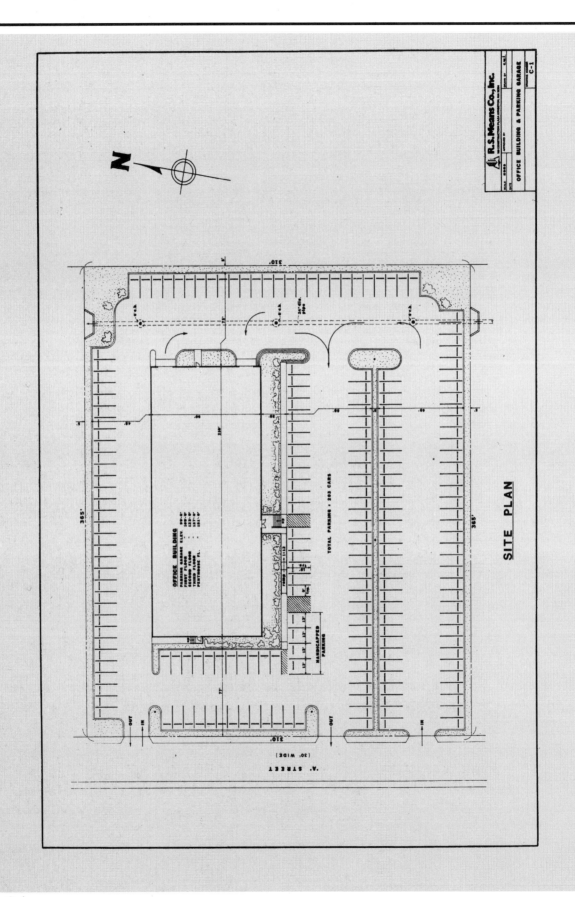

Figure 9.4

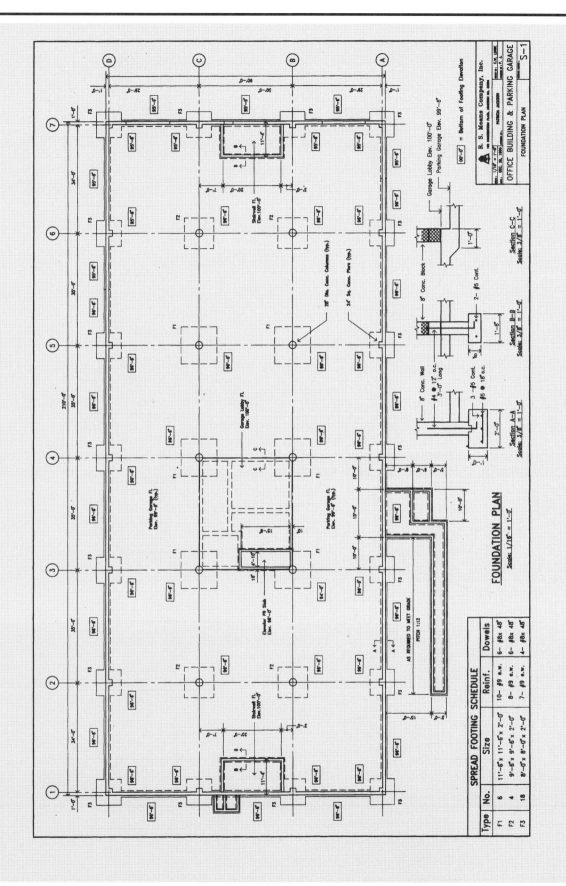

Figure 9.5a

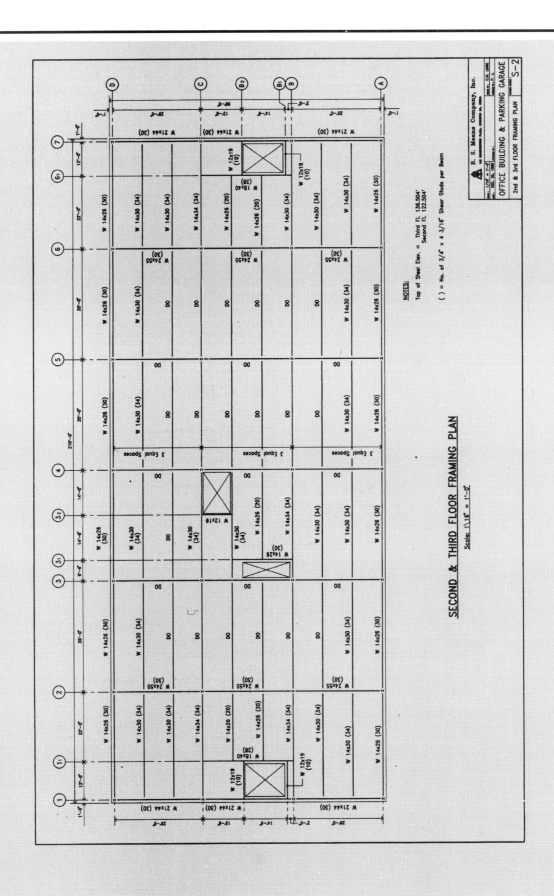

SECOND & THIRD FLOOR FRAMING PLAN

Scale: 1/16" = 1'-0"

NOTES:

Top of Steel Elev. = Third Fl. 136.504'
Second Fl. 122.504'

( ) = No. of 3/4" x 4 3/16" Shear Studs per Beam

OFFICE BUILDING & PARKING GARAGE

2nd & 3rd FLOOR FRAMING PLAN   S-2

R. S. Means Company, Inc.

Figure 9.5b

130

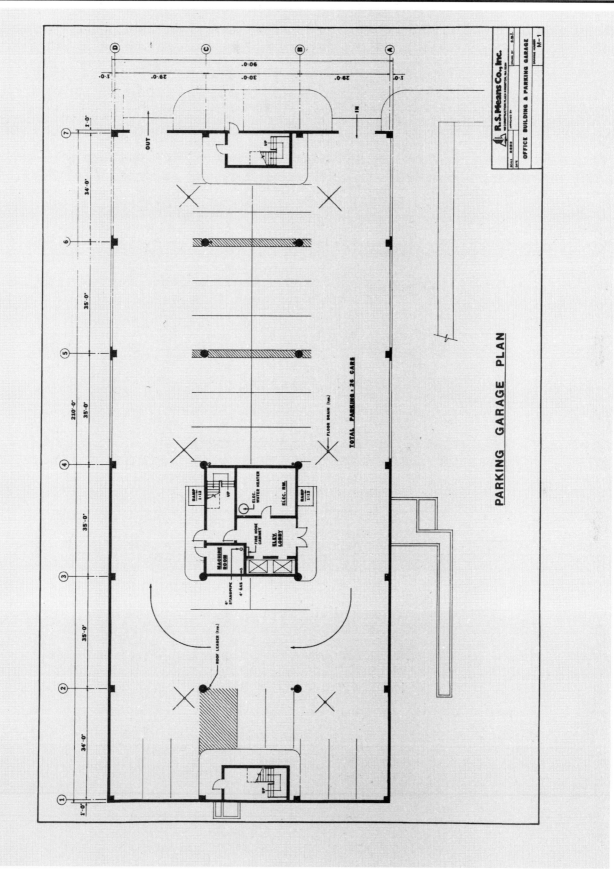

Figure 9.6

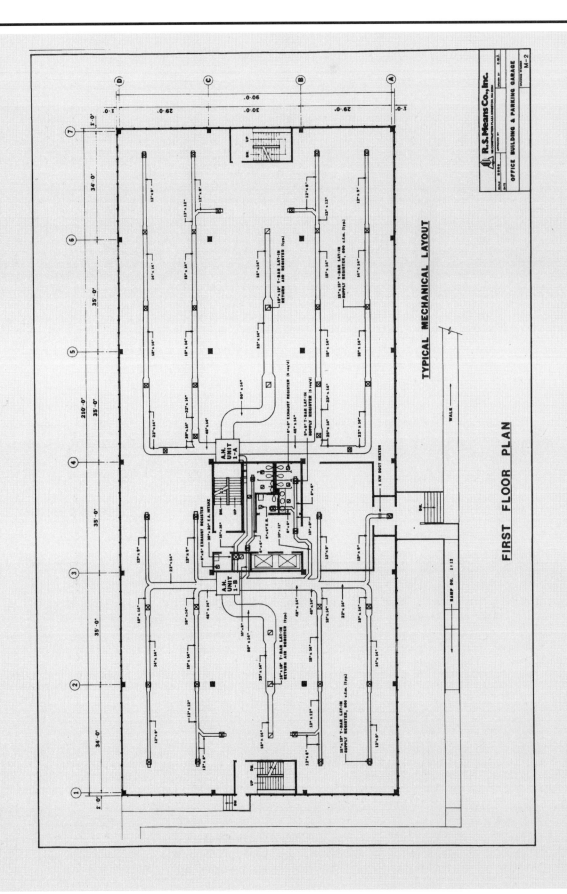

TYPICAL MECHANICAL LAYOUT

FIRST FLOOR PLAN

Figure 9.7

132

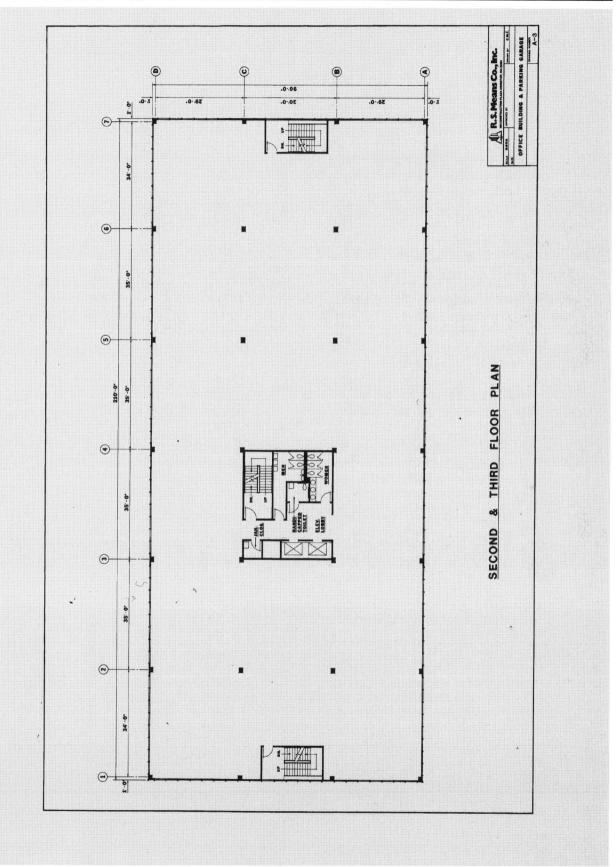

SECOND & THIRD FLOOR PLAN

Figure 9.8

133

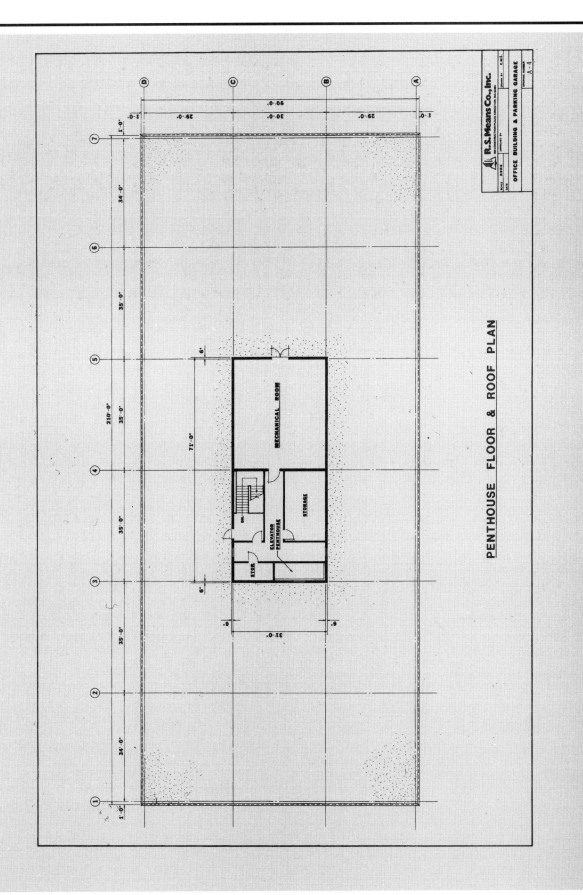

Figure 9.9

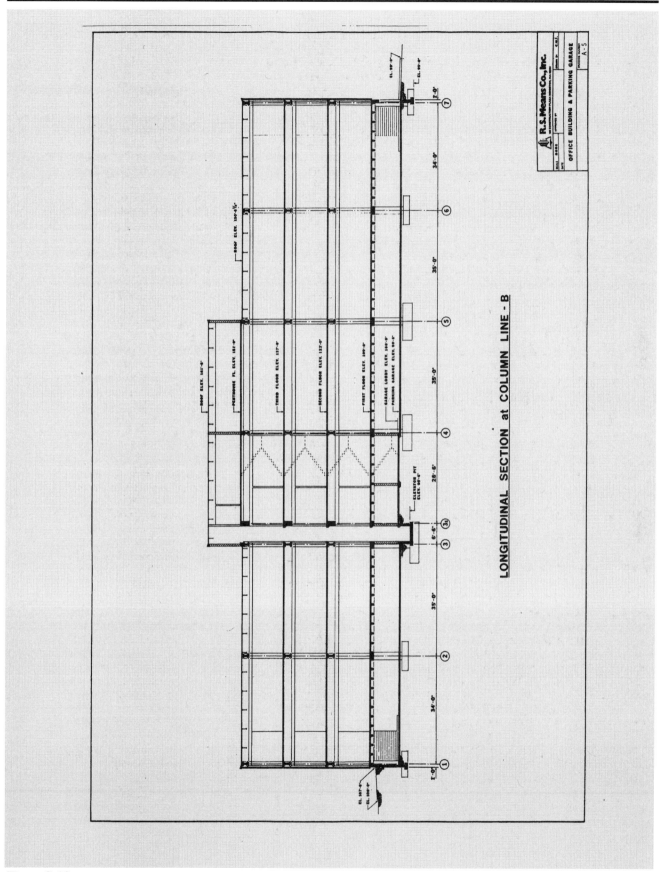

Figure 9.10

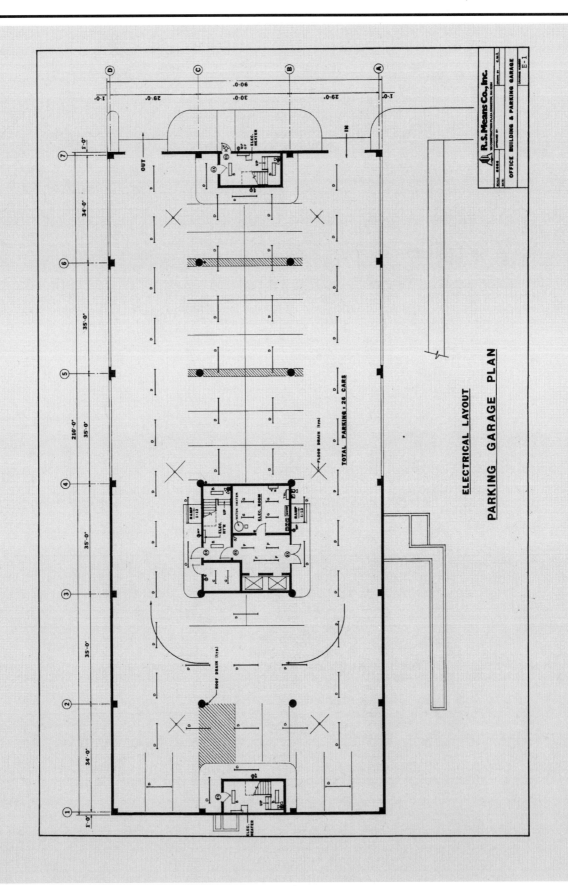

ELECTRICAL LAYOUT

PARKING GARAGE PLAN

Figure 9.11

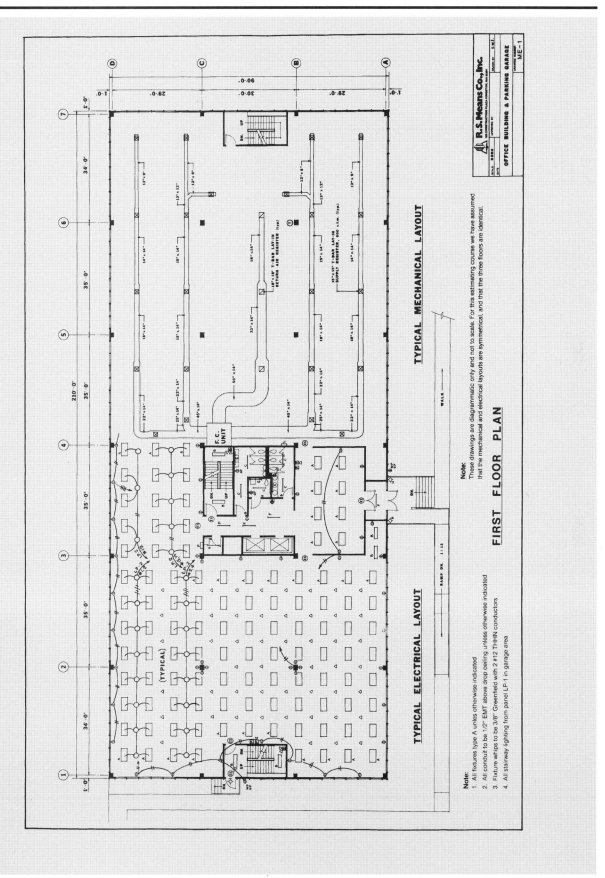

Figure 9.12

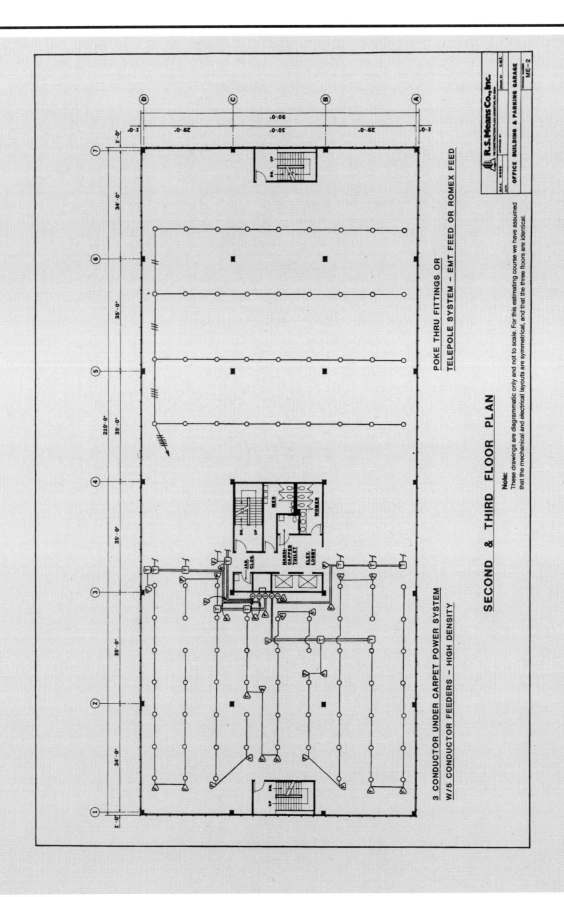

Figure 9.13

138

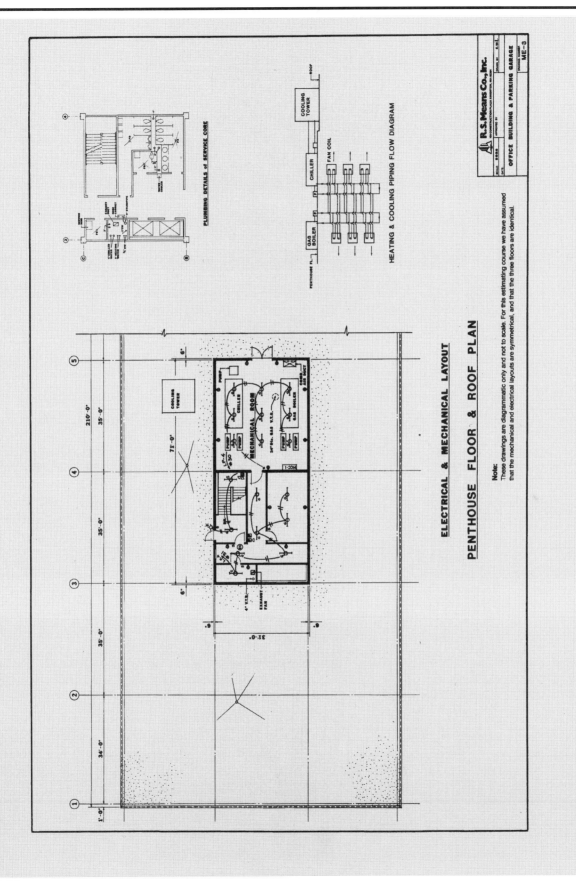

Figure 9.14

139

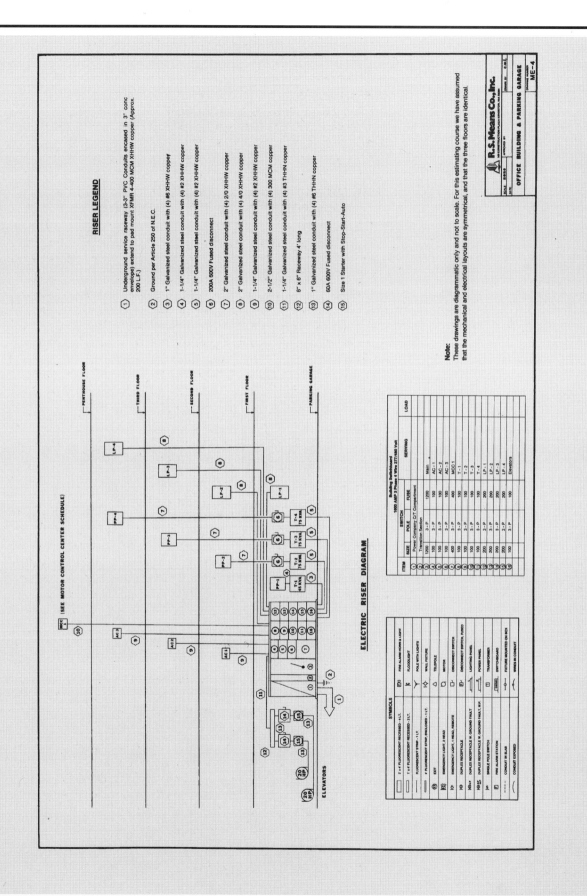

Figure 9.15

140

Professional Engineers, and others. These standardized documents usually include:

- Definitions
- Contract document descriptions
- Contractor's rights and responsibilities
- Architect-engineer's authority and responsibilities
- Owner's rights and responsibilities
- Variation from contract provisions
- Payment requirements and restrictions
- Requirements for the performance of the work
- Insurance and bond requirements
- Job conditions and operation

Since these documents are generic, additions, deletions and modifications unique to specific projects are often included in Supplementary General Conditions.

Estimated costs for Division 1 are often recorded on a standardized form or checklist similar to the Project Overhead Summary shown in Figures 9.16 and 9.17. Such preprinted forms or checklists are helpful to be sure that all requirements are included and priced. Many of the costs are dependent upon work in other divisions or on the total cost and/or time duration of the job. Project overhead costs should be determined throughout the estimating process and finalized when all other divisions have been estimated and a preliminary schedule established.

The following are brief discussions of various items that may be included as project overhead. The goal is to develop an approach to the estimate which will assure that all project requirements are included.

Personnel

Job site personnel may be included as either project overhead or office overhead, often depending upon the size of the project and the contractor's accounting methods. For example, if a project is large enough to require a full-time superintendent, then all costs for that person (time related) may be charged completely as project overhead. If the superintendent is responsible for a number of smaller jobs, then the expense may be either included in office overhead, or proportioned for each job. The same principles may apply to the salaries of field engineers, project managers, and time keepers.

If there is no full-time field engineer on the project, then costs may be incurred for establishing grades and setting stakes, building layout, producing as-built drawings, shop drawings or maintenance records. With regard to surveying and layout of the building and roads, it is important to determine who is the responsible party—the owner, general contractor or appropriate subcontractor. This responsibility should be established, if not done so in the contract documents, so that costs may be properly allocated.

Depending upon the size of the project, a carpenter and/or a laborer may be assigned to the job on a full-time basis for miscellaneous work. In such cases, workers are directly responsible to the job superintendent for various tasks; the costs of this work would not be attributable to any specific division and would most appropriately be included as project overhead.

Equipment

As discussed in Chapter 4, equipment costs may be recorded as project overhead or included in each appropriate division. Some equipment, however,

PROJECT OVERHEAD SUMMARY

PROJECT: Office Building   ESTIMATE NO:

LOCATION: ARCHITECT: DATE:

QUANTITIES BY: PRICES BY: EXTENSIONS BY: CHECKED BY:

DESCRIPTION	QUANTITY	UNIT	MATERIAL/EQUIP. UNIT	MATERIAL/EQUIP. TOTAL	LABOR UNIT	LABOR TOTAL	TOTAL COST UNIT	TOTAL COST TOTAL
Job Organization: Superintendent								
Project Manager								
Timekeeper & Material Clerk								
Clerical								
Safety, Watchman & First Aid								
Travel Expense: Superintendent								
Project Manager								
Engineering: Layout								
Inspection / Quantities								
Drawings								
CPM Schedule								
Testing: Soil								
Materials								
Structural								
Equipment: Cranes								
Concrete Pump, Conveyor, Etc.								
Elevators, Hoists								
Freight & Hauling								
Loading, Unloading, Erecting, Etc.								
Maintenance								
Pumping								
Scaffolding								
Small Power Equipment / Tools								
Field Offices: Job Office, Trailer								
Architect / Owner's Office								
Temporary Telephones								
Utilities								
Temporary Toilets								
Storage Areas & Sheds								
Temporary Utilities: Heat								
Light & Power								
PAGE TOTALS								

Figure 9.16

DESCRIPTION	QUANTITY	UNIT	MATERIAL/EQUIP.		LABOR		TOTAL COST	
			UNIT	TOTAL	UNIT	TOTAL	UNIT	TOTAL
Totals Brought Forward								
Winter Protection: Temp. Heat/Protection								
Snow Plowing								
Thawing Materials								
Temporary Roads								
Signs & Barricades: Site Sign								
Temporary Fences								
Temporary Stairs, Ladders & Floors								
Photographs								
Clean Up								
Dumpster								
Final Clean Up								
Continuous - One Laborer								
Punch List								
Permits: Building								
Misc.								
Insurance: Builders Risk - Additional Rider								
Owner's Protective Liability								
Umbrella								
Unemployment Ins. & Social Security								
(See Estimate Summary)								
Bonds								
Performance (See Estimate Summary)								
Material & Equipment								
Main Office Expense (See Estimate Summary)								
Special Items								
Totals:								

Figure 9.17

143

will be used by more than one trade; examples are personnel or material hoists and cranes. The allocation of costs in these cases should be according to company practice.

Testing

Various tests may be required by the owner, local building officials or as specified in the contract documents. Depending upon the stage of design development and the role of the contractor, soil borings and/or percolation tests may be required. Usually this type of testing is the responsibility of the designer or engineer, paid for directly by the owner. Most testing during construction is required to verify conformance of the materials and methods to the requirements of the specifications. The most common testing is as follows:

Soil Compaction: Soil compaction is usually specified as a percentage of maximum density. Strict compaction methods are required under stabs on grade and at backfill of foundation walls and footings. Testing may be required every day (or possibly every lift). Soil samples may have to be lab tested for cohesiveness, permeability and/or water content. Watering may be required to achieve the specified compaction.

Concrete: Concrete tests may be required at two stages: slump tests during placement and compression tests after a specified curing time. Slump tests, indicating relative water and cement content, may be required for each truck load, after placement of a certain number of cubic yards, each day or per building section. Compression tests are performed on samples placed in cylinders usually after 7 and 28 days of curing time. The cylinders are tested by outside laboratories. If the concrete samples fail to meet design specifications, core drilling of in-place building samples for further testing may be required.

Miscellaneous Testing: Other testing may be necessary based on the type of construction and owner or architect/engineer requirements. Core samples of asphalt paving are often required to verify specified thickness. Steel connection and weld testing may be specified for critical structural points. Masonry absorption tests may be also required.

The costs of testing installed materials may be included—in different ways—as project overhead. One method is separately itemizing the costs for each individual test; or, a fixed allowance can be made based on the size and type of project. Budget costs derived from this second method are shown in Figure 9.18. A third approach is to include a percentage of total project costs.

Temporary Services

Required temporary services may or may not be included in the specifications. A typical statement in the specifications is: "Contractor shall supply all material, labor, equipment, tools, utilities and other items and services required for the proper and timely performance of the work and completion of the project." As far as the owner and designer are concerned, such a statement eliminates a great deal of ambiguity. To the estimator, this means many items that must be estimated. Temporary utilities, such as heat, light, power and water are a major consideration. The estimator must not only account for anticipated monthly (or time related) costs, but should also be sure that installation and removal costs are included, whether by the appropriate subcontractor or the general contractor.

01300 | Administrative Requirements

		01320	Construction Progress Documents	CREW	DAILY OUTPUT	LABOR-HOURS	UNIT	2002 BARE COSTS				TOTAL INCL O&P	
								MAT.	LABOR	EQUIP.	TOTAL		
200	0600		Rule of thumb, CPM scheduling, small job ($10 Million)				Job					.05%	200
	0650		Large job ($50 Million +)									.03%	
	0700		Including cost control, small job									.08%	
	0750		Large job				↓					.04%	

		01321	Construction Photos										
500	0010		**PHOTOGRAPHS** 8" x 10", 4 shots, 2 prints ea., std. mounting				Set	276			276	305	500
	0100		Hinged linen mounts					300			300	330	
	0200		8" x 10", 4 shots, 2 prints each, in color					320			320	355	
	0300		For I.D. slugs, add to all above					3.89			3.89	4.28	
	0500		Aerial photos, initial fly-over, 6 shots, 1 print ea., 8" x 10"					665			665	730	
	0550		11" x 14" prints					735			735	810	
	0600		16" x 20" prints					950			950	1,050	
	0700		For full color prints, add					40%				40%	
	0750		Add for traffic control area				↓	273			273	300	
	0900		For over 30 miles from airport, add per				Mile	4.96			4.96	5.45	
	1000		Vertical photography, 4 to 6 shots with										
	1010		different scales, 1 print each				Set	1,025			1,025	1,125	
	1500		Time lapse equipment, camera and projector, buy					3,575			3,575	3,925	
	1550		Rent per month				↓	530			530	585	
	1700		Cameraman and film, including processing, B.&W.				Day	1,250			1,250	1,375	
	1720		Color				"	1,250			1,250	1,375	

01400 | Quality Requirements

		01450	Quality Control	CREW	DAILY OUTPUT	LABOR-HOURS	UNIT	2002 BARE COSTS				TOTAL INCL O&P	
								MAT.	LABOR	EQUIP.	TOTAL		
500	0010		FIELD TESTING										500
	0015		For concrete building costing $1,000,000, minimum				Project					5,100	
	0020		Maximum									41,000	
	0050		Steel building, minimum									5,100	
	0070		Maximum									16,000	
	0100		For building costing, $10,000,000, minimum									32,500	
	0150		Maximum				↓					52,000	
	0200		Asphalt testing, compressive strength Marshall stability, set of 3				Ea.					165	
	0220		Density, set of 3									95	
	0250		Extraction, individual tests on sample									150	
	0300		Penetration									45	
	0350		Mix design, 5 specimens									200	
	0360		Additional specimen									40	
	0400		Specific gravity									45	
	0420		Swell test									70	
	0450		Water effect and cohesion, set of 6									200	
	0470		Water effect and plastic flow									70	
	0600		Concrete testing, aggregates, abrasion, ASTM C 131									150	
	0650		Absorption, ASTM C 127									46	
	0800		Petrographic analysis, ASTM C 295									850	
	0900		Specific gravity, ASTM C 127									55	
	1000		Sieve analysis, washed, ASTM C 136									65	
	1050		Unwashed									65	
	1200		Sulfate soundness				↓					125	

Figure 9.18

The time of year may also have an impact on the cost of temporary services. Snow removal costs must be anticipated in climates where such conditions are likely. If construction begins during a wet season, or there is a high ground water table, then dewatering may be necessary. Usually the boring logs give the contractor a feet for the probable ground water elevation. Logs should be examined for the time of year in which the borings were taken. If high infiltration rates are expected, wellpoints may be required. The pumping allowance is usually priced from an analysis of the expected duration and volume of water. This information dictates pump size, labor to install, and power to operate the pumps. This can be an expensive item, since the pumps may have to operate 168 hours per week during certain phases of construction.

An office trailer and/or storage trailers or containers are usually required and included in the specifications. Even if these items are owned by the contractor, costs should still be allocated to the job as part of project overhead. Telephone, utility and temporary toilet facilities are other costs in this category.

Depending upon the location and local environment, some security services may be required. In addition to security personnel or guard dogs, fences, gates, special fighting and alarms may also be needed. A guard shack with heat, power and telephone, can be an expensive temporary cost.

Temporary Construction

Temporary construction may also involve many items which are not specified in the construction documents. Temporary partitions, doors, fences and barricades may be required to delineate or isolate portions of the building or site. In addition to these items, railings, catwalks or safety nets may also be necessary for the protection of workers. Depending upon the project size, an OSHA representative may visit the site to assure that all such safety precautions are being observed.

Ramps, temporary stairs and ladders are often necessary during construction for access between floors. When the permanent stairs are installed, temporary wood fillers are needed in metal pan treads until the concrete fill is placed. Workers will almost always use a new, permanent elevator for access throughout the building. While this use is almost always restricted, precautionary measures must be taken to protect the doors and cab. Invariably, some damage occurs. Protection of any and all finished surfaces throughout the course of the project must be priced and included in the estimate.

Job Cleanup

An amount should always be carried in the estimate for cleanup of the grounds and the building, both during the construction process and upon completion. The cleanup can be broken down into three basic categories, and these can be estimated separately:

- Continuous (daily or otherwise) cleaning of the building and site.
- Rubbish handling and removal.
- Final cleanup.

Costs for continuous cleaning can be included as an allowance, or estimated by required man-hours. Rubbish handling should include barrels, a trash chute if necessary, dumpster rental and disposal fees. These fees vary depending upon the project, and a permit may also be required. Costs for final cleanup should be based upon past projects and may include subcontract costs for items such as the cleaning of windows and waxing of floors. Included in the costs for final cleanup may be an allowance for repair and minor damage to finished work.

Miscellaneous General Conditions

Many other items must be taken into account when costs are being determined for project overhead. Among the major considerations are:

1. *Scaffolding or Swing Staging* – It is important to determine who is responsible for rental, erection and dismantling of scaffolding. If a subcontractor is responsible, it may be necessary to leave the scaffolding in place long enough for use by other trades. Scaffolding is priced by the section or per hundred square feet.

2. *Small Tools* – An allowance, based on past experience, should be carried for small tools. This allowance should cover hand tools as well as small power tools for use by workers on the general contractor's payroll. Small tools have a habit of "walking" and a certain amount of replacement is necessary. Special tools like magnetic drills may be required for specific tasks.

3. *Permits* – Various types of permits may be required depending upon local codes and regulations. Following are some examples:
   a. General building permit
   b. Subtrade permits (mechanical, electrical, etc.)
   c. Street use permit
   d. Sidewalk use permit
   e. Permit to allow work on Sundays
   f. Rubbish burning permit (if allowed)
   g. Blasting permit

Both the necessity of the permit and the responsibility for acquiring it must be determined. If the work is being done in an unfamiliar location, local building officials should be consulted regarding unusual or unknown requirements.

4. *Insurance* – Insurance coverage for each project and locality—above and beyond normal, required operating insurance—should be reviewed to assure that coverage is adequate. The contract documents will often specify certain required policy limits. The need for specific policies or riders should be anticipated (for example, fire or XCU—explosion collapse, underground).

Other items commonly included in project overhead are: photographs, job signs, sample panels and materials for owner/architect approval, and an allowance for replacement of broken glass. For some materials, such as imported goods or custom fabricated items, both shipping costs and off site storage fees can be expected. An allowance should be included for anticipated costs pertaining to punchlist items. These costs are likely to be based on past experience.

Some project overhead costs can be calculated at the beginning of the estimate. Others will be included as the estimating process proceeds. Still other costs are estimated last since they are dependent upon the total cost and duration of the project. Because many of the overhead items are not directly specified, the estimator must use experience and visualize the construction process to assure that all requirements are met. It is not important when or where these items are included, but that they are included. One contractor may list certain costs as project overhead, while another contractor would allocate the same costs (and responsibility) to a subcontractor. Either way, the costs are recorded in the estimate.

Sample Estimate: Division 1

At this initial stage of the estimating process, it is best to list as many as possible the items that are considered overhead. From a thorough review of the construction documents, especially the General and Supplementary Conditions, the estimator should be aware of both the specified requirements and those which are implied but not directly stated. Items are added to the list throughout the estimate. Pricing occurs at the end because a majority of the costs are time related or dependent upon total project cost.

The allocation of work and equipment to be included in the particular subtrades should be decided at this time. For example, the estimator must decide if the temporary fence is to be installed by the general contractor's personnel or if the work is to be subcontracted. Responsibility for establishing grades and building layout could be given to an employed field engineer, an outside engineering firm, or the appropriate subcontractors. These choices should be made at the beginning of the estimate. Figures 9.19 and 9.20 are the Project Overhead Summary for the three-story office building. All items known at this point are listed. Additions to the list and appropriate pricing will occur later in the estimate. Note that certain lump sum (LS) items have been included. Prices for these items (insurance, for example) can be determined either from historical costs or from telephone quotations.

Division 2: Site Work

Prior to quantity takeoff, pricing, or even the site visit, the estimator must read the specifications carefully and examine the plans thoroughly. By knowing—before the site visit—what to expect and what work is required, the estimator is less likely to overlook features and should be able to save time by avoiding a second site visit. In addition to being familiar with project requirements, the estimator might also find a Job Site Analysis form helpful (as in Figures 2.3 and 2.4) to assure that all items are included.

For the other divisions in new building construction, the plans and specifications include almost all of the data required for complete and proper estimating and construction. Site work, however, is similar to remodeling or renovation work. Certain existing conditions and "unknowns" may not be shown on the plans. It is the responsibility of the estimator to discover these conditions and to include appropriate costs in the estimate.

The first step in preparing the site work estimate is to determine areas and limits from the plans (to be checked in the field) for:

- Limits of the total parcel
- Limits of the work area
- Areas of tree removal and clearing
- Stripping of topsoil
- Cut and fill
- Building and utility excavation
- Roads and walks
- Paving
- Landscaping

There are three basic methods to determine area from a plan. The first is with a planimeter—a device which, when rolled along a designated perimeter, will measure area. The resulting measurement must then be converted to the scale

PROJECT
OVERHEAD SUMMARY

PROJECT: Office Building

LOCATION: ARCHITECT: DATE: Jan-02

QUANTITIES BY: ABC PRICES BY: DEF   EXTENSIONS BY: DEF   CHECKED BY: GHI

DESCRIPTION	QUANTITY	UNIT	MATERIAL/EQUIP. UNIT	MATERIAL/EQUIP. TOTAL	LABOR UNIT	LABOR TOTAL	TOTAL COST UNIT	TOTAL COST TOTAL
Job Organization: Superintendent		Week			1355			
Project Manager								
Timekeeper & Material Clerk		Week			790			
Clerical								
Safety, Watchman & First Aid								
Travel Expense: Superintendent								
Project Manager								
Engineering: Layout (3 person crew)	10	Day			780			
Inspection / Quantities								
Drawings								
CPM Schedule								
Testing: Soil	1	LS		9750				
Materials								
Structural								
Equipment: Cranes								
Concrete Pump, Conveyor, Etc.								
Elevators, Hoists								
Freight & Hauling								
Loading, Unloading, Erecting, Etc.								
Maintenance								
Pumping								
Scaffolding								
Small Power Equipment / Tools	0.5	%						
Field Offices: Job Office, Trailer		Mo	300					
Architect / Owner's Office								
Temporary Telephones		Mo	135					
Utilities								
Temporary Toilets		Mo	152.00					
Storage Areas & Sheds								
Temporary Utilities: Heat								
Light & Power	567	CSF						
PAGE TOTALS								

Figure 9.19

DESCRIPTION	QUANTITY	UNIT	MATERIAL/EQUIP.		LABOR		TOTAL COST	
			UNIT	TOTAL	UNIT	TOTAL	UNIT	TOTAL
Totals Brought Forward								
Winter Protection: Temp. Heat/Protection	56700	SF	0.45		0.45			
Snow Plowing								
Thawing Materials								
Temporary Roads	750	SY	3.13		1.67			
Signs & Barricades: Site Sign								
Temporary Fences	1	LS	130.00					
Temporary Stairs, Ladders & Floors	1350	LF	2.26		3.75			
Photographs								
Clean Up								
Dumpster		Week	425.00					
Final Clean Up	56.7	MSF	6.03		37.00			
Continuous - One Laborer		Week			938			
Punch List	0.2	%						
Permits: Building	1	%						
Misc.								
Insurance: Builders Risk - Additional Rider	1	%						
Owner's Protective Liability								
Umbrella								
Unemployment Ins. & Social Security								
Bonds								
Performance								
Material & Equipment								
Main Office Expense								
Special Items								
Totals:								

Figure 9.20

150

of the drawing. Trigonometry can also be used to determine area. It is applied to the distances and bearings supplied by a survey. Sines and cosines are used to calculate coordinates, which are in turn used to calculate area. The planimeter may be used only if the drawing is to scale and the scale is known. The trigonometric method may be used on a drawing not to scale, but only if survey data is available. Both methods are accurate.

A third method, triangulation, is less accurate but can be used in either case. This method involves dividing the area into triangles, squares and rectangles and determining the area of each. Some dimensions must be scaled. An example is shown in Figure 9.21. Triangulation is quick, and depending upon the shape of the site, is accurate to within a few percent.

Soil borings and analyses, if available, should be reviewed prior to the site visit. Test pit and percolation data can also be very useful. Soil type and porosity, any encountered ledge, and the height of the water table will all have a bearing on the site work estimate.

The Site Visit

Once the estimator is familiar with the project requirements, the site visit should be used to confirm and verify information. The initial examination should cover such items as general access, the amount and kind of traffic, location of existing utilities, and site storage. These conditions cannot be quantified, but will all have an impact on site work costs. Boundary lines and stakes should be located, especially if the work area is close to property lines.

The site visit may be conducted according to the sequence in which the site work will be performed: clearing and grubbing, excavation, blasting (if necessary), utilities and drainage, backfill, roads, paving and site improvements.

Estimates for tree removal can be made by actually counting and sizing each tree to be removed. If the site is small or the number of trees few, this may be the easiest (and most accurate) method. If large quantities are involved, a random count method will suffice. The first step is to determine the total acreage to be cleared. Counts of each type and size of vegetation should be made for 16' on either side of a line, 340' long. These counts (at least 3) should be made at random. Each count will reflect relative density for one fourth of an acre. Due to rising lumber costs (especially hardwood) and the growing use of firewood, the estimator may find a company or individual to cut and remove the trees at no cost. The expense of stump and brush removal must still be included.

The estimator should determine a rough soil profile, either from test pits, soil borings, or with a shovel. The depth of loam, or topsoil is important. The specifications may state whether the topsoil is to be stripped and stockpiled for reuse on the site, or hauled away. If there is no room for topsoil storage on the site so that all is removed (and possibly sold), topsoil may have to be purchased later for the landscaping. Test pits may also be a good indicator of soil stability. Depending upon the site conditions and project requirements, shoring or sheet piling may be required at trenches and excavations.

The estimator should roughly determine the location of building utilities and other excavation areas and should visualize the excavation process. Stockpile areas should be located and a rough determination reached as to proper equipment size and type. The excavation procedure should be based on a key rule of site work only move the material once. At this stage, the estimator may or may not have an indication as to whether fill will be required or excess hauled away. These questions will be answered during cut and fill calculations.

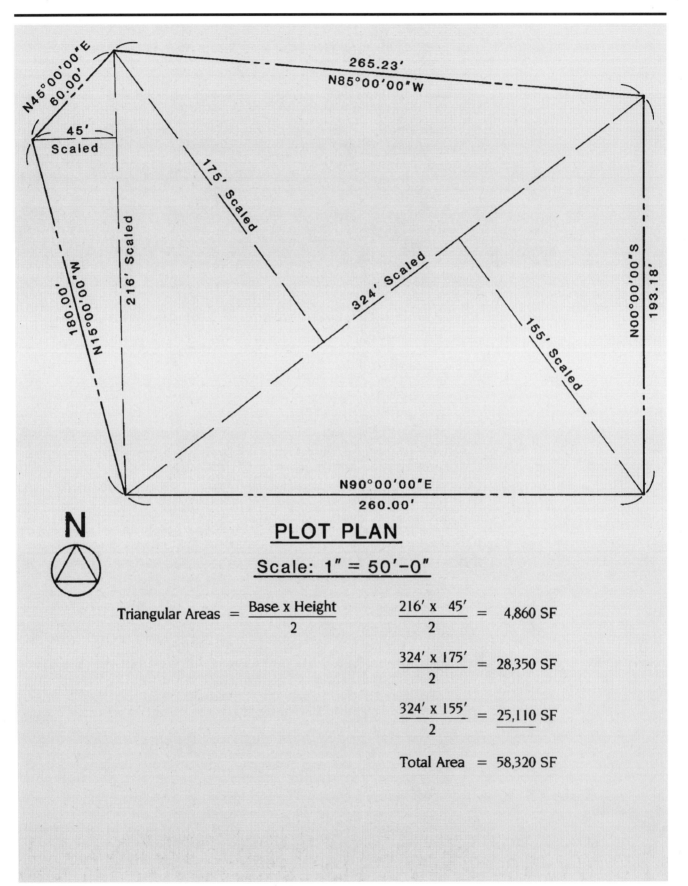

N45°00'00"E
60.00'

45'
Scaled

265.23'
N85°00'00"W

175' Scaled

216' Scaled

324' Scaled

155' Scaled

180.00'
N15°00'00"W

N00°00'00"S
193.18'

N90°00'00"E
260.00'

N

PLOT PLAN

Scale: 1" = 50'-0"

Triangular Areas $= \dfrac{\text{Base x Height}}{2}$

$\dfrac{216' \text{ x } 45'}{2} = 4{,}860$ SF

$\dfrac{324' \text{ x } 175'}{2} = 28{,}350$ SF

$\dfrac{324' \text{ x } 155'}{2} = 25{,}110$ SF

Total Area $= 58{,}320$ SF

Figure 9.21

152

If there is excess fill, haul distance to the dumping area (whether on site or off) will have a bearing not only on cost, but on the selection of equipment. Erosion control measures may be required to prevent soil runoff from stockpiles or excavation areas. If not specified, this possibility should be determined while at the site.

If ledge is evident, either visually or from borings, the estimator should be familiar with local regulations. Where blasting is permitted, items such as mats, overtime work and special permits can be very expensive. Local regulations may require that after exposure by excavation, the ledge be covered (backfilled) before blasting. Because of the fact that subsurface conditions may be unknown, blasting is usually included in the estimate as an allowance per cubic yard. This allowance must include all possibilities.

Existing utilities, whether on-site or off, overhead or underground, should be located and noted or verified on the site plan. The estimator should roughly note locations of the proposed utilities to determine if any problems or conflicts will arise due to existing conditions. For the same reasons, existing drainage conditions should be verified and compared to the proposed drainage.

If the specifications are not clear, the estimator should determine whether or not the excavated material is suitable for backfill. If not, the unsuitable material must be removed and new material purchased and delivered.

Roads, walks, paving, site improvements and landscaping should also be visualized to anticipate any unique conditions or requirements. The quality and detail of the site visit will be reflected in the accuracy and thoroughness of the site work estimate.

Earthwork

Responsibility for layout and staking of the building, grades, utilities and other excavation should be determined prior to estimating. If not, everyone will assume that the other guy is responsible.

There are two basic methods and a third variation of determining cuts and fills for the earthwork portion of a project. The first method is most appropriate for linear excavations (roads and large trenches). It involves determining the cross-sectional area of the excavation at appropriate locations. The area of the cross-sections can be determined by using a planimeter or by the methods shown in Figure 9.22. To calculate volume, two cross-sectional areas are averaged and multiplied by the distance between the sections. The selection and frequency of cross-section locations will have an effect on the calculations. Accuracy is improved if the sections are located so that topography is relatively uniform between each pair. Separate calculations must be made for cuts and for fill.

The second method is to superimpose a grid over the site plan and to determine the existing elevation at each intersection of the grid. The elevations at the four corners of each grid square are averaged. The difference between the proposed elevation and the existing average elevation for each square is calculated (whether cut or fill). All cuts and all fills are added separately to determine excavation quantities. The difference between the two is the amount that must be borrowed or hauled away. A quicker and slightly less accurate variation of this method can also be used. Instead of determining corner elevations and averaging, the estimator can visually determine the average elevations from the relationship of the grid squares to the contours.

A number of factors must be considered when calculating excavation quantities. If all dimensions (elevations and distances) are in feet, then the final

Level Cross-Section

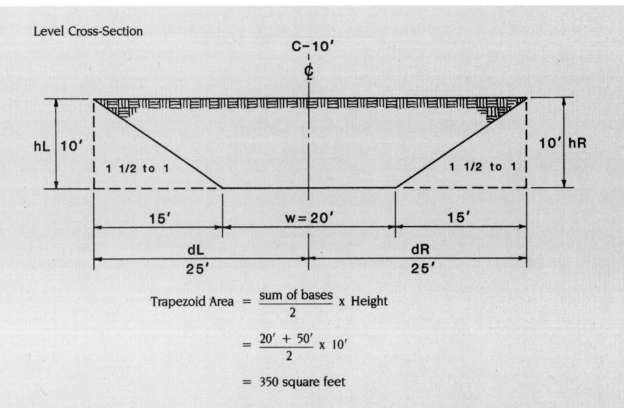

Trapezoid Area $= \dfrac{\text{sum of bases}}{2} \times \text{Height}$

$= \dfrac{20' + 50'}{2} \times 10'$

$= 350$ square feet

Three Level Cross Section

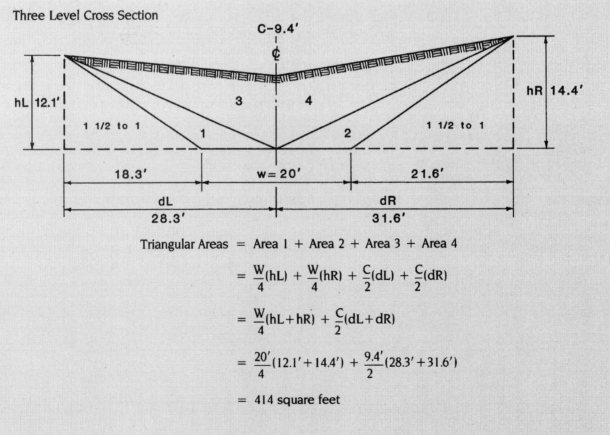

Triangular Areas $= \text{Area } 1 + \text{Area } 2 + \text{Area } 3 + \text{Area } 4$

$= \dfrac{W}{4}(hL) + \dfrac{W}{4}(hR) + \dfrac{C}{2}(dL) + \dfrac{C}{2}(dR)$

$= \dfrac{W}{4}(hL+hR) + \dfrac{C}{2}(dL+dR)$

$= \dfrac{20'}{4}(12.1'+14.4') + \dfrac{9.4'}{2}(28.3'+31.6')$

$= 414$ square feet

Figure 9.22

cubic foot quantities must be converted to cubic yards. When finish grade elevations are taken from a plan, the estimator must be sure to account for subbases and finish materials (gravel, paving, concrete, topsoil, etc.) which will be on top of the subgrade to reach the final elevations and grades.

Another major factor to consider is the difference between bank volume (material in a natural state), loose volume (after digging), and compacted volume. The relationship of swell and shrinkage is illustrated in Figure 9.23, with conversion tables shown in Figure 9.24. As shown, the volume of materials expands when disturbed or excavated. This same material can most often be compacted to less than the natural bank volume. An example of how these conversion tables are used is shown in the sample estimate. When pricing compaction, the estimator must consider the specified depth of lifts, or layers, which must each be compacted. The specifications should also be checked for required compaction testing.

After figuring bulk excavation, the estimator should take off the quantities for the following types of excavation, paying attention for such items as steps and haunches:

- Perimeter strip footings

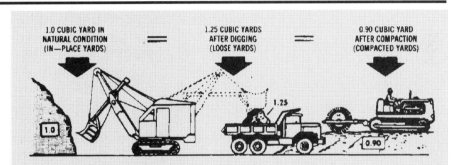

Approximate Material Characteristics*

Material	Loose (lb/cu yd)	Bank (lb/cu yd)	Swell (%)	Load Factor
Clay, dry	2,100	2,650	26	0.79
Clay, wet	2,700	3,575	32	0.76
Clay and gravel, dry	2,400	2,800	17	0.85
Clay and gravel, wet	2,600	3,100	17	0.85
Earth, dry	2,215	2,850	29	0.78
Earth, moist	2,410	3,080	28	0.78
Earth, wet	2,750	3,380	23	0.81
Gravel, dry	2,780	3,140	13	0.88
Gravel, wet	3,090	3,620	17	0.85
Sand, dry	2,600	2,920	12	0.89
Sand, wet	3,100	3,520	13	0.88
Sand and gravel, dry	2,900	3,250	12	0.89
Sand and gravel, wet	3,400	3,750	10	0.91

*Exact values will vary with grain size, moisture content, compaction, etc. Test to determine exact values for specific soils.

Figure 9.23

155

- Interior strip footings
- Spread footings
- Utility trenches, meter pits, manholes
- Special items (elevator pits, tunnels, etc.)
- Hand excavation

Caissons and Piles

Caissons, piles, and pressure-injected footings are specialty items usually taken off and priced by a subcontractor. The cost of caissons and piles, although priced by the linear foot, is dependent on the number of units to be installed. This is due to the fact that mobilization for the equipment can cost thousands of dollars before the first pile is driven. Typical mobilization costs are shown in Figure 9.25. The mobilization cost will have a lesser effect per unit, as the number of units required increases. When soliciting bids, the estimator should verify that all required costs are included. Pile caps, cutting, testing, and wet conditions will all add to the total cost.

Roads

Temporary roads are an item that may easily by underestimated or overlooked. Under dry conditions, watering, oiling, or calcium chloride treatment may be necessary to prevent dust. Under wet conditions, large quantities of crushed stone and trenching for drainage may be needed. Consideration must be given to the installation of utilities; this procedure may require removal and reconstruction of the temporary road.

As with the building layout, costs must also be included for the layout and grades of roads, curbs and associated utilities. Most specifications will state the requirements for work above rough grade: gravel base, subbase, finish paving courses, etc. Bulk cuts and fills for the roadwork should be determined as part of bulk excavation. Excavation for curbs must also be included. Handwork may be necessary.

Typical Soil Volume Conversion Factors				
Soil Type	Initial Soil Condition	Bank	Converted to: Loose	Converted to: Compacted
Clay	Bank	1.00	1.27	0.90
	Loose	0.79	1.00	0.71
	Compacted	1.11	1.41	1.00
Common earth	Bank	1.00	1.25	0.90
	Loose	0.80	1.00	0.72
	Compacted	1.11	1.39	1.00
Rock (blasted)	Bank	1.00	1.50	1.30
	Loose	0.67	1.00	0.87
	Compacted	0.77	1.15	1.00
Sand	Bank	1.00	1.12	0.95
	Loose	0.89	1.00	0.85
	Compacted	1.05	1.18	1.00

Figure 9.24

02450 | Foundation & Load Bearing Elements

02455 | Driven Piles

			DAILY	LABOR-		2002 BARE COSTS				TOTAL	
		CREW	OUTPUT	HOURS	UNIT	MAT.	LABOR	EQUIP.	TOTAL	INCL O&P	
220	1400	18" diameter, 7 ga.	B-19	480	.133	V.L.F.	21.50	4.04	3.93	29.47	34.50	**220**
	1500	For reinforcing steel, add				Lb.	.55			.55	.60	
	1700	For ball or pedestal end, add	B-19	11	5.818	C.Y.	83.50	176	171	430.50	565	
	1900	For lengths above 60', concrete, add	"	11	5.818	"	87	176	171	434	570	
	2000	For steel thin shell, pipe only				Lb.	.52			.52	.57	
	2200	Precast, prestressed, 50' long, 12" diam., 2-3/8" wall	B-19	720	.089	V.L.F.	10.20	2.69	2.62	15.51	18.40	
	2300	14" diameter, 2-1/2" wall		680	.094		13.40	2.85	2.77	19.02	22.50	
	2500	16" diameter, 3" wall		640	.100		18.55	3.03	2.95	24.53	28.50	
	2600	18" diameter, 3" wall	B-19A	600	.107		23.50	3.23	3.66	30.39	34.50	
	2800	20" diameter, 3-1/2" wall		560	.114		27	3.46	3.92	34.38	39.50	
	2900	24" diameter, 3-1/2" wall		520	.123		33	3.73	4.22	40.95	47	
	3100	Precast, prestressed, 40' long, 10" thick, square	B-19	700	.091		7	2.77	2.69	12.46	15.15	
	3200	12" thick, square		680	.094		8.85	2.85	2.77	14.47	17.40	
	3400	14" thick, square		600	.107		10.50	3.23	3.14	16.87	20	
	3500	Octagonal		640	.100		14	3.03	2.95	19.98	23.50	
	3700	16" thick, square		560	.114		16.70	3.46	3.37	23.53	27.50	
	3800	Octagonal		600	.107		16.80	3.23	3.14	23.17	27	
	4000	18" thick, square	B-19A	520	.123		19.95	3.73	4.22	27.90	32.50	
	4100	Octagonal	B-19	560	.114		19.90	3.46	3.37	26.73	31.50	
	4300	20" thick, square	B-19A	480	.133		25	4.04	4.57	33.61	39	
	4400	Octagonal	B-19	520	.123		22	3.73	3.62	29.35	34	
	4600	24" thick, square	B-19A	440	.145		35.50	4.41	4.99	44.90	52	
	4700	Octagonal	B-19	480	.133		31.50	4.04	3.93	39.47	46	
	4750	Mobilization for 10,000 L.F. pile job, add		3,300	.019			.59	.57	1.16	1.58	
	4800	25,000 L.F. pile job, add		8,500	.008			.23	.22	.45	.61	
350	0011	**PILING SPECIAL COSTS** pile caps, see Division 03310-240										**350**
	0500	Cutoffs, concrete piles, plain	1 Pile	5.50	1.455	Ea.		43.50		43.50	72.50	
	0600	With steel thin shell, add		38	.211			6.25		6.25	10.50	
	0700	Steel pile or "H" piles		19	.421			12.55		12.55	21	
	0800	Wood piles		38	.211			6.25		6.25	10.50	
	0900	Pre-augering up to 30' deep, average soil, 24" diameter	B-43	180	.267	L.F.		6.90	11.05	17.95	23	
	0920	36" diameter		115	.417			10.75	17.35	28.10	35.50	
	0960	48" diameter		70	.686			17.70	28.50	46.20	58.50	
	0980	60" diameter		50	.960			25	40	65	82	
	1000	Testing, any type piles, test load is twice the design load										
	1050	50 ton design load, 100 ton test				Ea.				14,000	15,000	
	1100	100 ton design load, 200 ton test								18,000	19,000	
	1150	150 ton design load, 300 ton test								22,500	24,000	
	1200	200 ton design load, 400 ton test								24,500	27,000	
	1250	400 ton design load, 800 ton test								28,350	31,500	
	1500	Wet conditions, soft damp ground										
	1600	Requiring mats for crane, add								40%	40%	
	1700	Barge mounted driving rig, add								30%	30%	
500	0010	**MOBILIZATION** Set up & remove, air compressor, 600 C.F.M. [R02455 -900]	A-5	3.30	5.455	Ea.		128	11.80	139.80	212	**500**
	0100	1200 C.F.M.	"	2.20	8.182			193	17.65	210.65	320	
	0200	Crane, with pile leads and pile hammer, 75 ton	B-19	.60	106			3,225	3,150	6,375	8,650	
	0300	150 ton	"	.36	177			5,375	5,225	10,600	14,400	
	0500	Drill rig, for caissons, to 36", minimum	B-43	2	24			620	995	1,615	2,050	
	0600	Up to 84"	"	1	48			1,250	2,000	3,250	4,100	
	0800	Auxiliary boiler, for steam small	A-5	1.66	10.843			255	23.50	278.50	420	
	0900	Large	"	.83	21.687			510	47	557	845	
	1100	Rule of thumb: complete pile driving set up, small	B-19	.45	142			4,300	4,200	8,500	11,600	
	1200	Large	"	.27	237			7,175	6,975	14,150	19,300	
850	0010	**PILES, STEEL** Not including mobilization or demobilization										**850**
	0100	Step tapered, round, concrete filled										

Figure 9.25

157

Bituminous paving is usually taken off and priced by the square yard, but it is sold by the ton. The average weight of bituminous concrete is 145 pounds per cubic foot. By converting, 1 square yard, 1" thick weighs approximately 110 pounds. This figure can be used to determine quantities (in tons) for purchase:

> Base course: 2" thick, 1,500 S.Y.
> 1,500 S.Y. × 2" × 110 lbs. = 330,000 lbs. = 165 tons
> Finish course: 1" thick, 1,500 S.Y.
> 1,500 S.Y. × 1" × 110 lbs. = 165,000 lbs. = 82.5 tons

Landscaping

If the required topsoil is stockpiled on site, equipment will be needed for hauling and spreading. Likewise if the stockpile is too large or too small, trucks will have to be mobilized for transport. If large trees and shrubs are specified, equipment may be required for digging and placing.

Landscaping, including lawn sprinklers, sod, seeding, etc., is usually done by subcontractors. The estimator should make sure that the subcontract includes a certain maintenance period during which all plantings are guaranteed. The required period will be most likely stated in the specifications and should include routine maintenance and replacement of dead plantings.

Estimating Hazardous Waste Remediation/Cleanup Work

Hazardous waste cleanup/site remediation work is more difficult to estimate than standard construction. The difficulty arises because all projects are unique and the hazardous materials are very different. In addition, the estimator must consider the cost of not only complying with applicable OSHA and other government regulations, but also the reduced worker productivity associated with these regulations. The cost factors associated with compliance include the following:

- safety training for management and construction personnel
- pre-employment and conclusion of the work medical exams
- site safety meetings
- shower facility and change room
- worker and equipment decontamination
- disposal of contaminated water
- suit up/off time
- personnel protective equipment
- nonproductive support workers for crew assistance and safety related monitoring
- work breaks due to varying ambient temperature levels

Productivity: The cost impact of these factors is dependent upon the types of hazardous waste and the remediation procedure. OSHA regulations and local authorities require different levels of personnel protective gear for different wastes. The levels of protection range from A to D. Level A protection includes a moon suit with air pack, while level D consists of standard construction clothing with the possible addition of a Tyvek suit, dust mask, or safety glasses. The productivity obtained from the work force varies dramatically depending on the level of protective gear compared to normal construction productivity. In general, D level protection allows a worker to be approximately 90% productive. Therefore, published construction data can be used to estimate the work.

The productivity of a C level worker with a full face mask and Tyvek suit is cut to approximately 75%. The B level protective gear of a full suit and air supplied from an outside source drops to approximately 35% efficiency. Level A protection consisting of a full moon suit with a self-contained breathing apparatus drops worker productivity to 20%. As the level of protection increases, additional nonproductive personnel are also required at the site to monitor and assure worker safety.

Disposal: Site remediation may consist of on-site encapsulation, and off-site disposal or destruction of the hazardous material by incineration or using an exotic treatment technique. Off-site disposal requires a complete chemical analysis before the material can be moved, acceptance of the material by the proposed recipient, and EPA manifesting (proper shipping documentation). Note: It is illegal to transport hazardous materials without the proper manifest.

When estimating off-site disposal, include all transportation costs, waiting time, and disposal fees. When estimating on-site incineration, the estimate must include the cost of trial burns to ensure destruction of the hazardous material and acceptance by the EPA of the stack emissions. The cost of treatment/disposal methods should be determined in conjunction with the manufacturer of the remediation equipement. Hazardous waste disposal is generally specified on a performance basis. Therefore, the process should be prequalified to ensure compliance with the specification.

Summary: Site remediation and hazardous waste cleanup should not be undertaken without complete knowledge of all applicable federal and state regulations. Those companies in the business have compliance departments because meeting all regulations is such an overriding concern in their work. The penalties for noncompliance are severe.

Sample Estimate: Division 2

Site work quantities may be derived from several different drawings. In this case, data are taken from the topographic survey, the finish grading plan and the site plan (Figures 9.2, 9.3 and 9.4). For smaller projects, these three plans may be combined as one. The detailed information that would normally be included in a complete set of plans and specifications is not provided for this example. For purposes of this demonstration, quantities as shown reflect realistic conditions and are based on normal construction techniques and materials.

For the site clearing portion of the estimate, areas to be cleared are measured from the topographic survey (Figure 9.2), but verification of area and determination of vegetation type must be done during the site visit. The estimate sheets for Division 2 are shown in Figures 9.26 through 9.29. Note that the units (acres and square yards) used for tree cutting and grubbing are different than those used for light clearing. The estimator should be sure to convert the required units for pricing. For the example, the whole site is to be cleared. The site dimensions are 310' × 365', or 113,150 square feet. Light clearing is priced by the acre, so the area is converted from 12,570 square yards to 2.6 acres. Prices for site clearing are taken from Figure 9.30.

Quantities for bulk excavation are derived using the alternative method as described above and as shown in Figure 9.31. In this example, a grid representing 20' × 20' squares is used. The estimator determines the average elevation change in each square and designates each as a cut (–) or a fill (+). The darkened square within the building is for excavation at the elevator pit. All cuts and fills are totalled separately and converted to cubic yards. An amount to

CONSOLIDATED ESTIMATE

PROJECT: Office Building
LOCATION:
TAKE OFF BY: ABC
CLASSIFICATION: Division 2
ARCHITECT:
QUANTITIES BY: ABC
PRICES BY: ABC

ESTIMATE NO:
DATE: Jan-02
CHECKED BY: GHI

EXTENSIONS BY: DEF

DESCRIPTION	SOURCE	QUANT.	UNIT	MATERIAL COST	MATERIAL TOTAL	LABOR COST	LABOR TOTAL	EQUIPMENT COST	EQUIPMENT TOTAL	SUBCONTRACT COST	SUBCONTRACT TOTAL	TOTAL COST	TOTAL TOTAL
Division 2: Site Work													
Site Cleaning													
Medium Trees -Cut	02230 200 0200	0.46	Acre			1725	794	1350	621				
Grub & Remove Stumps	02230 200 0250	0.46	Acre			650	299	1650	759				
Light Clearing	02230 220 0300	2.6	Acre			219	569	410	1066				
Earthwork													
Bulk Excavation	02315 400 1300	5430	CY			0.33	1792	0.66	3584				
Load Trucks (Loose)	02315 400 1550	3212	CY			0.54	1734	0.43	1381				
Haul (Loose)	02320 200 0400	3212	CY			1.11	3565	2.61	8383				
Compact Fill	02315 320 0300	2860	CY			0.26	744	0.59	1687				
Excess Excavation @Foundation	02315 410 5220	272	CY			0.43	117	1.29	351				
Backfill Compact @Foundation	02315 100 1300	302	CY			0.29	88	0.69	208				
@Foundation	02315 100 0600	302	CY			3.13	945	1.01	305				
Backfill Compact @Elevator Pit	02315 100 0010	30	CY			13.40	402	1.01	30				
@Elevator Pit	02315 100 0600	30	CY			3.13	94						
Footing Excavation, Spread	02315 900 0060	305	CY			2.19	668	1.08	329				
(Add to above 30%)	02315 900 2400	305	CY			0.66	201	0.32	98				
Footing Excavation, Continuous	02315 900 0060	85	CY			2.19	186	1.08	92				
Backfill @ Footings	02315 100 1900	390	CY			0.38	148	0.91	355				
Include Spread Excess													
Subtotals						$	12,347	$	19,250				

Figure 9.26

160

CONSOLIDATED ESTIMATE

PROJECT: Office Building	ESTIMATE NO:
LOCATION:	DATE: Jan-02
CLASSIFICATION: Division 2	CHECKED BY: GHI
ARCHITECT:	
TAKE OFF BY: ABC	QUANTITIES BY: ABC PRICES BY: EXTENSIONS BY:

DESCRIPTION	SOURCE	QUANT	UNIT	MATERIAL			LABOR			EQUIPMENT			SUBCONTRACT		TOTAL	
				COST	DEF TOTAL	TOTAL	COST	DEF TOTAL	TOTAL	COST	DEF TOTAL	TOTAL	COST	TOTAL	COST	TOTAL
Division 2: (Cont'd)																
Earthwork (cont'd)																
Utility Excavation																
24" Drain Excavation	02315 900 0300	230	CY				2.36	543		3.68		846				
Backfill & Compact	02315 100 0.66	287	CY				3.13	898		1.01		290				
Gas, Water, Sewer	02315 940 2600	240	LF				0.37	89		0.30		72				
Compaction	02315 940 3200	240	LF				0.19	46		0.15		36				
Sub Drain Gravel	02315 505 1300	46	CY	11.90	547		5.40	248		0.53		24				
Floor Slab Gravel, 6" Compacted	02315 505 0600	19000	SF	0.24	4560		0.14	2660		0.01		190				
Mobilization, Dozer, Loader Backhoe, Compactor	02305 250 0400	4	Ea.				57.00	228		201.00		804				
Site Drainage & Utilities																
Catch Basin	02630 200 1120	3	Ea.	540.00	1620		278.00	834		79.00		237				
Catch Basin Frames & Covers	02630 200 2100	3	Ea.	161.00	483		78.50	236		23.00		69				
Gas Service 3"	02550 464 1500	80	LF	1.80	144		2.86	229		1.16		93				
Water Service 6"	02510 800 1430	80	LF	12.25	980		5.90	472								
Gate Valve 6"	15110 200 2300	1	Ea.	795.00	795		268.00	268								
Sewer, 6" PVC	02530 780 2040	80	LF	2.48	198		1.83	146								
Foundation Drain	02620 280 3000	620	LF	2.00	1240		2.98	1848		0.45		279				
Drain Pipe 24" Concrete	02530 730 2040	310	LF	18.65	5782		11.90	3689		1.79		555				
Subtotals					$ 16,349			$ 12,433				$ 3,495				

Figure 9.27

CONSOLIDATED ESTIMATE

PROJECT: Office Building   CLASSIFICATION: Division 2
LOCATION: ARCHITECT:
TAKE OFF BY: ABC   QUANTITIES BY: ABC   PRICES BY: ABC   As Shown   EXTENSIONS BY:

ESTIMATE NO:
DATE: Jan-02
CHECKED BY: GHI

DESCRIPTION	SOURCE		QUANT	UNIT	MATERIAL COST	MATERIAL TOTAL	LABOR COST	LABOR TOTAL	EQUIPMENT COST	EQUIPMENT TOTAL	SUBCONTRACT COST	SUBCONTRACT TOTAL	TOTAL COST	TOTAL TOTAL
Division 2: (Cont'd)														
Roads & Walks														
Base Course 9"	02720	200 0200	8530	SY	4.55	38812	0.25	2133	0.55	4692				
Paving - Bituminous														
Binder	02740	300 0160	8530	SY	4.09	34888	0.46	3924	0.35	2986				
Wearing (@ $32/TON)	02740	300 0340	8530	SY	2.60	22178	0.33	2815	0.24	2047				
Concrete Curb														
Precast Straight	02770	215 0550	2050	LF	6.60	13530	2.04	4182	1.05	2153				
Precast Radius	02770	215 0600	240	LF	7.60	1824	4.39	1054	2.27	545				
Line Painting														
Stalls	02766	550 0800	203	Stall	2.64	536	3.61	733	1.01	205				
Arrows	02766	550 0620	150	SF	0.50	75	0.50	75	0.19	29				
Precast Bumpers	02840	700 1000	203	Ea.	30.00	6090	7.95	1614						
Concrete Sidewalks	02775	275 0310	1625	SF	1.15	1869	1.10	1788						
Concrete Steps	02778	280 0500	95	LF	4.46	424	11.40	1083	0.28	27				
from (03310-240-6850)														
Lawns & Plantings														
Landscaping	TELEPHONE QUOTE (Includes sales Tax)													
Sprinkler			15700	SF							0.80	12560		
Lawn, Top soil, Fert., Seed			15.7	MSF							595.00	9342		
Trees (60) & Shrubs (200),												16900		
Including Planting														
Subtotals						$ 120,225		$ 19,399		$ 12,682		$ 38,802		

Figure 9.28

CONSOLIDATED ESTIMATE

PROJECT: Office Building
LOCATION:
TAKE OFF BY: ABC

CLASSIFICATION: Division 2
ARCHITECT:
QUANTITIES BY: ABC   PRICES BY: DEF   EXTENSIONS BY: DEF

ESTIMATE NO:
DATE: Jan-02
CHECKED BY: GHI

SOURCE	DESCRIPTION	QUANT.	UNIT	MATERIAL		LABOR		EQUIPMENT		SUBCONTRACT		TOTAL	
				COST	TOTAL	COST	TOTAL	COST	TOTAL	COST	TOTAL	COST	TOTAL
	Division 2: (Cont'd)												
	Sheet 1 Subtotals					$	12,374	$	19,250				
	Sheet 2 Subtotals			$	16,349	$	12,433	$	3,495				
	Sheet 3 Subtotals			$	120,225	$	19,399	$	12,682	$	38,802		
	Division 2 Totals			$	136,574	$	44,206	$	35,427	$	38,802		

Figure 9.29

			CREW	DAILY OUTPUT	LABOR-HOURS	UNIT	MAT.	LABOR	EQUIP.	TOTAL	TOTAL INCL O&P	
		02225 | Selective Demolition						2002 BARE COSTS				
850	5040	Average	1 Carp	4	2	Ea.		60		60	93.50	850
	5080	Maximum	↓	2	4	↓		120		120	187	
		02230 | Site Clearing										
200	0010	**CLEAR AND GRUB** Cut & chip light, trees to 6" diam.	B-7	1	48	Acre		1,200	950	2,150	2,900	200
	0150	Grub stumps and remove	B-30	2	12			325	825	1,150	1,400	
	0200	Cut & chip medium, trees to 12" diam.	B-7	.70	68.571			1,725	1,350	3,075	4,150	
	0250	Grub stumps and remove	B-30	1	24			650	1,650	2,300	2,775	
	0300	Cut & chip heavy, trees to 24" diam.	B-7	.30	160			4,025	3,175	7,200	9,700	
	0350	Grub stumps and remove	B-30	.50	48			1,300	3,300	4,600	5,600	
	0400	If burning is allowed, reduce cut & chip				↓					40%	
	3000	Chipping stumps, to 18" deep, 12" diam.	B-86	20	.400	Ea.		12.50	6.80	19.30	26.50	
	3040	18" diameter		16	.500			15.60	8.50	24.10	33	
	3080	24" diameter		14	.571			17.85	9.70	27.55	37.50	
	3100	30" diameter		12	.667			21	11.30	32.30	44	
	3120	36" diameter		10	.800			25	13.55	38.55	52.50	
	3160	48" diameter	↓	8	1	↓		31	16.95	47.95	65.50	
	5000	Tree thinning, feller buncher, conifer										
	5080	Up to 8" diameter	B-93	240	.033	Ea.		1.04	1.68	2.72	3.42	
	5120	12" diameter		160	.050			1.56	2.52	4.08	5.15	
	5240	Hardwood, up to 4" diameter		240	.033			1.04	1.68	2.72	3.42	
	5280	8" diameter		180	.044			1.39	2.24	3.63	4.56	
	5320	12" diameter	↓	120	.067	↓		2.08	3.36	5.44	6.85	
	7000	Tree removal, congested area, aerial lift truck										
	7040	8" diameter	B-85	7	5.714	Ea.		145	99	244	330	
	7080	12" diameter		6	6.667			169	116	285	385	
	7120	18" diameter		5	8			202	139	341	465	
	7160	24" diameter		4	10			253	174	427	580	
	7240	36" diameter		3	13.333			335	231	566	775	
	7280	48" diameter	↓	2	20	↓		505	345	850	1,150	
220	0010	**CLEARING** Brush with brush saw	A-1	.25	32	Acre		750	243	993	1,450	220
	0100	By hand	"	.12	66.667			1,575	505	2,080	2,975	
	0300	With dozer, ball and chain, light clearing	B-11A	2	8			219	410	629	790	
	0400	Medium clearing		1.50	10.667			292	550	842	1,050	
	0500	With dozer and brush rake, light		10	1.600			43.50	82.50	126	158	
	0550	Medium brush to 4" diameter		8	2			54.50	103	157.50	197	
	0600	Heavy brush to 4" diameter	↓	6.40	2.500	↓		68.50	129	197.50	247	
	1000	Brush mowing, tractor w/rotary mower, no removal										
	1020	Light density	B-84	2	4	Acre		125	99.50	224.50	299	
	1040	Medium density		1.50	5.333			166	133	299	395	
	1080	Heavy density	↓	1	8	↓		250	199	449	595	
250	0010	**FELLING TREES & PILING** With tractor, large tract, firm										250
	0020	level terrain, no boulders, less than 12" diam. trees										
	0300	300 HP tractor, up to 400 trees/acre, 0 to 25% hardwoods	B-10M	.75	16	Acre		460	1,375	1,835	2,225	
	0340	25% to 50% hardwoods		.60	20			570	1,725	2,295	2,775	
	0370	75% to 100% hardwoods		.45	26.667			765	2,300	3,065	3,700	
	0400	500 trees/acre, 0% to 25% hardwoods		.60	20			570	1,725	2,295	2,775	
	0440	25% to 50% hardwoods		.48	25			715	2,150	2,865	3,475	
	0470	75% to 100% hardwoods		.36	33.333			955	2,875	3,830	4,600	
	0500	More than 600 trees/acre, 0 to 25% hardwoods		.52	23.077			660	2,000	2,660	3,175	
	0540	25% to 50% hardwoods		.42	28.571			820	2,450	3,270	3,950	
	0570	75% to 100% hardwoods	↓	.31	38.710	↓		1,100	3,325	4,425	5,375	
	0900	Large tract clearing per tree										
	1500	300 HP dozer, to 12" diameter, softwood	B-10M	320	.037	Ea.		1.07	3.23	4.30	5.20	
	1550	Hardwood	↓	100	.120	↓		3.43	10.35	13.78	16.60	

Figure 9.30

compensate for reduction in volume is added to the total fill yardage before net calculation. This is due to the decrease in volume of earth when compacted (see Figure 9.23). The costs for bulk excavation are taken from Figure 9.32. Compaction is not included and must be added separately.

Quantities for bulk excavation are based on finish grades. Excess excavation at the foundation to be backfilled must be included as shown in Figure 9.26. Calculations for compaction at the foundation and for loading and hauling excess fill are as follows:

Compaction at foundation:

Excavation (Bank C.Y.) = 272 C.Y.

$$\frac{1.0 \text{ Bank C.Y.}}{0.9 \text{ Compacted C.Y.}} \times 272 \text{ C.Y.} = 302 \text{ C.Y. required}$$

Load and haul excess fill:

Excess fill (Bank C.Y.) = 2,570 C.Y.

$$\frac{1.25 \text{ Loose C.Y.}}{1.0 \text{ Bank C.Y.}} \times 2,570 \text{ C.Y.} = 3,212 \text{ Loose C.Y.}$$

The ratios of bank to loose to compacted earth were derived from Figure 9.24. If the above calculations were not made, almost 700 cubic yards of loading and hauling may not have been included in the estimate. This represents 60-70 truckloads. The calculation for foundation compaction indicates that 302 cubic yards of bank material *when compacted* are required to replace 272 cubic yards of excavated bank material. Only backfill, (and not compaction) is included for the footings, because the costs are primarily for the spreading of excess material. Compaction is included with the foundation backfill and at the floor subbase.

Costs for asphaltic concrete pavement are taken from Figure 9.33. Note on the estimate sheet that the price per ton for the sample building (Figure 9.28) is different than that in *Means Building Construction Cost Data*. By simple conversion, prices in *Means Building Construction Cost Data* can be adjusted to local conditions:

Line No.: 02740-300-0851   Wearing Course

National average $28.50

Local cost (for sample project) $32.00

$$\frac{\$32.00}{\$28.50} = 1.12$$

Material cost per S.Y. × Conversion factor = Local cost per S.Y.

$2.32 per S.Y. × 1.12 = $2.60 per S.Y.

The landscaping and site improvements for the sample project are to be subcontracted. This relieves the general contractor, not only of the responsibility of pricing, but also of the problems of maintenance and potential plant replacement. The subcontract price, as received by telephone, is shown in Figure 9.34. The estimator must ask the appropriate questions to be sure that all work is included and all requirements met.

Division 3: Concrete

All cast in place for concrete concrete work involves the same basic elements: formwork, reinforcing, concrete, placement and finishing. There are two methods that can be used to estimate the concrete portion of a project. All of the above components may be grouped together to form systems or

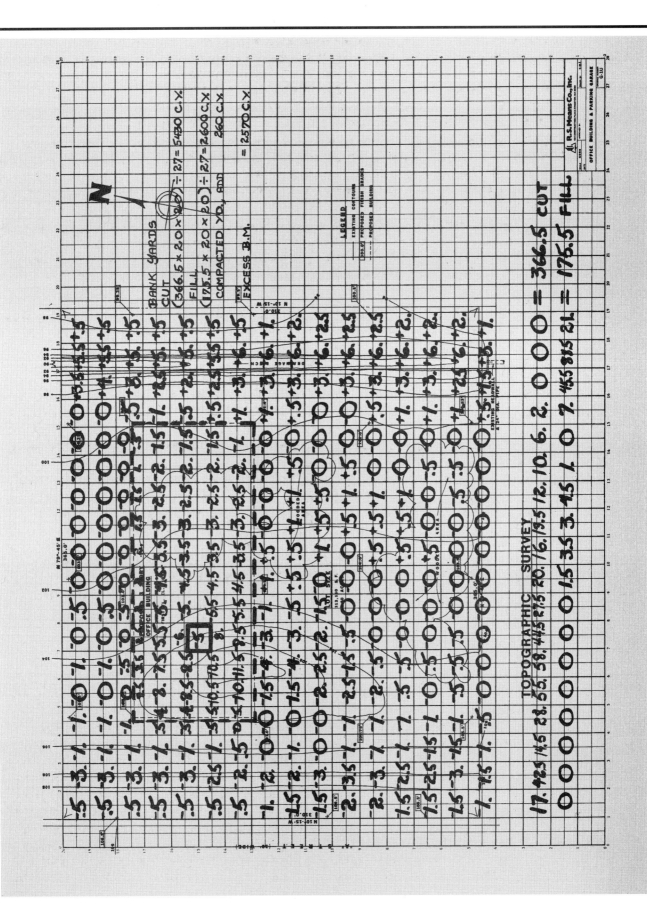

Figure 9.31

166

02315	Excavation and Fill		CREW	DAILY OUTPUT	LABOR-HOURS	UNIT	2002 BARE COSTS				TOTAL INCL O&P
							MAT.	LABOR	EQUIP.	TOTAL	
400	0010	**EXCAVATING, BULK BANK MEASURE** Common earth piled	R02315 -400									**400**
	0020	For loading onto trucks, add								15%	15%
	0050	For mobilization and demobilization, see division 02305-250	R02315 -450								
	0100	For hauling, see division 02320-200									
	0200	Backhoe, hydraulic, crawler mtd., 1 C.Y. cap. = 75 C.Y./hr.		B-12A	600	.027	C.Y.		.79	.94	1.73	2.22
	0250	1-1/2 C.Y. cap. = 100 C.Y./hr.		B-12B	800	.020			.59	.88	1.47	1.86
	0260	2 C.Y. cap. = 130 C.Y./hr.		B-12C	1,040	.015			.45	.93	1.38	1.71
	0300	3 C.Y. cap. = 160 C.Y./hr.		B-12D	1,280	.013			.37	1.64	2.01	2.36
	0310	Wheel mounted, 1/2 C.Y. cap. = 30 C.Y./hr.		B-12E	240	.067			1.97	1.33	3.30	4.44
	0360	3/4 C.Y. cap. = 45 C.Y./hr.		B-12F	360	.044			1.31	1.27	2.58	3.38
	0500	Clamshell, 1/2 C.Y. cap. = 20 C.Y./hr.		B-12G	160	.100			2.95	2.21	5.16	6.90
	0550	1 C.Y. cap. = 35 C.Y./hr.		B-12H	280	.057			1.69	2.51	4.20	5.30
	0950	Dragline, 1/2 C.Y. cap. = 30 C.Y./hr.		B-12I	240	.067			1.97	2.17	4.14	5.35
	1000	3/4 C.Y. cap. = 35 C.Y./hr.		"	280	.057			1.69	1.86	3.55	4.59
	1050	1-1/2 C.Y. cap. = 65 C.Y./hr.		B-12P	520	.031			.91	1.61	2.52	3.14
	1200	Front end loader, track mtd., 1-1/2 C.Y. cap. = 70 C.Y./hr.		B-10N	560	.021			.61	.47	1.08	1.45
	1250	2-1/2 C.Y. cap. = 95 C.Y./hr.		B-10O	760	.016			.45	.66	1.11	1.41
	1300	3 C.Y. cap. = 130 C.Y./hr.		B-10P	1,040	.012			.33	.66	.99	1.23
	1350	5 C.Y. cap. = 160 C.Y./hr.		B-10Q	1,280	.009			.27	.76	1.03	1.25
	1500	Wheel mounted, 3/4 C.Y. cap. = 45 C.Y./hr.		B-10R	360	.033			.95	.60	1.55	2.10
	1550	1-1/2 C.Y. cap. = 80 C.Y./hr.		B-10S	640	.019			.54	.43	.97	1.30
	1600	2-1/4 C.Y. cap. = 100 C.Y./hr.		B-10T	800	.015			.43	.46	.89	1.16
	1650	5 C.Y. cap. = 185 C.Y./hr.		B-10U	1,480	.008			.23	.53	.76	.93
	1800	Hydraulic excavator, truck mtd, 1/2 C.Y. = 30 C.Y./hr.		B-12J	240	.067			1.97	3.07	5.04	6.35
	1850	48 inch bucket, 1 C.Y. = 45 C.Y./hr.		B-12K	360	.044			1.31	2.41	3.72	4.63
	3700	Shovel, 1/2 C.Y. capacity = 55 C.Y./hr.		B-12L	440	.036			1.07	.80	1.87	2.50
	3750	3/4 C.Y. capacity = 85 C.Y./hr.		B-12M	680	.024			.69	.80	1.49	1.93
	3800	1 C.Y. capacity = 120 C.Y./hr.		B-12N	960	.017			.49	.72	1.21	1.53
	3850	1-1/2 C.Y. capacity = 160 C.Y./hr.		B-12O	1,280	.013			.37	.72	1.09	1.35
	3900	3 C.Y. cap. = 250 C.Y./hr.		B-12T	2,000	.008			.24	.65	.89	1.07
	4000	For soft soil or sand, deduct									15%	15%
	4100	For heavy soil or stiff clay, add									60%	60%
	4200	For wet excavation with clamshell or dragline, add									100%	100%
	4250	All other equipment, add									50%	50%
	4400	Clamshell in sheeting or cofferdam, minimum		B-12H	160	.100			2.95	4.39	7.34	9.30
	4450	Maximum		"	60	.267			7.85	11.70	19.55	25
	8000	For hauling excavated material, see div. 02320-200										
410	0010	**EXCAVATING, BULK, DOZER** Open site											**410**
	2000	75 H.P., 50' haul, sand & gravel		B-10L	460	.026	C.Y.		.75	.63	1.38	1.84
	2020	Common earth			400	.030			.86	.73	1.59	2.11
	2040	Clay			250	.048			1.37	1.17	2.54	3.37
	2200	150' haul, sand & gravel			230	.052			1.49	1.27	2.76	3.66
	2220	Common earth			200	.060			1.72	1.46	3.18	4.22
	2240	Clay			125	.096			2.75	2.33	5.08	6.75
	2400	300' haul, sand & gravel			120	.100			2.86	2.43	5.29	7.05
	2420	Common earth			100	.120			3.43	2.91	6.34	8.45
	2440	Clay			65	.185			5.30	4.48	9.78	13
	3000	105 H.P., 50' haul, sand & gravel		B-10W	700	.017			.49	.62	1.11	1.43
	3020	Common earth			610	.020			.56	.71	1.27	1.64
	3040	Clay			385	.031			.89	1.12	2.01	2.60
	3200	150' haul, sand & gravel			310	.039			1.11	1.40	2.51	3.23
	3220	Common earth			270	.044			1.27	1.60	2.87	3.70
	3240	Clay			170	.071			2.02	2.54	4.56	5.90
	3300	300' haul, sand & gravel			140	.086			2.45	3.09	5.54	7.15
	3320	Common earth			120	.100			2.86	3.61	6.47	8.35

Figure 9.32

02700 | Bases, Ballasts, Pavements & Appurtenances

02740 | Flexible Pavement

	Line	Description	CREW	DAILY OUTPUT	LABOR-HOURS	UNIT	MAT.	LABOR	EQUIP.	TOTAL	TOTAL INCL O&P	
300	0300	Wearing course, 1" thick R02065 -300	B-25B	10,575	.009	S.Y.	1.52	.24	.18	1.94	2.24	300
	0340	1-1/2" thick		7,725	.012		2.32	.33	.24	2.89	3.32	
	0380	2" thick		6,345	.015		3.12	.40	.29	3.81	4.36	
	0420	2-1/2" thick		5,480	.018		3.84	.46	.34	4.64	5.30	
	0460	3" thick		4,900	.020		4.58	.51	.38	5.47	6.25	
	0800	Alternate method of figuring paving costs										
	0810	Binder course, 1-1/2" thick	B-25	630	.140	Ton	27	3.60	2.69	33.29	38.50	
	0811	2" thick		690	.128		27	3.28	2.46	32.74	38	
	0812	3" thick		800	.110		27	2.83	2.12	31.95	36.50	
	0813	4" thick		850	.104		27	2.67	2	31.67	36.50	
	0850	Wearing course, 1" thick	B-25B	575	.167		28.50	4.37	3.24	36.11	41.50	
	0851	1-1/2" thick		630	.152		28.50	3.99	2.96	35.45	40.50	
	0852	2" thick		690	.139		28.50	3.65	2.70	34.85	39.50	
	0853	2-1/2" thick		745	.129		28.50	3.38	2.50	34.38	39	
	0854	3" thick		800	.120		28.50	3.14	2.33	33.97	38.50	
	1000	Pavement replacement over trench, 2" thick	B-37	90	.533	S.Y.	3.17	13.25	1.30	17.72	25.50	
	1050	4" thick		70	.686		6.30	17.05	1.67	25.02	35	
	1080	6" thick		55	.873		10	21.50	2.13	33.63	47	
315	0011	**PAVING** Asphaltic concrete, parking lots & driveways										315
	0020	6" stone base, 2" binder course, 1" topping	B-25C	9,000	.005	S.F.	1.21	.14	.16	1.51	1.73	
	0300	Binder course, 1-1/2" thick		35,000	.001		.32	.04	.04	.40	.46	
	0400	2" thick		25,000	.002		.41	.05	.06	.52	.59	
	0500	3" thick		15,000	.003		.64	.08	.10	.82	.94	
	0600	4" thick		10,800	.004		.83	.12	.13	1.08	1.25	
	0800	Sand finish course, 3/4" thick		41,000	.001		.18	.03	.04	.25	.29	
	0900	1" thick		34,000	.001		.22	.04	.04	.30	.35	
	1000	Fill pot holes, hot mix, 2" thick	B-16	4,200	.008		.42	.19	.11	.72	.87	
	1100	4" thick		3,500	.009		.61	.22	.13	.96	1.17	
	1120	6" thick		3,100	.010		.83	.25	.15	1.23	1.47	
	1140	Cold patch, 2" thick	B-51	3,000	.016		.48	.38	.05	.91	1.17	
	1160	4" thick		2,700	.018		.90	.43	.06	1.39	1.72	
	1180	6" thick		1,900	.025		1.41	.60	.08	2.09	2.58	

02750 | Rigid Pavement

	Line	Description	CREW	DAILY OUTPUT	LABOR-HOURS	UNIT	MAT.	LABOR	EQUIP.	TOTAL	TOTAL INCL O&P	
100	0010	**CONCRETE PAVEMENT** Including joints, finishing, and curing										100
	0020	Fixed form, 12' pass, unreinforced, 6" thick	B-26	3,000	.029	S.Y.	17.35	.78	.63	18.76	21	
	0100	8" thick		2,750	.032		24	.85	.68	25.53	28.50	
	0200	9" thick		2,500	.035		28	.93	.75	29.68	33	
	0300	10" thick		2,100	.042		30	1.11	.90	32.01	35.50	
	0400	12" thick		1,800	.049		32.50	1.30	1.05	34.85	38.50	
	0500	15" thick		1,500	.059		36	1.55	1.26	38.81	44	
	0510	For small irregular areas, add					100%					
	0700	Finishing, broom finish small areas	2 Cefi	120	.133	S.Y.		3.83		3.83	5.65	
	1000	Curing, with sprayed membrane by hand	2 Clab	1,500	.011	"	.44	.25		.69	.87	
	1650	For integral coloring, see div. 03310-220										

02766 | Pavement Markings

	Line	Description	CREW	DAILY OUTPUT	LABOR-HOURS	UNIT	MAT.	LABOR	EQUIP.	TOTAL	TOTAL INCL O&P	
550	0010	**LINES ON PAV'T** Acrylic waterborne, white or yellow, 4" wide	B-78	20,000	.002	L.F.	.13	.06	.02	.21	.25	550
	0200	6" wide		11,000	.004		.12	.10	.04	.26	.33	
	0500	8" wide		10,000	.005		.17	.11	.04	.32	.42	
	0600	12" wide		4,000	.012		.31	.29	.11	.71	.91	
	0620	Arrows or gore lines		2,300	.021	S.F.	.50	.50	.19	1.19	1.53	
	0640	Temporary paint, white or yellow		15,000	.003	L.F.	.15	.08	.03	.26	.32	
	0660	Removal	1 Clab	300	.027			.63		.63	.97	
	0680	Temporary tape	2 Clab	1,500	.011		1.46	.25		1.71	2	

Figure 9.33

168

TELEPHONE QUOTATION

PROJECT Office Building

DATE

FIRM QUOTING Landscaping, Inc.

TIME

PHONE (   )

ADDRESS

BY

ITEM QUOTED Landscaping & Sprinklers

RECEIVED BY EBW

WORK INCLUDED	AMOUNT OF QUOTATION
Automatic sprinkler system	1 2 5 6 0
Top soil ⎫	
Seeding ⎬	9 3 4 2
Trees ⎫	
Shrubs ⎬	1 6 9 0 0
(Area ≈ 15,700 SF or 1745 SY	

DELIVERY TIME	**TOTAL BID**	3 8 8 0 2

DOES QUOTATION INCLUDE THE FOLLOWING: If ☐ NO is checked, determine the following:

STATE & LOCAL SALES TAXES	☑ YES	☐ NO	MATERIAL VALUE
DELIVERY TO THE JOB SITE	☒ YES	☐ NO	WEIGHT
COMPLETE INSTALLATION	☒ YES	☐ NO	QUANTITY
COMPLETE SECTION AS PER PLANS & SPECIFICATIONS	☒ YES	☐ NO	DESCRIBE BELOW

EXCLUSIONS AND QUALIFICATIONS

Maintenance : Routine   60 Days

Guarantee: Sprinkler   1 Year
              Dead Replacement   90 Days

ADDENDA ACKNOWLEDGEMENT	**TOTAL ADJUSTMENTS**	
	ADJUSTED TOTAL BID	

ALTERNATES

ALTERNATE NO.	
ALTERNATE NO.	
ALTERNATE NO.	
ALTERNATE NO.	
ALTERNATE NO.	
ALTERNATE NO.	
ALTERNATE NO.	

Figure 9.34

assemblies, such as an 8" foundation wall, 24" × 24" column, and a 1 × 2' strip footing. Prices for such systems are shown in Figure 9.3 5. A more accurate, but time consuming method is to quantify and price the individual components of the concrete work.

Note that the costs in Figure 9.35 are primarily shown per cubic yard of concrete. Prices may also be separated for minimum, average or maximum reinforcing. Care should be taken when using this pricing method to assure that the proposed system closely matches the priced system. For example, one foundation wall system cost cannot be developed to accurately price different size walls. An 8" wall will require 81 square feet of formwork contact area for every cubic yard of concrete. In contrast, a 12" wall will require 54 square feet of forms. The ratio of square feet of form contact area to concrete volume is different for the two wall thicknesses. Reinforcing quantities may also differ per cubic yard. If systems are used for concrete pricing, all variables must be considered.

In the charts below, SFCA is square feet of formwork contact area.

Wall	SFCA/S.F. Wall	SFCA/C.Y. Concrete
8" wall	2.0	81.0
12" wall	2.0	54.0
16" wall	2.0	40.5

Columns	SFCA/L.F. Column	SFCA/C.Y. Concrete
12" square	4.0	108.0
16" square	5.3	81.0
24" square	8.0	54.0

Unless the estimator has prices for all possible cast in place concrete system variations, the most accurate estimating method is to determine the costs for each component of the concrete work.

Cast in Place Concrete

Before beginning the quantity takeoff, the estimator should carefully review the specifications for requirements that may not be indicated on the plans. Such items may include concrete type (regular or lightweight), concrete strengths, and curing time. Specified additives—such as colors and air entrainment—must also be included in the takeoff, along with protection during curing. In addition to the specifications and plans, details and sections showing concrete work will usually be provided. Reinforcing requirements and accessories will be indicated on these drawings.

When estimating concrete work by individual component, it might be possible to apply the dimensions and quantities calculated for one component to another. It is helpful to take off quantities for all components of one type of concrete work at the same time. For example, the square feet of form contact area may also be the same area that will require finishing.

An example of this quantity takeoff method is shown in Figures 9.36 and 9.37. Footing, Column, and Beam Schedules, which are normally included on foundation plans, are shown in Figure 9.41. Note in Figure 9.37 that volume (to determine quantity of concrete) is not converted to cubic yards for each system until total cubic feet are determined. This eliminates the effect of rounding errors and helps to assure that such conversions are done consistently and accurately. If concrete is taken off and priced in this way, the quantities and prices of a given project can be used to develop historical costs for future projects.

03200 | Concrete Reinforcement

		03240	Fibrous Reinforcing	CREW	DAILY OUTPUT	LABOR-HOURS	UNIT	MAT.	LABOR	EQUIP.	TOTAL	TOTAL INCL O&P	
								2002 BARE COSTS					
300	0010		FIBROUS REINFORCING										300
	0100		Synthetic fibers, add to concrete				Lb.	3.79			3.79	4.17	
	0110		1-1/2 lb. per C.Y.				C.Y.	5.85			5.85	6.45	
	0150		Steel fibers, add to concrete				Lb.	.44			.44	.48	
	0155		25 lb. per C.Y.				C.Y.	11			11	12.10	
	0160		50 lb. per C.Y.					22			22	24	
	0170		75 lb. per C.Y.					34			34	37.50	
	0180		100 lb. per C.Y.					44			44	48.50	

03300 | Cast-In-Place Concrete

		03310	Structural Concrete		CREW	DAILY OUTPUT	LABOR-HOURS	UNIT	MAT.	LABOR	EQUIP.	TOTAL	TOTAL INCL O&P	
									2002 BARE COSTS					
200	0010		CONCRETE, FIELD MIX FOB forms 2250 psi	R03310 -080				C.Y.	68			68	75	200
	0020		3000 psi					"	71			71	78	
220	0010		CONCRETE, READY MIX Regular weight	R03310 -040										220
	0020		2000 psi					C.Y.	65			65	71.50	
	0100		2500 psi	R03310 -050					58			58	64	
	0150		3000 psi CN						69			69	76	
	0200		3500 psi						71			71	78	
	0300		4000 psi						74			74	81.50	
	0350		4500 psi						75.50			75.50	83	
	0400		5000 psi CN						77			77	84.50	
	0411		6000 psi						88			88	96.50	
	0412		8000 psi						143			143	158	
	0413		10,000 psi						203			203	224	
	0414		12,000 psi						246			246	270	
	1000		For high early strength cement, add						10%					
	1010		For structural lightweight with regular sand, add						25%					
	2000		For all lightweight aggregate, add						45%					
	3000		For integral colors, 2500 psi, 5 bag mix											
	3100		Red, yellow or brown, 1.8 lb. per bag, add					C.Y.	14.55			14.55	16	
	3200		9.4 lb. per bag, add						78.50			78.50	86.50	
	3400		Black, 1.8 lb. per bag, add						17.25			17.25	19	
	3500		7.5 lb. per bag, add						72.50			72.50	79.50	
	3700		Green, 1.8 lb. per bag, add						34.50			34.50	38	
	3800		7.5 lb. per bag, add						165			165	181	
240	0010		CONCRETE IN PLACE Including forms (4 uses), reinforcing	R03310 -010										240
	0050		steel, including finishing unless otherwise indicated											
	0300		Beams, 5 kip per L.F., 10' span	R03310 -100	C-14A	15.62	12.804	C.Y.	223	385	46	654	905	
	0350		25' span		"	18.55	10.782		209	325	38.50	572.50	790	
	0500		Chimney foundations, industrial, minimum	R04210 -055	C-14C	32.22	3.476		140	100	1.11	241.11	310	
	0510		Maximum		"	23.71	4.724		163	136	1.51	300.51	395	
	0700		Columns, square, 12" x 12", minimum reinforcing		C-14A	11.96	16.722		242	505	60	807	1,125	
	0720		Average reinforcing			10.13	19.743		340	595	71	1,006	1,400	
	0740		Maximum reinforcing			9.03	22.148		420	670	79.50	1,169.50	1,600	
	0800		16" x 16", minimum reinforcing			16.22	12.330		193	375	44.50	612.50	845	
	0820		Average reinforcing			12.57	15.911		305	480	57	842	1,175	
	0840		Maximum reinforcing			10.25	19.512		420	590	70	1,080	1,475	
	0900		24" x 24", minimum reinforcing			23.66	8.453		163	256	30.50	449.50	620	
	0920		Average reinforcing			17.71	11.293		255	340	40.50	635.50	865	

Figure 9.35

171

No matter what format estimators use to compile and list quantities for concrete work, each of these quantities must be computed. Once the format has been established, it should be used throughout the estimate and on all future estimates, establishing consistency and minimizing errors and omissions.

Figure 9.38 graphically illustrates the relationship of the different components of concrete work based on a complete, concrete-framed structure. The way in which the components relate will vary for different and individual types of concrete work, but this pie chart suggests the relative importance of each component.

Formwork: As seen in Figure 9.38, for a complete concrete framed building, formwork can account for more than one third of all concrete costs. For some individual concrete systems using beams, columns, and walls, job-built formwork can account for 50% – 60%, or more, of the costs.

Prefabricated, modular forms can help to significantly reduce formwork costs. This is because these forms can be used over and over again, in some cases, hundreds of times. The initial purchase price of prefabricated forms is higher than job-built, but the effective higher cost decreases with each reuse. These prefabricated forms are only useful and cost effective for repetitive, standard, modular concrete work. To illustrate how the cost of formwork (whether prefabricated or job-built) decreases with each use, please refer to Figure 9.39. This table shows the costs per 100 SFCA (square feet of contact area) for job-built foundation wall forms, 8' high. Costs for the first use of the forms include construction, erection, stripping and cleaning. Costs for reuse include the same erection, stripping, and cleaning costs plus an allowance for 10% material replacement. The difference in cost between the first use and the reuse is the bulk of initial materials and construction labor. The effective costs of the formwork based on the number of uses are calculated below. Costs are from Figure 9.39.

$$\frac{\text{Cost of First Use} + (\text{Cost per Reuse} \times \text{No. of Reuses})}{\text{No. of Uses}} = \frac{\text{Cost per Use}}{(\text{for No. of Uses})}$$

Using the costs in Figure 9.39:

Two Uses:

$$\frac{\$380.12 + (\$263.16 \times 1)}{2} = \$321.64 \text{ per use for two uses}$$

Three Uses:

$$\frac{\$380.12 + (\$263.16 \times 2)}{3} = \$302.15 \text{ per use for three uses}$$

Four Uses:

$$\frac{\$380.12 + (\$263.16 \times 3)}{4} = \$292.40 \text{ per use for four uses}$$

The estimator should expect to reuse forms within a project as well as from one project to the next. To determine the optimum number of forms required, a preliminary schedule (as described in Chapter 5) can be extremely useful. For a building with hundreds of similar columns, only a certain percentage of that number of forms need be built. The appropriate number of forms is determined by the sequence of construction and by the curing time before the forms can be stripped and reused. (See Chapter 5 and Figures 5.4 and 5.5.) Curing time is often the limiting factor when determining the required number and reuse of forms.

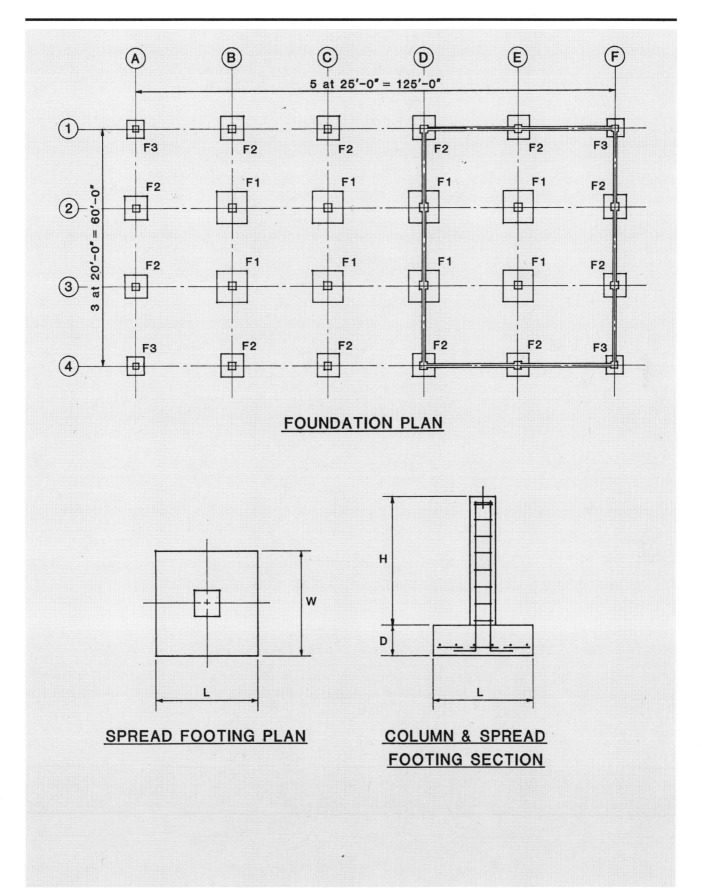

FOUNDATION PLAN

SPREAD FOOTING PLAN

**COLUMN & SPREAD
FOOTING SECTION**

Figure 9.36

173

PROJECT **Sample Project**

ESTIMATE NO.

LOCATION ARCHITECT DATE

TAKE OFF BY EXTENSIONS BY: CHECKED BY:

DESCRIPTION	NO.	DIMENSIONS				Concrete Volume	UNIT	Form Area	UNIT	Finished Area	UNIT	MISC.	UNIT
Spread Footings							CF		SF		SF		
F-1	8	8	8	1.83		937		469		512			
F-2	12	6	6	1.33		575		383		432			
F-3	4	4.5	4.5	1		81		72		81			
Totals						1593	CF	924	SF	1025	SF		
						59	CY						
Wall Footings													
Column Line 1,4	2	38.75	2	1		155		155		155			
Column Line D,F	2	43.5	2	1		174		174		174			
Totals						329	CF	329	SF	329	SF		
						13	CY						
Walls													
Column Line 1,4	2	51.5	1	11		1133		2266		1133			
Column Line D,F	2	61.5	1	11		1353		2706		1353			
Pilasters	10	1.5	1.5	11		83		440		55			
Corners	10 x 8											80	Ea.
Brick Shelf													
Column Line 1		51.5	.33	1		(17)		52					
Column Line F		61.5	.33	1		(21)		62					
Totals						2531	CF	5526	SF	2541	SF		
						94	CY						
Set Anchor Bolts												48	Ea.
Keyway	2	39										78	
	2	44										88	
												166	LF
Columns C-1	8	2	2	12		384		768					
C-2	12	1.67	1.67	12		400		960		ALL			
C-3	4	1.33	1.33	12		85		256		↓			
Totals						869	CF	1984	SF	1984	SF		
						32	CY						

Figure 9.37

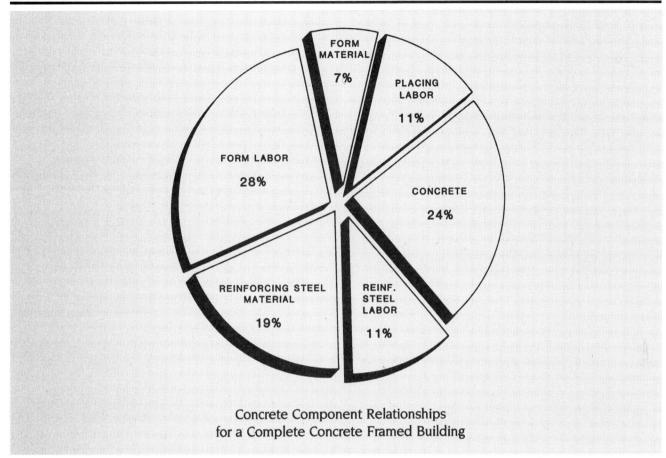

Concrete Component Relationships
for a Complete Concrete Framed Building

Figure 9.38

FOUNDATION WALL, 8' High (Line 03110-455-2000)			First Use			Reuse		
Item	Cost	Unit	Quantity	Material	Installation	Quantity	Material	Installation
5/8" exterior plyform	$864.00	M.S.F.	110 S.F.	$ 95.04		11.0 S.F.	$ 9.50	
Lumber	531.00	M.B.F.	140 B.F.	74.34		14.0 B.F.	7.43	
Accessories			Allow	25.00		Allow	25.00	
Make up, crew C-2	29.24	L.H.	5.0 L.H.		$146.20	1.0 L.H.		$ 29.24
Erect and strip			6.5 L.H.		190.06	6.5 L.H.		190.06
Clean and move			1.5 L.H.		43.86	1.5 L.H.		43.86
Bare Total per 100 S.F.C.A.			13.0 L.H.	$194.38	$380.12	9.0 L.H.	$41.93	$263.16

Figure 9.39

175

Reinforcing: The size and quantities of reinforcing steel can vary greatly with the type of structure, design load, concrete strength, and type of steel. Steel material prices should always be verified by local suppliers. Labor costs will also vary depending upon the size of reinforcing. Reinforcing is priced and purchased by weight. While a smaller and lighter bar may be installed faster than a larger, heavier bar, more tonnage is installed faster with the heavier reinforcing. For example, in slabs and footings, a rodman can install a ton of heavy bars in about nine hours, while a ton of light bars requires about fifteen hours.

Specified sizes for reinforcing are shown in sections and details. Quantities are measured from the plans as linear feet for each size; this figure must then be converted to weight. Conversion factors are given in Figure 9.40. Accessories, dowels, and splicing should be priced separately. If post-tensioning is specified, the estimator must be sure to include all requirements, coatings, wrappings, etc. Grouting after tensioning may also be required.

When the plans show the complete concrete design but are not sufficient to serve as placing drawings, suppliers will often provide the detailing service by preparing reinforcing steel placing drawings. This service should not be substituted for and does not take the place of structural engineering services. In conjunction with the drawings, the suppliers may produce material lists to aid the estimator. For large projects, the Concrete Reinforcing Steel Institute (CRSI) may provide these services to all prospective bidders.

Welded wire fabric for slabs should be measured allowing a 10% overlap. (Material costs for welded wire fabric, as presented in *Means Building Construction Cost Data* include a 10% overlap.) Weights, sizes, and new designations of welded wire fabric are shown in Figure 9.40.

Because of the quantity and weight involved, handling costs for reinforcing must also be added. Unloading, sorting, and crane placement costs can add 5% to 7% to the cost of reinforcing.

Placing Concrete: There are many ways to transport concrete from the truck to the installation locations. The most economical method is by direct chute because little labor or equipment is required from truck to forms; gravity does the work. The construction of an elaborate chute system, while an expensive initial cost, may be less costly overall than alternative methods.

The estimator should visualize the process and use experience to determine the most economical approach. The number of forms, size of the pours, and size of the finishing crews will all have a direct bearing on the choice of proper placement method. If the size of the pour is large and adequate finishing workers and laborers are available, pumping concrete is the fastest, but most expensive method. If a crane is available on-site, placement with the crane and a bucket may be the most feasible. Conveyors, wheelbarrows, and even hand carried buckets may be used depending upon the type of work involved. Motorized carts may also be used in conjunction with each of the above mentioned methods. Temporary ramps and runways will be required for these carts. Costs are presented in *Means Building Construction Cost Data* for each method of placing concrete for different types of structures. Only an analysis of each individual project will determine the most feasible and economical method of concrete placement.

Finishing Concrete: Requirements for concrete finishes are usually found in the specifications portion of the construction documents. For floors, walks, and other horizontal surfaces, not only must the type of finish be included

		Nominal Dimensions - Round Sections		
Bar Size Designation	Weight Pounds Per Foot	Diameter Inches	Cross-Sectional Area-Sq. Inches	Perimeter Inches
#3	.376	.375	.11	1.178
#4	.668	.500	.20	1.571
#5	1.043	.625	.31	1.963
#6	1.502	.750	.44	2.356
#7	2.044	.875	.60	2.749
#8	2.670	1.000	.79	3.142
#9	3.400	1.128	1.00	3.544
#10	4.303	1.270	1.27	3.990
#11	5.313	1.410	1.56	4.430
#14	7.650	1.693	2.25	5.320
#18	13.600	2.257	4.00	7.090

Reinforcing Bars

Common Stock Styles of Welded Wire Fabric

	New Designation	Old Designation	Steel Area Per Foot				Approximate Weight Per 100 Sq. Ft.	
			Longitudinal		Transverse			
	Spacing - Cross Sectional Area (IN.)-(Sq. IN. 100)	Spacing Wire Gauge (IN.)-(AS & W)	IN.	CM	IN.	CM	LB	KG
Rolls	6 x 6-W1.4 x W1.4	6 x 6-10 x 10	0.028	0.071	0.028	0.071	21	9.53
	6 x 6-W2.0 x W2.0	6 x 6-8 x 8 (1)	0.040	0.102	0.040	0.102	29	13.15
	6 x 6-W2.9 x W2.9	6 x 6-6 x 6	0.058	0.147	0.053	0.147	42	19.05
	6 x 6-W4.0 x W4.0	6 x 6-4 x 4	0.080	0.203	0.080	0.203	58	26.31
	4 x 4-W1.4 x W1.4	4 x 4-10 x 10	0.042	0.107	0.042	0.107	31	14.06
	4 x 4-W2.0 x W2.0	4 x 4-8 x 8 (1)	0.060	0.152	0.060	0.152	43	19.50
	4 x 4-W2.9 x W2.9	4 x 4-6 x 6	0.087	0.221	0.087	0.221	62	28.12
	4 x 4-W4.0 x W4.0	4 x 4-4 x 4	0.120	0.305	0.120	0.305	85	38.56
Sheets	6 x 6-W2.9 x W2.9	6 x 6-6 x 6	0.058	0.147	0.058	0.147	42	19.05
	6 x 6-W4.0 x W4.0	6 x 6-4 x 4	0.080	0.203	0.080	0.203	58	26.31
	6 x 6-W5.5 x W5.5	6 x 6-2 x 2 (2)	0.110	0.279	0.110	0.279	80	36.29
	6 x 6-W4.0 x W4.0	4 x 4-4 x 4	0.120	0.305	0.120	0.305	85	38.56

NOTES
1. Exact W-number size for 8 gauge is W2.1
2. Exact W-number size for 2 gauge is W5.4

Figure 9.40

(broom, trowel, etc.), but also any required surface treatments and additives. Such treatments may include hardeners, colors, and abrasives. Costs for floor finishing will also depend upon the allowable tolerances for floor leveling. Typical tolerances may allow a 1/8" variation in 10'. Closer required tolerances will require more labor and finishing costs.

Wall finishing may vary from patching voids and tie holes to creating hammered or sandblasted finishes. Architectural finishes may also be achieved using special form surface materials, such as boards or strips.

Spread Footings: Plans for a building with spread footings (as in Figure 9.36) generally include a footing schedule that lists, counts, and defines each type and size of spread footing. Reinforcing data is usually included. An example footing schedule is shown in Figure 9.41; it includes information on reinforcing requirements. When the schedule is provided, it eliminates the repetition of detailing on the plans and simplifies the estimator's task. Dimensions can be transferred directly onto the Quantity Sheet (see Figure 9.37). Using this form and the footing schedule, the procedure is as follows:

- Write down the number of each size and the dimensions of each of the spread footings shown on the plans.
- Extend each of these lines to obtain the Volume, Form Area, and Finish Area.
- Total the extensions.
- Count the total number of footings on the plan to make sure none has been left out.

When working with these plans for spread footings, take off related items such as anchor bolts, templates, base plates, and reinforcing steel. Reinforcing steel may be transferred to a separate sheet. These other items should be computed and totalled on the same Quantity Sheet from which they were derived in order to avoid errors and omissions.

Continuous Wall Footings and Grade Beams: Wall footings transmit loads directly to the soil. Grade Beams are self-supporting structural members that carry wall loads across unacceptable soil to spread footings, caissons, or support piles. The sequence for quantity takeoff is the same as discussed above. Again, dimensions should be recorded and extended on a Quantity Sheet.

Wall footings are stepped when grade changes necessitate changes in footing elevation. Figure 9.42 illustrates a stepped footing. Notice that in plan view, the footing may easily be mistaken as continuous. Where stepped footings occur, extra formwork (2 × Area A plus Area B) and extra concrete (W × Area A) are required in addition to the quantities for the continuous footing. Accessories should also be taken off at the same time using the pertinent dimensions and derived quantities.

Pile Caps: Pile Caps are basically spread footings on piles. The shape is not always rectangular, so that the form area is determined by the perimeter × depth. Concrete volume should be not deducted for the protrusion of the piles into the cap.

Piers: Piers are used to extend the bearing of the superstructure down to the footing. They vary in length to accommodate sub-surface conditions, original site grades and frost penetration. Piers are often included on the Footing Schedule.

Walls: Walls are separated by height during takeoff. Generally this is done in 4' increments such as under 4', 4' to 8', and 8' to 12'. Different forming and placing costs may be used for each of these heights.

- Using a quantity sheet, define each wall. If all walls have footings, the wall footing schedule can be used as a basis.
- Write in the dimensions for each of these walls.
- From these dimensions, compute the Volume and Form Area for each wall, dividing them into convenient modular categories. Brick shelves and slab seats should be listed and deducted from concrete volume.
- Total the extensions.
- Upon examination of the wall sections and details, list all associated items: beams seats, anchor bolts, base plates, chamfers, pilasters, rustications, architectural finishes, openings, and reinforcing steel. Where pilasters occur, figure the wall forming straight through and then add the area of the pilasters as a separate item. Costs are different for walls and pilasters. The additional materials will be used in waste, bracing, and framing.
- Building a perimeter wall with many angles and corners will take more time and more material to form than a wall with four square corners. Note the number of corners on the Quantity Sheet so that an allowance for material and labor can be made when pricing the wall forming.

Footing Schedule			
Ident.	No.	Size	Reinforcing
F-1	8	8'-0" x 8'0" x 1'-10"	9-#6 e.w.
F-2	12	6'-0" x 6'0" x 1'-4"	8-#5 e.w.
F-3	4	4'-6" x 4'6" x 1'-0"	5-#5 e.w.

Column Schedule			
Ident.	No.	Size	Reinforcing
C-1	8	24" x 24" x 12'	8-#11 ties #4 @ 22"
C-2	12	20" x 20" x 12'	8-#9 ties #3 @ 18"
C-3	4	16" x 16" x 12'	8-#9 ties #3 @ 16"

Beam Schedule			
Ident.	No.	Size	Reinforcing
B-1 Ext.	1	1' x 2'-6" x 60'	See Re-Bar Schedule
B-2 Ext.	2	1' x 2'-4" x 75'	
B-3 Int.	2	8" x 1'-6" x 60'	
B-4 Int.	2	6" x 1'-6" x 125'	

Figure 9.41

The above procedure will be identical for foundation walls and retaining walls. If retaining walls are battered, be sure to note this. Formwork erection time may be up to 20% slower as a result. For special architectural effects, the use of form liners, rustication strips, retarders, and sandblasting must be considered.

Columns: Columns are commonly listed in a schedule similar to that used for spread footings. An example is shown in Figure 9.41. Quantities are derived using the same procedure discussed above. When estimating columns, the following associated items must be considered:

- Height Measurement from floor slab below to underside of beam or slab.
- Multiple use of forms.
- Chamfer strips.
- Finishing requirements.
- Capitals.
- Reinforcing steel.
- Clamps, inserts, and placing pockets.

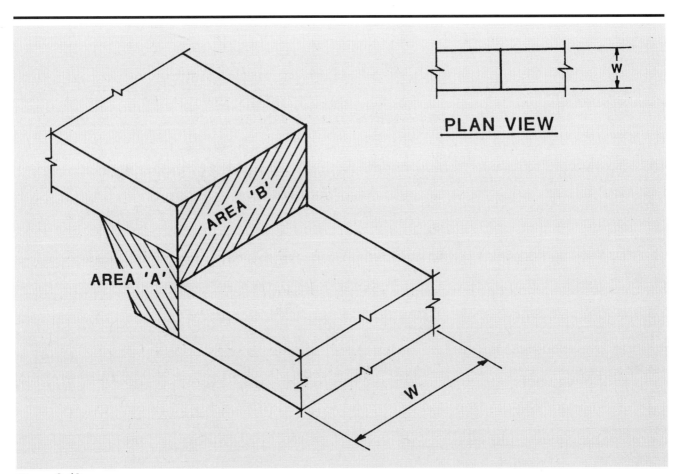

Figure 9.42

Beams: As with footings and columns, cast in place beams are also often listed on a schedule. If not otherwise designated, interior and exterior beams should be listed separately. The following principles should be used when estimating beams:

- The depths given in schedules represent design depths. Slab thickness must be deducted when calculating volume and contact area.
- Exterior and interior beams should be listed separately. An exterior beam requires forming for the full design depth on one side. The access and bracing of this exterior form makes it more costly to erect and strip.
- Formwork or volume should not be deducted for columns.
- Beams and slabs are usually placed simultaneously. Beam quantities should be separated by floor to be included with slab placement.
- The support method should be determined—whether the formwork is hung from beams, supported with adjustable horizontal shores, or shored from the structure below. Reshoring costs must be included if required.
- Finishing requirements and design will affect the multiple use of forms.

Calculations for Formed, Cast in Place Concrete:

Spread Footings and Pile Caps:

Volume $-$ Length (L) $\times$ Width W $\times$ Depth (D)

Form Area $-$ (2L + 2W) $\times$ D

Finish Area $-$ L $\times$ W

Continuous Footings:

Volume $-$ Length (L) $\times$ Width (W) $\times$ Depth (D)

Form Area $-$ 2L $\times$ D

Finish Area $-$ L $\times$ W

Rectangular Columns:

Volume $-$ Length (L) $\times$ Width (W) $\times$ Height (H)

Form Area $-$ (2L + 2W) $\times$ H

Finish Area $-$ (2L + 2W) $\times$ H

Round Columns:

Volume $-$ 3.146 $\times$ Radius (R) $\times$ R $\times$ Height (H)

Form Area $-$ 3.146 $\times$ Diameter (D) $\times$ H

Finish Area $-$ 3.146 $\times$ D $\times$ H

Walls:

Volume $-$ Length (L) $\times$ Height (H) $\times$ Thickness (T)

Form Area $-$ L $\times$ H $\times$ 2

Finish Area $-$ L $\times$ H (for each side)

Beams:

Exterior:

Volume $-$ L $\times$ W $\times$ (Design depth (d) $-$ Slab thickness (t))

Form Area $-$ (L $\times$ W) + (L $\times$ d) + (L $\times$ (d $-$ t))

Finish Area $-$ Same as form area

Interior:

Volume — Length × Width × (d − t)

Form Area — (L × W) + 2 (L × (d − t))

Finish Area — Same as form area

Slabs: Concrete stabs can be divided into two categories: slabs on grade including sidewalks and concrete paving, and elevated slabs. Several items must be considered when estimating slabs on grade.

- Granular Base
- Fine Grading
- Vapor Barrier
- Edge Forms and Bulkheads
- Expansion Joints
- Contraction Joints
- Screeds
- Welded Wire Fabric and Reinforcing
- Concrete Material
- Finish and Topping
- Curing and Hardeners
- Control Joints
- Concrete "Outs"
- Haunches
- Drops

Quantities for most of the items above can be derived from basic dimensions: length, width, and slab thickness. When these quantities are calculated, some basic rules can be used:

- Allow about 25% compaction for granular base.
- Allow 10% overlap for vapor barrier and welded wire fabric.
- Allow 5% extra concrete for waste.
- No deductions should be made for columns or "outs" under 10 S.F.
- If screeds are separated from forming and placing costs, figure 1 L.F. of screed per 10 S.F. of finish area.

Elevated slabs, while similar to slabs on grade, require special consideration. For all types of elevated slabs, edge forms should be estimated carefully. Stair towers, elevator shafts, and utility chases all require edge forms. The appropriate quantities for each of these "outs" should be deducted from the concrete volume and finish area. Edge form materials and installation will vary depending upon the slab type and thickness. Special consideration should be given to pipe chases that require concrete placement after the pipes are in position. This is done in order to maintain the fireproofing integrity. Hand mixing and placing may be necessary.

When estimating beam-supported slabs, beams should not normally be deducted from the floor form area. The extra costs for bracing and framing at the beams will account for the difference. When concrete for the beams and slab is placed at the same time, the placement costs will be the same for both.

Slab placement costs should be increased based on the beam volume. Separate costs for beam concrete placement should not be included. For hung slabs, separation of costs for beams and slab is even more difficult. These costs may be treated as one item.

Metal decking for elevated slabs is usually installed by the steel erector. When determining concrete quantities, measure the slab thickness to the middle of the corrugation.

Cast in place concrete joist and dome (or waffle) slabs are more difficult to estimate. Pans and domes are shored with either open or closed deck forming. For concrete joist systems, quantities for concrete can be calculated from Figure 9.43. Volume of beams must be added separately. Volume of waffle slabs, because of the varying sizes of solid column heads, should be estimated differently. Total volume from top of slab to bottom of pan should be calculated. The volume of the actual number of domes (voids) is then deducted. Volumes for standard domes are listed in Figure 9.44.

Stairs: When taking off cast in place stairs, a careful investigation should be made to ensure that all inserts, such as railing pockets, anchors, nosings, and reinforcing steel have been included. Also indicate on your quantity sheet any special tread finishes; these can sometimes represent a considerable cost.

Stairs cast on fill are most easily estimated and recorded by the square foot. The calculation is slant length times width. Because of the commonly used ratios adopted by designers, this method works for high or low risers with wide or narrow treads. Shored cast in place stairs can be estimated in the same way—slant length times width.

	Concrete Quantities (CF concrete/SF floor) for Single & Multiple Span Concrete Joist Construction					
	Width					
	20″ Forms			**30″ Forms**		
Depth (Rib/Slab)	5″ Rib 25″ O.C.	6″ Rib 26″ O.C.	7″ Rib 27″ O.C.	5″ Rib 35″ O.C.	6″ Rib 36″ O.C.	7″ Rib 37″ O.C.
8″/3″	0.40	0.42	—	0.36	0.37	—
8″/4½″	0.53	0.55	—	0.48	0.50	—
10″/3″	0.45	0.47	—	0.39	0.41	—
10″/4½″	0.57	0.61	—	0.51	0.53	—
12″/3″	0.49	0.52	—	0.42	0.45	—
12″/4½″	0.62	0.65	—	0.55	0.57	—
14″/3″	0.54	0.57	—	0.45	0.48	—
14″/4½″	0.66	0.69	—	0.58	0.61	—
16″/3″	—	0.63	0.66	—	0.52	0.55
16″/4½″	—	0.75	0.79	—	0.65	0.68
20″/3″	—	0.75	0.79	—	0.61	0.64
20″/4½″	—	0.87	0.91	—	0.74	0.77

Figure 9.43

Treads and landings of prefabricated metal pan stairs should be identified on the takeoff sheet. The cost of hand placing the fill concrete should be included in Division 3.

For all cast in place, formed concrete work, the following factors can affect the number of pans to rent, the S.F. of forms to build, the number of shores to provide, etc.:

- Placement rate of concrete for crew.
- Concrete pump or crane and bucket rental cost per pour.
- Finishing rate for concrete finishers.
- Curing time before stripping and reshoring.
- Forming and stripping time.
- Number of reuses of forms.

Experience (or consultation with an experienced superintendent) will tell the estimator which combination of these six items will limit the quantity of forms to be built, or the number of pans to be rented.

Precast Concrete

Typical precast units are estimated and purchased by the square foot or linear foot. The manufacturer will most often deliver precast units. Costs for delivery are sometimes separate from the purchase price and must be included in the estimate. Most manufacturers also have the capability to erect these units, and, because of experience, are best suited to the task.

Each type and size of precast unit should be separated on the quantity or estimate sheet. Typical units are shown in Figure 9.45. The estimator should verify the availability of specified units with local suppliers. For types that are seldom used, casting beds may have to be custom built, thus increasing cost.

Connections and joint requirements will vary with the type of unit, from welding or bolting embedded steel to simple grouting. Toppings must also be included, whether lightweight or regular concrete on floors, insulating concrete for roofs. The specifications and construction details must be examined carefully.

Sample Estimate: Division 3

When performing the estimate for the sample division building project, a number of principles mentioned above are applied. The most important is common sense, together with the ability to visualize the construction process. Most information for the quantity takeoff would normally be supplied on

	Standard Dome sizes and Volumes									
	19" Domes				30" Domes					
Depth	6"	8"	10"	12"	8"	10"	12"	14"	16"	20"
Volume (CF per Dome)	1.09	1.41	1.90	2.14	3.85	4.78	5.53	6.54	7.44	9.16

Figure 9.44

R03410-030   Prestressed Precast Concrete Structural Units

See also R03410-090 for post-tensioned prestressed concrete.

Type	Location	Depth	Span in Ft.		Live Load Lb. per S.F.
Double Tee 8' to 10'	Floor	28" to 34"	60 to 80		50 to 80
	Roof	12" to 24"	30 to 50		40
	Wall	Width 8'	Up to 55' high		Wind
Multiple Tee 8'	Roof	8" to 12"	15 to 40		40
	Floor	8" to 12"	15 to 30		100
Plank or	Roof		Roof	Floor	40 for Roof
	or	4" 6" 8" 10" 12"	13 22 26 33 42	12 18 25 29 32	100 for Floor
	Floor				
Single Tee 8' to 10	Roof	28" 32" 36" 48"	40 80 100 120		40
AASHO Girder	Bridges	Type 4 5 6	100 110 125		Highway
Box Beam 4'	Bridges	15" 27" 33"	40 to 100		Highway

The majority of precast projects today utilize double tees rather than single tees because of speed and ease of installation. As a result casting beds at manufacturing plants are normally formed for double tees. Single tee projects will therefore require an initial set up charge to be spread over the individual single tee costs.

For floors, a 2" to 3" topping is field cast over the shapes. For roofs, insulating concrete or rigid insulation is placed over the shapes.

Member lengths up to 40' are standard haul, 40' to 60' require special permits and lengths over 60' must be escorted. Over width and/or over length can add up to 100% on hauling costs.

Large heavy members may require two cranes for lifting which would increase erection costs by about 45%. An eight man crew can install 12 to 20 double tees, or 45 to 70 quad tees or planks per day.

Grouting of connections must also be included.

Several system buildings utilizing precast members are available. Heights can go up to 22 stories for apartment buildings. Optimum design ratio is 3 S.F. of surface to 1 S.F. of floor area.

Figure 9.45

drawings, details, and in the specifications. Even when all data is provided, however, the estimator must be familiar with standard materials and practices.

Concrete work is composed of many components, most of which are based on similar dimensions. It makes sense to organize all takeoff data on a Quantity Sheet as shown in Figure 9.37. A partial takeoff for Division 3 of the sample project is shown in Figure 9.46. Note that every quantity that is to be priced has been converted to the appropriate units and delineated on the sheet. As the quantities are transferred to the pricing sheets, colored pencil check marks will ensure inclusion of all items. The estimate sheets for Division 3 are shown in Figures 9.47 to 9.51.

Most quantities are derived from the plan in Figure 9.5a. Heights are determined from the building section in Figure 9.10. Dimensioned details are provided in complete plans and specifications. The spread footing quantities are easily obtained from the drawings. Care must be exercised when estimating the continuous footings. Note on Figure 9.46 that the complete perimeter is used for a gross quantity and that a deduction is included where the spread footings interrupt the continuous footing. The keyway and dowel supports, as shown in Section A-A in Figure 9.5a, continue across the spread footings. Thus, the linear dimension of the keyway and dowel supports will be greater than that of the continuous footing. (Note that the keyway and dowel supports are included in more than the perimeter footings.) The steps in the perimeter footing between column lines 5 and 6 (Figure 9.5a) are not listed separately due to their insignificant effect on the total cost. This is a judgment call of the type that should only be made based on experience.

Quantities for the perimeter wall are calculated with no deduction for the pilasters. Such a deduction should be more than compensated by extra costs for framing and bracing at the connections. Pilaster formwork should be recorded separately. Costs used for placing the pilaster concrete are the same as for the perimeter wall because placement for both will occur simultaneously.

Reinforcing quantities are taken from Reference Table R03310-010 in *Means Building Construction Cost Data*. These figures are based on pounds per cubic yard of concrete. When detailed information is available, actual linear quantities and counts of splices and accessories of each size and type should be taken off and priced for greater accuracy. Note that the costs used for welded wire fabric as presented in *Means Building Construction Cost Data* (Figure 9.52) include a 10% overlap so that quantities in this case are derived from actual dimensions.

Concrete material costs for the waffle slab include a cost for high early strength. The 2nd and 3rd floor elevated slabs include percentages for high early strength and for lightweight aggregate. The appropriate percentages are obtained from Figure 9.53 and added to the unit costs before entry on the estimate sheet. Costs for concrete are readily available from local suppliers and should always be verified.

Figures 9.53 (bottom) and 9.54 show costs for complete concrete systems in place. These systems are most often priced per cubic yard of concrete. The costs are based on averages for each type and size of system as listed. These figures provide an excellent checklist to quickly compare costs of similar systems estimated by individual components for possible omissions or duplications. For the sample estimate such a comparison can be made. The

QUANTITY SHEET

PROJECT: Office Building
LOCATION:
TAKE OFF BY:

Division 3
ARCHITECT:
EXTENSIONS BY:

DESCRIPTION	NO.	DIMENSIONS			Forms	UNIT	Volume	UNIT	Reinf.	UNIT	Finish	UNIT
Spread Footings	6	11.5'	11.5'	2'	552	SFCA	1587	CF				
	4	9.5'	9.5'	2'	304		722					
	18	8'	8'	2'	1152		2304					
					2008	SFCA	4613	CF				
Dowel Supports	28 Ea.						171	CY	8892	Lbs.		
									4.45	T		
Continuous Footings												
Perimeter	1	600'	2'	1'	1200	SFCA	1200	CF	2600	Lbs.		
									1.30	T		
Deduct Spr. Footings	(1	8'	2'	1'	-288		-288)					
Ramp		166'	1.5'	0.67'	498		166					
Stoop		25'	1.5'	0.67'	75		25					
Stairs	2	40'	1.5'	0.67'	240		81					
Core		160'	1.5'	0.67'	480		161					
					2205	SFCA	1345	CF				
							50	CY				
Dowel Supports		1031	LF									
Keyway		791	LF									
Walls - Perimeter												
Deduct Stairs / Core												
Pit		600'	1'	10'	12000	SFCA	4020	CF			6000	SF
Ramp & Stoop		42'	0.83'	4'	336		140					
Deduct - Doors		191'	1'	-4'	1528		512					
Openings	(2	10'	1'	10'	-400		-134)					
	(6	31'	1'	-4.5'	-1674		-561)					
					11790	SFCA	3977	CF	18988	Lbs.		
							148	CY	9.49	T		
Box Openings		466	LF		81	LF						
Columns 28" diam.	9	30" diam.	9'		93	LF	398	CF			636	
	1	30" diam.	12'				59				94	
											730	SF
Pilasters 3 Sides	14	2'	1.67'	9'	673	SFCA	457	CF	8908	Lbs.		
							17	CY	4.45	T		
2 Sides	4	1.67'	1.67'	9'	120		421	CF				
							100					
					793	SFCA	521	CF	9956	Lbs.	793	SF
							19	CY	4.98	T		
Chamfer	32	288	LF	9'								

Figure 9.46

187

CONSOLIDATED ESTIMATE

PROJECT: Office Building
LOCATION:
TAKE OFF BY: ABC
CLASSIFICATION: Division 3
ARCHITECT: ABC
QUANTITIES BY: ABC
PRICES BY: As Shown
EXTENSIONS BY:

ESTIMATE NO:
DATE: Jan-02
CHECKED BY: GHI

DESCRIPTION	SOURCE			QUANT.	UNIT	MATERIAL		DEF	LABOR		DEF	EQUIPMENT		SUBCONTRACT		TOTAL	
						COST	TOTAL		COST	TOTAL		COST	TOTAL	COST	TOTAL	COST	TOTAL
Division 3: Concrete																	
Form Work																	
Spread Footing	03110	430	5150	2008	SFCA	0.56	1124		2.19	4398							
Dowel Supports	03110	430	6100	18	Ea.	13.80	248		45.50	819							
	03110	430	6150	10	Ea.	24.00	240		53.50	535							
Continuous Footings	03110	430	0150	2205	SFCA	0.74	1632		1.87	4123							
Dowel Supports	03110	430	0500	1031	LF	0.80	825		1.82	1876							
Keyway (excl. stairs & core)	03110	430	1500	791	LF	0.19	150		0.45	356							
Walls:																	
Pit (10")	03110	455	2000	336	SFCA	1.83	615		4.68	1572							
Ramp & Stoops (8")	03110	455	2000	1528	SFCA	1.83	2796		4.68	7151							
Perimeter (12" x 10")	03110	455	2550	9926	SFCA	0.60	5956		3.55	35237							
Deduct Openings																	
Garage																	
Above Grade																	
Box Openings	03110	455	0150	466	LF	1.84	857		5.00	2330							
Columns - 28" Diam., 9@ 9' 1 @12'	03110	410	1850	93	LF	13.55	1260		7.25	674							
Pilasters	03110	455	8600	793	SFCA	0.72	571		5.20	4124							
Subtotals							$ 16,275			$ 63,196							

Figure 9.47

188

CONSOLIDATED ESTIMATE

PROJECT: Office Building
LOCATION:
TAKE OFF BY: ABC
CLASSIFICATION: Division 3
ARCHITECT: As Shown
QUANTITIES BY: ABC   PRICES BY: ABC   EXTENSIONS BY: DEF
ESTIMATE NO:
DATE: Jan-02
CHECKED BY: GHI

DESCRIPTION	SOURCE	QUANT.	UNIT	MATERIAL COST	MATERIAL TOTAL	LABOR COST	LABOR TOTAL	EQUIPMENT COST	EQUIPMENT TOTAL	SUBCONTRACT COST	SUBCONTRACT TOTAL	TOTAL COST	TOTAL TOTAL
Division 3: (Cont'd)													
Form Work (Cont'd)													
Chamfer 3/4"	03150 160 2200	288	LF	0.43	124	0.46	132						
4" Slab Edge Form	03110 445 3000	65	LF	0.41	27	1.51	98						
@ Pit & Openings													
Waffle Slab (1st Floor)	03110 420 4500	18220	LF	4.70	85634	3.47	63223						
		18900											
Deduct, Stairs		-600											
Shaft		-80											
Opening Edge Forms	03110 420 5000	273	SFCA	3.38	923	7.40	2020						
Perimeter Edge Forms	03110 420 7000	600	SFCA	0.38	228	1.82	1092						
Perimeter Work Deck	03110 420 8000	200	LF	11.00	2200	10.10	2020						
Bulkhead Forms	03110 420 6000	5000	LF	1.65	8250	2.81	14050						
Reinforcing Steel													
Footings	03210 600 0500	5.75	TON	535.00	3076	520.00	2990						
Walls	03210 600 0700	9.49	TON	535.00	5077	365.00	3464						
Columns	03210 600 0250	9.43	TON	550.00	5187	475.00	4479						
Waffle Slab	03210 600 0400	23.59	TON	595.00	14036	385.00	9082						
					$ 124,761		$ 102,651						

Figure 9.48

189

CONSOLIDATED ESTIMATE

PROJECT: Office Building	CLASSIFICATION: Division 3
LOCATION:	ARCHITECT:
TAKE OFF BY: ABC	QUANTITIES BY: ABC PRICES BY: ABC

ESTIMATE NO:	
DATE: Jan-02	
CHECKED BY: GHI	
EXTENSIONS BY:	

SOURCE	DESCRIPTION	QUANT	UNIT	MATERIAL COST (DEF)	MATERIAL TOTAL	LABOR COST	LABOR TOTAL	EQUIPMENT COST (DEF)	EQUIPMENT TOTAL	SUBCONTRACT COST	SUBCONTRACT TOTAL	TOTAL COST	TOTAL
	Division 3: (Cont'd)												
	Reinforcing Steel												
	WWF: 6 x 6 10/10												
	Elevated Slab	37800	SF										
	Garage Slab	18900											
	Penthouse Floor	2100											
	Deduct - Stair	-1200											
	Elevator	-360											
		57240	SF										
03220 200 0100	Total WWF	572.4	CSF	7.00	4007	15.65	8958						
	Cast in Place Concrete												
	Spreading Footings												
03310 220 0150	Concrete - Incl. 5% Waste	180	CY	69.00	12420								
03310 700 2600	Placing	180	CY			9.85	1773	0.59	106				
	Continuous Footings												
03310 220 0150	Concrete - Incl. 5% Waste	52	CY	69.00	3588								
03310 700 1900	Placing	52	CY			9.85	512	0.59	31				
	Walls												
03310 220 0300	Concrete - Incl. 5% Waste	233	CY	74.00	17242								
03310 700 5100	Placing	233	CY			14.75	3437	6.85	1596				
03350 350 0010	Finishing	6000	SF	0.03	180	0.43	2580						
	Columns												
03310 220 0150	Concrete - Incl. 5% Waste	18	CY	74.00	1332								
03310 700 1900	Placing	18	CY			11.60	209	5.40	97				
03350 350 0010	Finishing	730	SF	0.03	22	0.43	314						
					$ 38,791		$ 17,783		$ 1,830				

Figure 9.49

CONSOLIDATED ESTIMATE

PROJECT: Office Building
LOCATION:
TAKE OFF BY: ABC
CLASSIFICATION: Division 3
ARCHITECT:
QUANTITIES BY: ABC
PRICES BY: ABC
ESTIMATE NO:
DATE: Jan-02
CHECKED BY: GHI
EXTENSIONS BY:
PRICES BY: DEF DEF

DESCRIPTION	SOURCE			QUANT	UNIT	MATERIAL COST	MATERIAL TOTAL	LABOR COST	LABOR TOTAL	EQUIPMENT COST	EQUIPMENT TOTAL	SUBCONTRACT COST	SUBCONTRACT TOTAL	TOTAL COST	TOTAL TOTAL
Division 3: (Cont'd)															
Cast in Place Concrete															
Pilasters															
Concrete - Incl. 5% Waste	03310	220	0300	20	CY	74.00	1480								
Placing	03310	700	4950	180	CY			16.20	2916	7.55	1359				
Finishing	03350	350	0010	793	SF	0.03	24	0.43	341						
Slab on Grade															
Concrete - Incl. 5% Waste	03310	220	0150	348	CY	69.00	24012								
Placing	03310	700	4600	348	CY			7.15	2488	0.43	150				
Finishing	03350	300	0150	18210	SF			0.36	6556						
Pit Slab															
Concrete - Incl. 5% Waste	03310	220	0150	3	CY	69.00	207								
Placing	03310	700	4600	3	CY			7.15	21	0.43	1				
Finishing	03350	300	0150	80	SF			0.36	29						
Waffle Slab															
Concrete - Incl. 5% Waste	03310	220	0300	555	CY	74.00	41070								
Add10% for high strength	03310	220	1000	555	CY	7.40	4107								
Placing	03310	700	1400	555	CY			11.60	6438	5.40	2997				
Finishing	03350	300	0150	18900	SF			0.36	6804						
5-1/2" Elevated Slab															
4" Lt. Wt. Concrete - Incl. 5%	03310	220	0300	462	CY	74.00	34188								
Add 25% for struct lightweight	03310	220	1010	462	CY	18.50	8547								
Placing	03310	700	1400	462	CY			11.60	5359	5.40	2495				
Finishing	03350	300	0150	37800	SF			0.36	13608						
Curing - 3 floors	03390	200	0300	567	CSF	4.85	2750	3.95	2240						
						$	116,385	$	46,800	$	7,002				

Figure 9.50

191

CONSOLIDATED ESTIMATE

PROJECT: Office Building
LOCATION:
TAKE OFF BY: ABC

CLASSIFICATION: Division 3
ARCHITECT:
QUANTITIES BY: ABC
PRICES BY: As Shown
EXTENSIONS BY: DEF

ESTIMATE NO:
DATE: Jan-02
CHECKED BY: GHI

SOURCE	DESCRIPTION	QUANT	UNIT	MATERIAL DEF		LABOR DEF		EQUIPMENT DEF		SUBCONTRACT		TOTAL	
				COST	TOTAL	COST	TOTAL	COST	TOTAL	COST	TOTAL	COST	TOTAL
	Division 3: (Cont'd)												
	Cast in Place Concrete (Cont'd)												
03150 660 1750	Stair Treads & Landings	1436	SF	3.75	5385	2.30	3303						
	Subtotals				$ 5,385		$ 3,303						
	Sheet 1 Subtotals				$ 16,275		$ 63,196						
	Sheet 1 Subtotals				$ 124,761		$ 102,651						
	Sheet 1 Subtotals				$ 38,791		$ 17,783		$ 1,830				
	Sheet 1 Subtotals				$ 116,385		$ 46,800		$ 7,002				
	Sheet 1 Subtotals				$ 5,385		$ 3,303						
	Division 3 Total				$ 312,367		$ 240,339		$ 8,832				

Figure 9.51

03200 | Concrete Reinforcement

03210 | Reinforcing Steel

			CREW	DAILY OUTPUT	LABOR-HOURS	UNIT	2002 BARE COSTS MAT.	LABOR	EQUIP.	TOTAL	TOTAL INCL O&P	
700	1700	Straight bars, #10 & #11 — R03210 -070	C-5	140	.400	Ea.	17.20	13.25	5.20	35.65	46.50	700
	1750	#14 bars		130	.431		20.50	14.30	5.55	40.35	52	
	1800	#18 bars		75	.747		31.50	25	9.65	66.15	86	
	2100	#11 to #18 & #14 to #18 transition		75	.747		33.50	25	9.65	68.15	88	
	2400	Bent bars, #10 & #11		105	.533		29.50	17.70	6.90	54.10	69	
	2500	#14		90	.622		39	20.50	8.05	67.55	85.50	
	2600	#18		70	.800		56	26.50	10.35	92.85	116	
	2800	#11 to #14 transition		75	.747		41	25	9.65	75.65	96	
	2900	#11 to #18 & #14 to #18 transition		70	.800		56	26.50	10.35	92.85	116	

03220 | Welded Wire Fabric

			CREW	DAILY OUTPUT	LABOR-HOURS	UNIT	MAT.	LABOR	EQUIP.	TOTAL	TOTAL INCL O&P	
200	0010	**WELDED WIRE FABRIC** ASTM A185 — R03220 -030										200
	0050	Sheets										
	0100	6 x 6 - W1.4 x W1.4 (10 x 10) 21 lb. per C.S.F.	2 Rodm	35	.457	C.S.F.	7	15.65		22.65	34	
	0200	6 x 6 - W2.1 x W2.1 (8 x 8) 30 lb. per C.S.F.		31	.516		8.95	17.70		26.65	40	
	0300	6 x 6 - W2.9 x W2.9 (6 x 6) 42 lb. per C.S.F.		29	.552		11.85	18.90		30.75	45	
	0400	6 x 6 - W4 x W4 (4 x 4) 58 lb. per C.S.F.		27	.593		16.80	20.50		37.30	53	
	0500	4 x 4 - W1.4 x W1.4 (10 x 10) 31 lb. per C.S.F.		31	.516		10.85	17.70		28.55	42	
	0600	4 x 4 - W2.1 x W2.1 (8 x 8) 44 lb. per C.S.F.		29	.552		13.20	18.90		32.10	46.50	
	0650	4 x 4 - W2.9 x W2.9 (6 x 6) 61 lb. per C.S.F.		27	.593		20	20.50		40.50	56.50	
	0700	4 x 4 - W4 x W4 (4 x 4) 85 lb. per C.S.F.		25	.640		26.50	22		48.50	66	
	0750	Rolls										
	0800	2 x 2 - #14 galv., 21 lb/C.S.F., beam & column wrap	2 Rodm	6.50	2.462	C.S.F.	17.45	84.50		101.95	161	
	0900	2 x 2 - #12 galv. for gunite reinforcing	"	6.50	2.462		19.45	84.50		103.95	164	
	1000	Specially fabricated heavier gauges in sheets	4 Rodm	50	.640			22		22	37	
	1010	Material only, minimum				Ton	510			510	560	
	1020	Average					705			705	775	
	1030	Maximum					920			920	1,000	

03230 | Stressing Tendons

			CREW	DAILY OUTPUT	LABOR-HOURS	UNIT	MAT.	LABOR	EQUIP.	TOTAL	TOTAL INCL O&P	
600	0010	**PRESTRESSING STEEL** Post-tensioned in field — R03410 -090										600
	0100	Grouted strand, 50' span, 100 kip	C-3	1,200	.053	Lb.	2.58	1.67	.14	4.39	5.75	
	0150	300 kip		2,700	.024		2.11	.74	.06	2.91	3.61	
	0300	100' span, 100 kip		1,700	.038		2.58	1.18	.10	3.86	4.88	
	0350	300 kip		3,200	.020		2.47	.62	.05	3.14	3.81	
	0500	200' span, 100 kip		2,700	.024		2.58	.74	.06	3.38	4.13	
	0550	300 kip		3,500	.018		2.47	.57	.05	3.09	3.70	
	0800	Grouted bars, 50' span, 42 kip		2,600	.025		.46	.77	.06	1.29	1.83	
	0850	143 kip		3,200	.020		.44	.62	.05	1.11	1.57	
	1000	75' span, 42 kip		3,200	.020		.46	.62	.05	1.13	1.60	
	1050	143 kip		4,200	.015		.40	.48	.04	.92	1.26	
	1200	Ungrouted strand, 50' span, 100 kip	C-4	1,275	.025		1.92	.87	.05	2.84	3.64	
	1250	300 kip		1,475	.022		2.10	.75	.05	2.90	3.63	
	1400	100' span, 100 kip		1,500	.021		1.95	.74	.04	2.73	3.45	
	1450	300 kip		1,650	.019		2.10	.67	.04	2.81	3.49	
	1600	200' span, 100 kip		1,500	.021		1.95	.74	.04	2.73	3.45	
	1650	300 kip		1,700	.019		2.10	.65	.04	2.79	3.45	
	1800	Ungrouted bars, 50' span, 42 kip		1,400	.023		.28	.79	.05	1.12	1.70	
	1850	143 kip		1,700	.019		.28	.65	.04	.97	1.44	
	2000	75' span, 42 kip		1,800	.018		.28	.62	.04	.94	1.39	
	2050	143 kip		2,200	.015		.28	.51	.03	.82	1.18	
	2220	Ungrouted single strand, 100' slab, 25 kip		1,200	.027		2.09	.93	.06	3.08	3.92	
	2250	35 kip		1,475	.022		1.95	.75	.05	2.75	3.47	

Figure 9.52

component costs for spread footings from the estimate sheets of Division 3 are added:

Spread Footings:

	Bare Costs			
	Material	Labor	Equipment	Total
Formwork	$ 1,124	$4,398		$ 5,522
Reinf.	2,369	2,302		4,671
Concrete	12,420			12,420
Placing		1,773	$106	1,879
Total	$15,913	$8,473	$106	$24,492
$/CY (171 CY)	93.06	49.55	0.62	143.23

Reinforcing costs, separated for the spread footings only, are obtained from the appropriate quantities, from Figure 9.46, and the unit costs from Figure 9.49. Comparison to line 3850 in Figure 9.54 verifies that costs for the building's spread footings are good. Similar crosschecks can be made throughout the estimate to help prevent gross errors. Quantities and prices for all components of each of the concrete systems have been separately itemized. By organizing the estimate in this way, costs and quantities can be compared to historical figures and used as a comparison for future projects.

Division 4: Masonry

Masonry is generally estimated by the piece (brick, block, etc.) or by wall area (per square foot). Quantities based on square feet of surface area are a function of:

- Size of the masonry unit.
- Bond (pattern or coursing).
- Thickness of mortar joints.
- Thickness of wall.

Similar to other divisions, masonry estimates should be performed in as detailed a manner as is practical. The masonry specifications will contain pertinent information which should be noted before proceeding with the quantity takeoff. Such items as mortar types, joint reinforcing, and cleaning requirements should all be described and defined. Typical mortar types (mixtures) most commonly specified are shown in Figure 9.55. Required mortar additives (coloring, anti-hydro, bonding agents, etc.) will be specified and should be noted.

It is important that mortar specifications be followed carefully. What little money may be saved by "skimping" on the mortar is not really worth the risks involved.

When all requirements in the specifications have been noted, the plans must be carefully reviewed. Drawings and details of masonry work are usually interspersed throughout the plans. All sheets should be checked. Walls and partitions of different masonry types and sizes are estimated separately. Requirements which must be listed and estimated individually include:

- Number and type of masonry units
- Bonding patterns
- Special coursing
- Openings
- Lintels and accessories
- Joint size (mortar quantities) and finish
- Grouting (cores, pilasters, door jambs, and bond beams, incl. reinforcing)
- Joint reinforcing

03240	Fibrous Reinforcing	CREW	DAILY OUTPUT	LABOR-HOURS	UNIT	2002 BARE COSTS				TOTAL INCL O&P	
						MAT.	LABOR	EQUIP.	TOTAL		
300	0010	**FIBROUS REINFORCING**										**300**
	0100	Synthetic fibers, add to concrete				Lb.	3.79			3.79	4.17
	0110	1-1/2 lb. per C.Y.				C.Y.	5.85			5.85	6.45
	0150	Steel fibers, add to concrete				Lb.	.44			.44	.48
	0155	25 lb. per C.Y.				C.Y.	11			11	12.10
	0160	50 lb. per C.Y.					22			22	24
	0170	75 lb. per C.Y.					34			34	37.50
	0180	100 lb. per C.Y.					44			44	48.50

03310	Structural Concrete		CREW	DAILY OUTPUT	LABOR-HOURS	UNIT	2002 BARE COSTS				TOTAL INCL O&P	
							MAT.	LABOR	EQUIP.	TOTAL		
200	0010	**CONCRETE, FIELD MIX** FOB forms 2250 psi	R03310 -080				C.Y.	68			68	75	**200**
	0020	3000 psi					"	71			71	78
220	0010	**CONCRETE, READY MIX** Regular weight	R03310 -040										**220**
	0020	2000 psi					C.Y.	65			65	71.50
	0100	2500 psi						58			58	64
	0150	3000 psi CN	R03310 -050					69			69	76
	0200	3500 psi						71			71	78
	0300	4000 psi						74			74	81.50
	0350	4500 psi						75.50			75.50	83
	0400	5000 psi CN						77			77	84.50
	0411	6000 psi						88			88	96.50
	0412	8000 psi						143			143	158
	0413	10,000 psi						203			203	224
	0414	12,000 psi						246			246	270
	1000	For high early strength cement, add						10%				
	1010	For structural lightweight with regular sand, add						25%				
	2000	For all lightweight aggregate, add						45%				
	3000	For integral colors, 2500 psi, 5 bag mix										
	3100	Red, yellow or brown, 1.8 lb. per bag, add					C.Y.	14.55			14.55	16
	3200	9.4 lb. per bag, add						78.50			78.50	86.50
	3400	Black, 1.8 lb. per bag, add						17.25			17.25	19
	3500	7.5 lb. per bag, add						72.50			72.50	79.50
	3700	Green, 1.8 lb. per bag, add						34.50			34.50	38
	3800	7.5 lb. per bag, add						165			165	181
240	0010	**CONCRETE IN PLACE** Including forms (4 uses), reinforcing	R03310 -010										**240**
	0050	steel, including finishing unless otherwise indicated										
	0300	Beams, 5 kip per L.F., 10' span	R03310 -100	C-14A	15.62	12.804	C.Y.	223	385	46	654	905
	0350	25' span		"	18.55	10.782		209	325	38.50	572.50	790
	0500	Chimney foundations, industrial, minimum	R04210 -055	C-14C	32.22	3.476		140	100	1.11	241.11	310
	0510	Maximum		"	23.71	4.724		163	136	1.51	300.51	395
	0700	Columns, square, 12" x 12", minimum reinforcing		C-14A	11.96	16.722		242	505	60	807	1,125
	0720	Average reinforcing			10.13	19.743		340	595	71	1,006	1,400
	0740	Maximum reinforcing			9.03	22.148		420	670	79.50	1,169.50	1,600
	0800	16" x 16", minimum reinforcing			16.22	12.330		193	375	44.50	612.50	845
	0820	Average reinforcing			12.57	15.911		305	480	57	842	1,175
	0840	Maximum reinforcing			10.25	19.512		420	590	70	1,080	1,475
	0900	24" x 24", minimum reinforcing			23.66	8.453		163	256	30.50	449.50	620
	0920	Average reinforcing			17.71	11.293		255	340	40.50	635.50	865

Figure 9.53

03310 | Structural Concrete

	Line	Description	Ref.	CREW	DAILY OUTPUT	LABOR-HOURS	UNIT	MAT.	LABOR	EQUIP.	TOTAL	TOTAL INCL O&P	
									2002 BARE COSTS				
240	0940	Maximum reinforcing	R03310-010	C-14A	14.15	14.134	C.Y.	345	425	50.50	820.50	1,100	240
	1000	36" x 36", minimum reinforcing			33.69	5.936		150	179	21.50	350.50	470	
	1020	Average reinforcing	R03310-100		23.32	8.576		237	259	31	527	705	
	1040	Maximum reinforcing			17.82	11.223		325	340	40.50	705.50	935	
	1100	Columns, round, tied, 12" diameter, minimum reinforcing	R04210-055		20.97	9.537		199	288	34	521	710	
	1120	Average reinforcing			15.27	13.098		320	395	47	762	1,025	
	1140	Maximum reinforcing			12.11	16.515		430	500	59.50	989.50	1,325	
	1200	16" diameter, minimum reinforcing			31.49	6.351		218	192	23	433	570	
	1220	Average reinforcing			19.12	10.460		345	315	37.50	697.50	920	
	1240	Maximum reinforcing			13.77	14.524		475	440	52	967	1,275	
	1300	20" diameter, minimum reinforcing			41.04	4.873		209	147	17.50	373.50	480	
	1320	Average reinforcing			24.05	8.316		325	251	30	606	785	
	1340	Maximum reinforcing			17.01	11.758		435	355	42	832	1,075	
	1400	24" diameter, minimum reinforcing			51.85	3.857		182	117	13.85	312.85	400	
	1420	Average reinforcing			27.06	7.391		299	223	26.50	548.50	710	
	1440	Maximum reinforcing			18.29	10.935		415	330	39.50	784.50	1,025	
	1500	36" diameter, minimum reinforcing			75.04	2.665		173	80.50	9.55	263.05	330	
	1520	Average reinforcing			37.49	5.335		261	161	19.15	441.15	560	
	1540	Maximum reinforcing			22.84	8.757		375	265	31.50	671.50	860	
	1900	Elevated slabs, flat slab, 125 psf Sup. Load, 20' span		C-14B	38.45	5.410		153	163	18.65	334.65	445	
	1950	30' span			50.99	4.079		140	123	14.05	277.05	365	
	2100	Flat plate, 125 psf Sup. Load, 15' span			30.24	6.878		163	208	23.50	394.50	530	
	2150	25' span			49.60	4.194		134	127	14.45	275.45	365	
	2300	Waffle const., 30" domes, 125 psf Sup. Load, 20' span			37.07	5.611		197	169	19.35	385.35	505	
	2350	30' span			44.07	4.720		180	142	16.30	338.30	440	
	2500	One way joists, 30" pans, 125 psf Sup. Load, 15' span			27.38	7.597		231	229	26	486	645	
	2550	25' span			31.15	6.677		216	201	23	440	580	
	2700	One way beam & slab, 125 psf Sup. Load, 15' span			20.59	10.102		186	305	35	526	725	
	2750	25' span			28.36	7.334		169	221	25.50	415.50	565	
	2900	Two way beam & slab, 125 psf Sup. Load, 15' span			24.04	8.652		174	261	30	465	635	
	2950	25' span			35.87	5.799		148	175	20	343	460	
	3100	Elevated slabs including finish, not											
	3110	including forms or reinforcing											
	3150	Regular concrete, 4" slab		C-8	2,613	.021	S.F.	.94	.56	.26	1.76	2.18	
	3200	6" slab			2,585	.022		1.45	.57	.26	2.28	2.76	
	3250	2-1/2" thick floor fill			2,685	.021		.64	.55	.25	1.44	1.82	
	3300	Lightweight, 110# per C.F., 2-1/2" thick floor fill			2,585	.022		.74	.57	.26	1.57	1.97	
	3400	Cellular concrete, 1-5/8" fill, under 5000 S.F.			2,000	.028		.49	.74	.34	1.57	2.04	
	3450	Over 10,000 S.F.			2,200	.025		.39	.67	.31	1.37	1.79	
	3500	Add per floor for 3 to 6 stories high			31,800	.002			.05	.02	.07	.09	
	3520	For 7 to 20 stories high			21,200	.003			.07	.03	.10	.15	
	3800	Footings, spread under 1 C.Y.		C-14C	38.07	2.942	C.Y.	100	84.50	.94	185.44	244	
	3850	Over 5 C.Y.			81.04	1.382		92	40	.44	132.44	164	
	3900	Footings, strip, 18" x 9", plain			41.04	2.729		91	78.50	.87	170.37	225	
	3950	36" x 12", reinforced			61.55	1.820		92.50	52.50	.58	145.58	185	
	4000	Foundation mat, under 10 C.Y.			38.67	2.896		124	83.50	.93	208.43	268	
	4050	Over 20 C.Y.			56.40	1.986		110	57	.64	167.64	212	
	4200	Grade walls, 8" thick, 8' high		C-14D	45.83	4.364		126	130	15.65	271.65	360	
	4250	14' high			27.26	7.337		149	219	26.50	394.50	540	
	4260	12" thick, 8' high			64.32	3.109		111	93	11.15	215.15	279	
	4270	14' high			40.01	4.999		119	149	17.95	285.95	385	
	4300	15" thick, 8' high			80.02	2.499		104	74.50	8.95	187.45	241	
	4350	12' high			51.26	3.902		106	117	14	237	315	
	4500	18' high			48.85	4.094		116	122	14.70	252.70	335	
	4520	Handicap access ramp, railing both sides, 3' wide		C-14H	14.58	3.292	L.F.	139	98	2.44	239.44	310	
	4525	5' wide			12.22	3.928		146	117	2.91	265.91	345	

Figure 9.54

- Wall ties
- Flashing, reglets, weepholes
- Control joints
- Cleaning (water or chemical)
- Scaffolding and equipment

Each type of specified masonry unit should be identified on a takeoff sheet, not only by the kind of unit (brick, block, etc.) but also by type of construction (wall, arch, foundation, etc.). The square feet and/or linear feet of each listing should be measured from the plans and elevations. It should be remembered that exterior measurements of a building's perimeter include an overlap of the thickness of the walls for every outside corner. This is accepted accuracy and should be included, since corners tend to require more cutting and waste.

When measuring exterior dimensions at inside corners, however, the estimator must add to the actual dimensions to account for the extra material and labor involved to construct the inside corners.

Brick Mortar Mixes*					
Type	Portland Cement	Hydrated Lime	Sand** (maximum)**	Strength	Use
M	1	1/4	3-3/4	High	General use where high strength is required, especially good compressive strength; work that is below grade and in contact with earth.
S	1	1/2	4-1/2	High	Okay for general use, especially good where high lateral strength is desired.
N	1	1	6	Medium	General use when masonry is exposed above grade; best to use when high compressive and lateral strengths are not required.
0	1	2	9	Low	Do not use when masonry is exposed to severe weathering; acceptable for non-loadbearing walls of solid units and interior non-loadbearing partitions of hollow units.

*The water used should be of the quality of drinking water. Use as much as is needed to bring the mix to a suitably plastic and workable state.
**The sand should be damp and loose. A general rule for sand content is that it should not be less than 2-1/4 or more than 3 times the sum of the cement and lime volumes.

Figure 9.55

Openings or "outs" should be deleted from overall quantity only if greater than two square feet. Lintels should be listed separately at this time by type and size (steel angle, precast, etc.). Responsibility for purchase of steel lintels should be established. Often, loose lintels are supplied by the steel fabricator and only installed by the masonry contractor.

When the overall square foot quantities of masonry work have been determined, the number of units may be calculated. Figure 9.56 lists different types of masonry units and corresponding quantities. Note that quantities per square foot as shown are based on certain joint sizes. If different joint sizes are specified, quantities should be adjusted. In Figure 9.56, quantities are listed not only by the type of unit but also by bond and coursing. Typical brick and block types and joints, bonds and coursing are shown in Figures 9.57 to 9.59. The descriptions of different bonds in Figure 9.58 include suggested waste allowance percentages which should be added to the quantities before pricing.

It is difficult to develop accurate unit price data for masonry due to fluctuation of both material prices and labor productivity. The estimator should always call suppliers to verify material costs and availability. All unit prices for brick masonry in *Means Building Construction Cost Data*, (Figure 9.60) include the material cost per thousand pieces in the line description. If local costs are different, the material unit costs should be adjusted.

Productivity is affected by the type and complexity of work and can be severely affected by weather. Assume that a crew (including masons, helpers, and appropriate equipment) can install approximately 600 standard bricks per man per day. This productivity rate assumes fairly straight runs (few openings) and good weather. For complicated work (many openings and corners), assume the same crew can install 400 bricks per man per day. Even for a simple building elevation as shown in Figure 9.61, both rates would apply. The high productivity portions (HP) are those areas, between rows of windows, with no openings. Low productivity portions (LP) are those areas which include the window jambs, headers, and sills.

Total wall area	1,920 S.F.
Deduct openings	− 376 S.F.
Total masonry area	1,544 S.F.
Standard brick (Fig. 9.56)	× 6.75 brick/S.F.
Total bricks	10,422 bricks
5% waste (Running bond, Fig. 9.58)	+ 521
	10,943 bricks

56% of the wall area is high productivity and 44% is low productivity.

56% of 10,943	= 6,128 bricks (HP)
44% of 10,943	= 4,815 bricks (LP)

HP area: $\dfrac{6{,}128 \text{ bricks}}{600 \text{ bricks/day}}$ = 10 days

LP area: $\dfrac{4{,}815 \text{ bricks}}{400 \text{ bricks/day}}$ = 12 days

From quick, simple calculations, the estimator has the material quantity and the time required. A material cost is obtained from a supplier and the cost per day for the crew is multiplied by the time durations. Total costs for the work are obtained. Unit costs are determined by dividing the total costs—material or labor or both—by the appropriate units—square feet or per thousand brick.

Temperature and humidity variations can have a significant impact on masonry work. Not only are the general working conditions affected, but the drying time and workability of mortar have an effect on productivity. Figure 9.62 is a graphic representation of the relative effects of temperature and humidity. The numbers in the grid are factors to be multiplied by a maximum productivity. The following procedure can be used to derive this theoretical maximum (based on ideal weather conditions, 75 degrees F., 60% humidity—see Figure

Brick Quantities

| Running Bond | | | | | | For Other Bonds Standard Size Add to S.F. Quantities in Table to Left | | |
| Number of Brick per S.F. of Wall - Single Wythe with 3/8" Joints | | | | C.F. of Mortar per M Bricks, Waste Included | | | | |
Type Brick	Nominal Size (incl. mortar) L H W	Modular Coursing	Number of Brick per S.F.	3/8" Joint	1/2" Joint	Bond Type	Description	Factor
Standard	8 x 2-2/3 x 4	3C = 8"	6.75	10.3	12.9	Common	full header every fifth course	+20%
Economy	8 x 4 x 4	1C = 4"	4.50	11.4	14.6		full header every sixth course	+16.7%
Engineer	8 x 3-1/5 x 4	5C = 16"	5.63	10.6	13.6	English	full header every second course	+50%
Fire	9 x 2-1/2 x 4-1/2	2C = 5"	6.40	550 # Fireclay	–	Flemish	alternate headers every course	+33.3%
Jumbo	12 x 4 x 6 or 8	1C = 4"	3.00	23.8	30.8		every sixth course	+5.6%
Norman	12 x 2-2/3 x 4	3C = 8"	4.50	14.0	17.9	Header = W x H exposed		+100%
Norwegian	12 x 3-1/5 x 4	5C = 16"	3.75	14.6	18.6	Rowlock = H x W exposed		+100%
Roman	12 x 2 x 4	2C = 4"	6.00	13.4	17.0	Rowlock stretcher = L x W exposed		+33.3%
SCR	12 x 2-2/3 x 6	3C = 8"	4.50	21.8	28.0	Soldier = H x L exposed		–
Utility	12 x 4 x 4	1C = 4"	3.00	15.4	19.6	Sailor = W x L exposed		-33.3%

Concrete Block Quantities

| Concrete Blocks Nominal Size (incl. 3/8" joint) | Approximate Weight per S.F. | | Blocks per 100 S.F. | Mortar per M block | |
	Standard	Lightweight		Partitions	Back up
2" x 8" x 16"	20 PSF	15 PSF	113	16 C.F.	36 C.F.
4"	30	20		31	51
6"	42	30		46	66
8"	55	38		62	82
10"	70	47		77	97
12"	85	55		92	112

Glass Block Quantities

| | Per 100 S.F. | | Per 1000 Block | | | | |
Size	No. of Block	Mortar 1/4" Joint	Asphalt Emulsion	Caulk	Expansion Joint	Panel Anchors	Wall Mesh
6" x 6"	410 ea.	5.0 C.F.	.17 gal.	1.5 gal.	80 L.F.	20 ea.	500 L.F.
8" x 8"	230	3.6	.33	2.8	140	36	670
12" x 12"	102	2.3	.67	6.0	312	80	1000
Approximate quantity per 100 S.F.			.07 gal.	0.6 gal.	32 L.F.	9 ea.	51, 68, 102 L.F.

Figure 9.56

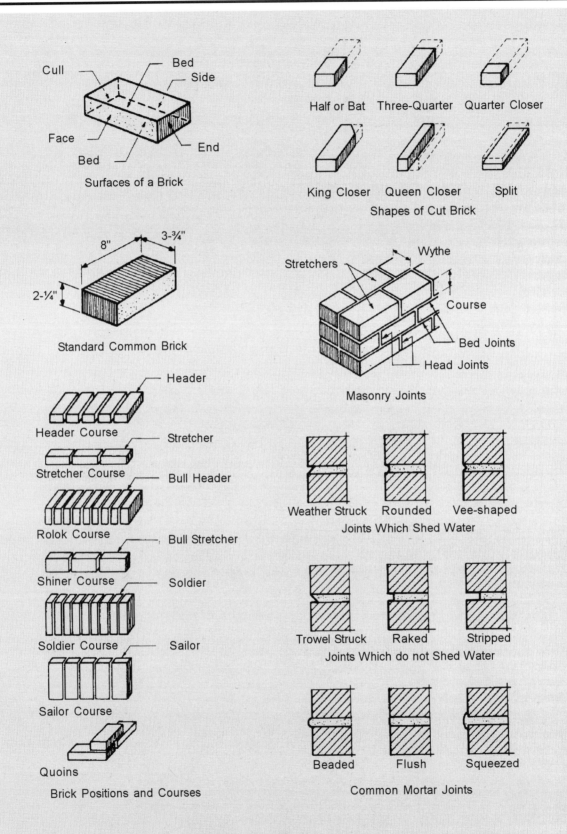

Surfaces of a Brick

Cull — Bed — Side — Face — Bed — End

Shapes of Cut Brick

Half or Bat   Three-Quarter   Quarter Closer
King Closer   Queen Closer   Split

Standard Common Brick

8"   3-¾"   2-¼"

Masonry Joints

Stretchers — Wythe — Course — Bed Joints — Head Joints

Brick Positions and Courses

Header — Header Course
Stretcher — Stretcher Course
Bull Header — Rolok Course
Bull Stretcher — Shiner Course
Soldier — Soldier Course
Sailor — Sailor Course
Quoins

Joints Which Shed Water

Weather Struck   Rounded   Vee-shaped

Joints Which do not Shed Water

Trowel Struck   Raked   Stripped

Common Mortar Joints

Beaded   Flush   Squeezed

Figure 9.57

200

Running or Stretcher Bond	The face brick are all stretchers and are tied to the backing by metal or reinforcing. Waste – 5%.
Common or American Bond	Every sixth course of stretcher bond is usually a header course. Waste – 4%.
Flemish Bond	Each course has alternate headers and stretchers with the alternate headers centered over the stretcher. Waste – 3 to 5%.
English Bond	Consists of alternate headers and stretchers with the vertical joints in the header and stretcher aligning or breaking over each other. Waste – 8 to 15%.
Stack Bond	Has no overlapping of units since all vertical joints are aligned. Usually this pattern is bonded to the backing with rigid steel ties. Waste – 3%.
English Cross or Dutch Bond	Built up of interlocking crosses. This wall consists of two headers and a stretcher forming a cross. Waste – 8%.

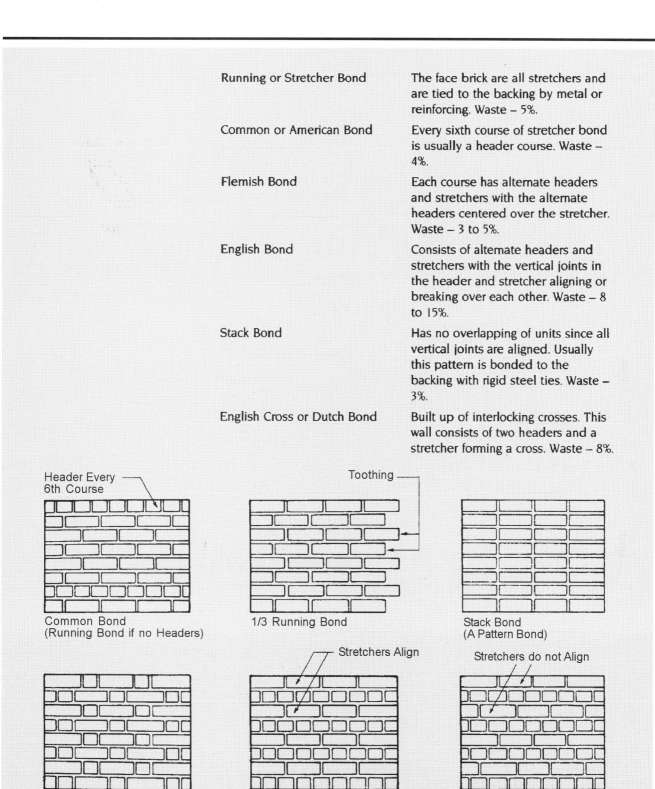

Figure 9.58

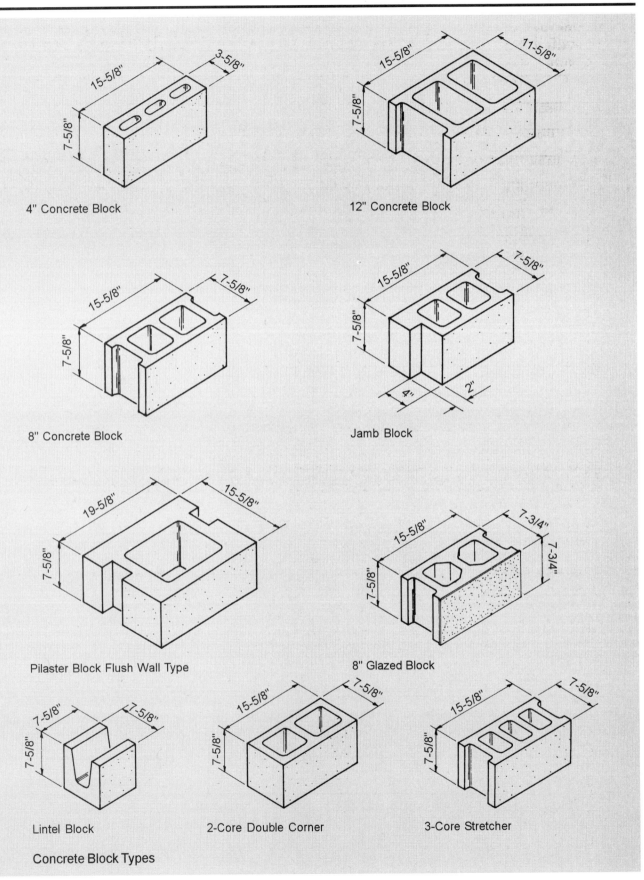

4" Concrete Block

12" Concrete Block

8" Concrete Block

Jamb Block

Pilaster Block Flush Wall Type

8" Glazed Block

Lintel Block

2-Core Double Corner

3-Core Stretcher

Concrete Block Types

Figure 9.59

04050 | Basic Masonry Materials & Methods

R01540 -100

04090		Masonry Accessories	CREW	DAILY OUTPUT	LABOR-HOURS	UNIT	2002 BARE COSTS MAT.	LABOR	EQUIP.	TOTAL	TOTAL INCL O&P	
420	0500	8" x 8" units, 8" thick				S.F.	.66			.66	.73	420
	0550	12" thick				↓	.82			.82	.90	
650	0010	**PARGETING** Regular Portland cement, 1/2" thick	D-1	2.50	6.400	C.S.F.	13.20	173		186.20	280	650
	5100	Waterproof Portland cement	"	2.50	6.400	"	14.55	173		187.55	281	
700	0010	**SCAFFOLDING & SWING STAGING** See division 01540										700
850	0010	**VENT BOX** See division 04090-860										850
860	0010	**VENT BOX** Extruded aluminum, 4" deep, 2-3/8" x 8-1/8"	1 Bric	30	.267	Ea.	30.50	8.15		38.65	46	860
	0050	5" x 8-1/8"		25	.320		42.50	9.75		52.25	61.50	
	0100	2-1/4" x 25"		25	.320		85	9.75		94.75	109	
	0150	5" x 16-1/2"		22	.364		44	11.10		55.10	65.50	
	0200	6" x 16-1/2"		22	.364		85	11.10		96.10	111	
	0250	7-3/4" x 16-1/2"	↓	20	.400		80	12.20		92.20	107	
	0400	For baked enamel finish, add					35%					
	0500	For cast aluminum, painted, add					60%					
	1000	Stainless steel ventilators, 6" x 6"	1 Bric	25	.320		91	9.75		100.75	115	
	1050	8" x 8"		24	.333		96	10.15		106.15	122	
	1100	12" x 12"		23	.348		112	10.60		122.60	139	
	1150	12" x 6"		24	.333		96	10.15		106.15	122	
	1200	Foundation block vent, galv., 1-1/4" thk, 8" high, 16" long, no damper	↓	30	.267		18	8.15		26.15	32.50	
	1250	For damper, add				↓	6			6	6.60	
900	0010	**WALL PLUGS** For nailing to brickwork, 26 ga., galvanized, plain	1 Bric	10.50	.762	C	25	23		48	63	900
	0050	Wood filled	"	10.50	.762	"	84	23		107	128	

04200 | Masonry Units

04210		Clay Masonry Units	CREW	DAILY OUTPUT	LABOR-HOURS	UNIT	2002 BARE COSTS MAT.	LABOR	EQUIP.	TOTAL	TOTAL INCL O&P	
100	0010	**COMMON BUILDING BRICK** C62, TL lots, material only										100
	0020	Standard, minimum				M	270			270	297	
	0050	Average (select)	↓			"	315			315	345	
120	0010	**BRICK VENEER** Scaffolding not included, truck load lots										120
	0015	Material costs incl. 3% brick and 25% mortar waste										
	0020	Standard, select common, 4" x 2-2/3" x 8" (6.75/S.F.)	D-8	1.50	26.667	M	360	740		1,100	1,525	
	0050	Red, 4" x 2-2/3" x 8", running bond		1.50	26.667		400	740		1,140	1,575	
	0100	Full header every 6th course (7.88/S.F.)		1.45	27.586		400	765		1,165	1,625	
	0150	English, full header every 2nd course (10.13/S.F.)		1.40	28.571		395	790		1,185	1,650	
	0200	Flemish, alternate header every course (9.00/S.F.)		1.40	28.571		395	790		1,185	1,650	
	0250	Flemish, alt. header every 6th course (7.13/S.F.)		1.45	27.586		400	765		1,165	1,625	
	0300	Full headers throughout (13.50/S.F.)		1.40	28.571		395	790		1,185	1,650	
	0350	Rowlock course (13.50/S.F.)		1.35	29.630		395	820		1,215	1,675	
	0400	Rowlock stretcher (4.50/S.F.)		1.40	28.571		400	790		1,190	1,675	
	0450	Soldier course (6.75/S.F.)		1.40	28.571		400	790		1,190	1,675	
	0500	Sailor course (4.50/S.F.)		1.30	30.769		400	850		1,250	1,750	
	0601	Buff or gray face, running bond, (6.75/S.F.)		1.50	26.667		400	740		1,140	1,575	
	0700	Glazed face, 4" x 2-2/3" x 8", running bond		1.40	28.571		1,500	790		2,290	2,875	
	0750	Full header every 6th course (7.88/S.F.)		1.35	29.630		1,475	820		2,295	2,850	
	1000	Jumbo, 6" x 4" x 12",(3.00/S.F.)		1.30	30.769		1,250	850		2,100	2,675	
	1051	Norman, 4" x 2-2/3" x 12" (4.50/S.F.)	↓	1.45	27.586	↓	735	765		1,500	1,975	

R04210 -120

R04210 -120 R04210 -180 R04210 -500

Figure 9.60

BUILDING ELEVATION

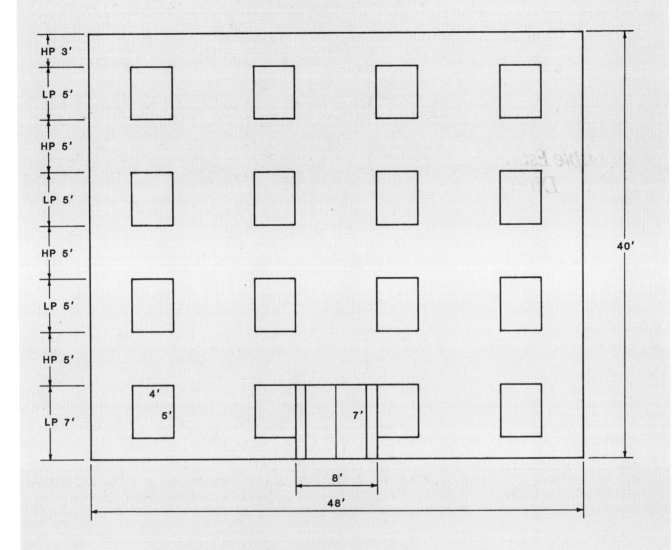

HP – HIGH PRODUCTIVITY
LP – LOW PRODUCTIVITY

Figure 9.61

204

9.62). For a given month, assume the productivity of a crew to be 500 units per day per mason. A record of the local, average temperature and humidity for that month is obtained from the National Weather Service, for example, 60 degrees F., 75% relative humidity. The factor for those conditions is 0.765, from Figure 9.62.

$$\frac{500 \text{ units/day}}{0.765} = 654 \text{ units/day (theoretical maximum)}$$

This theoretical maximum serves as a practical example for using the chart to predict productivity based on anticipated weather conditions.

Sample Estimate: Division 4

For the masonry portion sample building project, assume that good historical records are available. The following data have been developed from past jobs:

Productivity:

Regular block – 150 S.F. block/mason/day
8" × 16" × 8" – 170 block/mason/day

Split face block – 125 S.F. block/mason/day
8" × 16" × 6" – 142 block/mason/day

Mortar – 62 C.F./M block
(Figure 9.56) 10 C.F./mason/day

Bare Costs:

Mason	$30.50/hr., $244/day
Mason's helper	$23.50/hr., $188/day
Mortar	$3.86/C.F.
Mixer	$80.00/day
Scaffolding	$2750/month
Forklift	$232.20/day
Masonry saw	$44.50/day

Even though the mixer and scaffolding are owned, costs are charged to each job. A crew of four masons and two helpers will be available and used for this job. Costs are developed as follows:

Material:

Regular block:
9,430 S.F. × 113 block/100 S.F. = 10,656 block
(10,656 + 5% waste) x $1.05 ea.
11,189 block × $1.05 = $11,748

Split face:
2,040 S.F. × 113 block/100 S.F. = 2,305 block
(2,305 + 5% waste) × $2.08 ea.
2,420 block × $2.08 = $5,034

BRICK PRODUCTIVITY TABLE

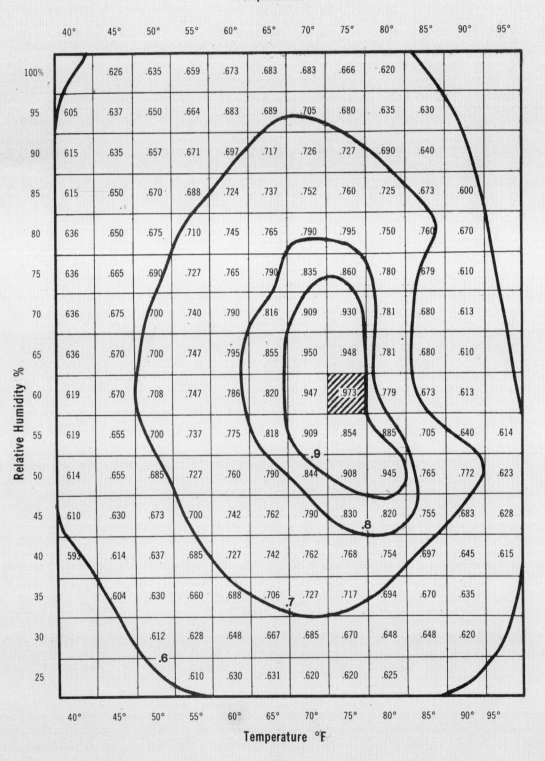

Figure 9.62

206

Mortar:

    62 C.F./M block × 12,961 block = 804 C.F.

    804 C.F. × $3.86 C.F. = $3,103

Reinforcing:

    4,008 Lb. × $0.28/Lb. = $1,122

Lintels:

    Labor costs included below.

    Material costs included in Division 5.

Total Material Costs:

Regular Block	$11,748
Split Face Block	5,034
Mortar	3,103
Reinforcing	1,122
	$21,007

Labor

Regular block:

$$\frac{9,430 \text{ S.F.}}{150 \text{ S.F./mason/day} \times 4 \text{ masons}} = 16 \text{ days}$$

Split block:

$$\frac{2,040 \text{ S.F.}}{125 \text{ S.F./mason/day} \times 4 \text{ masons}} = 4 \text{ days}$$

4 masons × $244/day × 20 days = $19,520

2 helpers × $188/day × 20 days = $7,520

1 helper × $ 35.25/hour × 20 hours = $705 (overtime)

Total Labors Costs:

Masons	$19,520
Helpers	8,225
	$27,745

Equipment:

Mixer:

$80.00/day × 20 days = $1,600

Scaffolding:

24 sections for 1 month @) $295/CSF × 10 CSF = $2,750

Crane:

4 hours × $128.70/hour = $515

Forklift:

10 days × $232.20/day = $2,322

Masonry saw:

10 days × $44.50/day = $445

Total Equipment Costs:

Mixer	$1,600
Scaffolding	2,750
Crane	515
Forklift	2,322
Masonry saw	445
	$4,882

The costs are entered as shown in Figure 9.63. Unit costs for materials should be obtained from or verified by local suppliers. The most accurate prices are those developed from recent data. A thorough and detailed cost control and accounting system is important for the development of such costs. Overtime costs for one helper are for mixing mortar before the start of the workday.

Division 5: Metals

The metals portion of a building project, and the corresponding estimate should be broken down into basic components: structural metals, metal joists, metal decks, miscellaneous metals, ornamental metals, expansion control, and fasteners. The items in a building project are shown on many different sheets of the drawings and may or may not be thoroughly listed in the specifications. This is especially true of the miscellaneous metals that are listed under Division 5. A complete and detailed review of the construction documents is therefore necessary, noting all items and requirements. Most structural steel work is subcontracted to specialty fabricators and erectors. However, the estimator for the general contractor may perform a takeoff to assure that all specific work is included. Pricing based on total weight (tonnage) can be used to verify subcontractor prices.

Structural Metals

The various structural members should be identified and separately listed by type and size. For example:

	Size	Length	Quantity
Columns:	W8 × 67	12'-6"	16
	W12 × 58	12'-6"	8
Beams:	W14 × 30	30'-0"	10
	W14 × 30	28'-0"	6
	W14 × 26	30'-0"	16
	W24 × 55	30'-0"	12

The quantities may be converted to weight (tonnage) for pricing. This figure is based on weight per linear foot. Base plates, leveling plates, anchor bolts, and other accessories should be taken off and listed at this time. Connections should also be noted. Costs for structural steel in *Means Building Construction Cost Data* (Figures 9.64 and 9.65) are given for both individual members and for complete projects including bolted connections. If specified, costs for high strength steel and or high strength bolts must be added separately (Figure 9.65). Welded connections should be listed by fillet size and length and priced separately. Light gauge and special framing for items such as hanging lintels, fascias, and parapets, should be taken off and priced separately. The amount of this type of metal work will vary depending upon design and project requirements.

CONSOLIDATED ESTIMATE

PROJECT: Office Building
LOCATION:
TAKE OFF BY: ABC
CLASSIFICATION: Division 4
ARCHITECT:
QUANTITIES BY: ABC
PRICES BY: As Shown
EXTENSIONS BY:

ESTIMATE NO:
DATE: Jan-02
CHECKED BY: GHI

DESCRIPTION	SOURCE	QUANT	UNIT	MATERIAL DEF	COST	TOTAL	LABOR DEF	COST	TOTAL	EQUIPMENT DEF	COST	TOTAL	SUBCONTRACT COST	TOTAL	TOTAL COST	TOTAL
Division 4: Masonry																
Regular Block 8" x 16" x 8" (based on $1.05/Ea.)	04420 220 1150	11189	Ea.		1.05	11748										
Split Face Block 8" x 16" x 8" (based on $2.08/Ea.)	04420 240 6150	2420	Ea.		2.08	5034										
Mortar	04060 500 2100	804	CF		3.86	3103										
Reinforcing	04080 200 0010	4008	Lb.		0.28	1122										
Labor: Masons		80	Day					244.00	19520							
Mason Helper		40	Day					188.00	7520							
Overtime		20	Hour					35.25	705							
Equipment: Mixer		20	Day								80.00	1600				
Scaffolding (24 Sections)	01540 750 0090	10	CSF		245.00	2450		30.00	300							
Crane	01590 600 2500	4	Hour								128.70	515				
Forklift	01590 400 2040	10	Day								232.20	2322				
Masonry Saw	01590 400 6000	10	Day								44.50	445				
Division 4 Total					$	23,458		$	28,045		$	4,882				

Figure 9.63

For development of a budget cost or verification of a subcontractor bid, the estimator can apply average allowance percentages to gross tonnage for:

Base plates	2 to 3%	
Column splices & Beam connections	8 to 10%	
Total allowance	10 to 13%	of main members

Erection costs can be determined in two ways. The first method is to base labor and equipment costs on gross tonnage, as shown in Figure 9.65. The second method is to base costs on the number of pieces. A typical crew and crane can set 35 to 60 pieces per day on the average. Based on normal sizes of beams, girders and columns, this is equivalent to approximately 25 tons per day. The installation rate (number of pieces per day) will not only vary due to crew size, speed, and efficiency, but will also be affected by job conditions (relative access or mobility) and the weight of the pieces. A crew may set more smaller, lighter pieces (but less total tonnage) per day than larger or heavier pieces.

Metal Joists and Decks

The estimator will frequently find H-Series Open Web Joists used for the direct support of floor and roof decks in buildings. When long clear spans are desired, the designer will specify often the Longspan Steel Joist LH-Series; these are suitable for the direct support of floors and roof decks. Deep Longspan Steel Joists DLH-Series may be specified for the direct support of roof decks. Some plans will not indicate each joist in place, but rather the designation and spacing. The overall length of each span will be dimensioned. Required bridging may or may not be shown, as this may be defined in the specifications. The quantity takeoff sheet should indicate the joist designation, quantity, length, and weight per lineal foot for total tonnage. Extensions of bottom chords, bridging (whether diagonal or horizontal), and any other related items must also be tabulated. The weight of joists per linear foot is usually not specified on the drawings, but is shown in manufacturers' catalogues.

Metal roof deck or form deck (sometimes termed centering) is often used on open web joist construction. Metal deck used on structural steel framing are manufactured in open type, cellular or composite configuration. The finish may be either galvanized painted or black.

Metal joists are most often priced by the ton. Installation costs may be determined in the same way as for structural steel. Metal deck is taken off and priced by the square foot based on type (open, cellular, composite) gauge and finish.

Miscellaneous and Ornamental Metals

A thorough review of all drawings and specifications is necessary to ensure that all miscellaneous metal requirements are included. Listed below are typical miscellaneous and ornamental metals and the appropriate takeoff units. All items should be listed and priced separately.

				DAILY	LABOR-			2002 BARE COSTS			TOTAL	
	05120	**Structural Steel**	CREW	OUTPUT	HOURS	UNIT	MAT.	LABOR	EQUIP.	TOTAL	INCL O&P	
600	0100	Double panel convex roof, spans to 200'	E-2	960	.058	S.F.	7.95	1.94	1.48	11.37	13.75	**600**
	0200	Double panel arched roof, spans to 300'	↓	760	.074	↓	12.25	2.44	1.87	16.56	19.80	
640	0010	**STRUCTURAL STEEL MEMBERS**										**640**
	0020	Shop fabricated for 1-2 story bldg., bolted conn's., 100 tons										
	0100	W 6 x 9 R05120-210	E-2	600	.093	L.F.	6.05	3.10	2.37	11.52	14.60	
	0120	x 16		600	.093		10.80	3.10	2.37	16.27	19.80	
	0140	x 20		600	.093		13.50	3.10	2.37	18.97	23	
	0300	W 8 x 10		600	.093		6.75	3.10	2.37	12.22	15.35	
	0320	x 15		600	.093		10.10	3.10	2.37	15.57	19.05	
	0350	x 21		600	.093		14.15	3.10	2.37	19.62	23.50	
	0360	x 24		550	.102		16.15	3.38	2.58	22.11	26.50	
	0370	x 28		550	.102		18.85	3.38	2.58	24.81	29.50	
	0500	x 31		550	.102		21	3.38	2.58	26.96	31.50	
	0520	x 35		550	.102		23.50	3.38	2.58	29.46	34.50	
	0540	x 48		550	.102		32.50	3.38	2.58	38.46	44	
	0600	W 10 x 12		600	.093		8.10	3.10	2.37	13.57	16.85	
	0620	x 15		600	.093		10.10	3.10	2.37	15.57	19.05	
	0700	x 22		600	.093		14.80	3.10	2.37	20.27	24.50	
	0720	x 26		600	.093		17.50	3.10	2.37	22.97	27	
	0740	x 33		550	.102		22	3.38	2.58	27.96	33	
	0900	x 49		550	.102		33	3.38	2.58	38.96	45	
	1100	W 12 x 14		880	.064		9.45	2.11	1.62	13.18	15.85	
	1300	x 22		880	.064		14.80	2.11	1.62	18.53	21.50	
	1500	x 26		880	.064		17.50	2.11	1.62	21.23	24.50	
	1520	x 35		810	.069		23.50	2.29	1.75	27.54	32	
	1560	x 50		750	.075		33.50	2.48	1.90	37.88	43.50	
	1580	x 58		750	.075		39	2.48	1.90	43.38	49.50	
	1700	x 72		640	.087		48.50	2.90	2.22	53.62	61	
	1740	x 87		640	.087		58.50	2.90	2.22	63.62	72	
	1900	W 14 x 26		990	.057		17.50	1.88	1.44	20.82	24	
	2100	x 30		900	.062		20	2.06	1.58	23.64	27.50	
	2300	x 34		810	.069		23	2.29	1.75	27.04	31	
	2320	x 43		810	.069		29	2.29	1.75	33.04	38	
	2340	x 53		800	.070		35.50	2.32	1.78	39.60	45.50	
	2360	x 74		760	.074		50	2.44	1.87	54.31	61.50	
	2380	x 90		740	.076		60.50	2.51	1.92	64.93	73	
	2500	x 120		720	.078		81	2.58	1.97	85.55	95.50	
	2700	W 16 x 26		1,000	.056		17.50	1.86	1.42	20.78	24	
	2900	x 31		900	.062		21	2.06	1.58	24.64	28.50	
	3100	x 40		800	.070		27	2.32	1.78	31.10	35.50	
	3120	x 50		800	.070		33.50	2.32	1.78	37.60	43	
	3140	x 67	↓	760	.074		45	2.44	1.87	49.31	56	
	3300	W 18 x 35	E-5	960	.083		23.50	2.81	1.57	27.88	32.50	
	3500	x 40		960	.083		27	2.81	1.57	31.38	36	
	3520	x 46		960	.083		31	2.81	1.57	35.38	40.50	
	3700	x 50		912	.088		33.50	2.96	1.66	38.12	44	
	3900	x 55		912	.088		37	2.96	1.66	41.62	48	
	3920	x 65		900	.089		44	3	1.68	48.68	55	
	3940	x 76		900	.089		51	3	1.68	55.68	63.50	
	3960	x 86		900	.089		58	3	1.68	62.68	70.50	
	3980	x 106		900	.089		71.50	3	1.68	76.18	85.50	
	4100	W 21 x 44		1,064	.075		29.50	2.53	1.42	33.45	38.50	
	4300	x 50		1,064	.075		33.50	2.53	1.42	37.45	43	
	4500	x 62		1,036	.077		42	2.60	1.46	46.06	52	
	4700	x 68		1,036	.077		46	2.60	1.46	50.06	56.50	
	4720	x 83	↓	1,000	.080	↓	56	2.70	1.51	60.21	68	

Figure 9.64

05120 | Structural Steel

		CREW	DAILY OUTPUT	LABOR-HOURS	UNIT	2002 BARE COSTS				TOTAL INCL O&P
						MAT.	LABOR	EQUIP.	TOTAL	
680	0010	**STRUCTURAL STEEL PROJECTS** Bolted, unless noted otherwise R05080-310									680
	0200	Apartments, nursing homes, etc., 1 to 2 stories	E-5	10.30	7.767	Ton	1,225	262	147	1,634	1,975
	0300	3 to 6 stories R05090-510	"	10.10	7.921		1,250	267	149	1,666	2,000
	0400	7 to 15 stories	E-6	14.20	9.014		1,275	305	116	1,696	2,075
	0500	Over 15 stories R05120-210	"	13.90	9.209		1,325	310	119	1,754	2,125
	0700	Offices, hospitals, etc., steel bearing, 1 to 2 stories **CN**	E-5	10.30	7.767		1,225	262	147	1,634	1,975
	0800	3 to 6 stories R05120-220	E-6	14.40	8.889		1,250	300	114	1,664	2,025
	0900	7 to 15 stories		14.20	9.014		1,275	305	116	1,696	2,075
	1000	Over 15 stories R05120-230	↓	13.90	9.209		1,325	310	119	1,754	2,125
	1100	For multi-story masonry wall bearing construction, add						30%			
	1300	Industrial bldgs., 1 story, beams & girders, steel bearing	E-5	12.90	6.202		1,225	209	117	1,551	1,850
	1400	Masonry bearing	"	10	8		1,225	270	151	1,646	1,975
	1500	Industrial bldgs., 1 story, under 10 tons,									
	1510	steel from warehouse, trucked	E-2	7.50	7.467	Ton	1,475	248	190	1,913	2,275
	1600	1 story with roof trusses, steel bearing	E-5	10.60	7.547		1,450	254	142	1,846	2,200
	1700	Masonry bearing	"	8.30	9.639		1,450	325	182	1,957	2,375
	1900	Monumental structures, banks, stores, etc., minimum	E-6	13	9.846		1,225	330	127	1,682	2,075
	2000	Maximum	"	9	14.222		2,025	480	183	2,688	3,275
	2200	Churches, minimum	E-5	11.60	6.897	Ton	1,150	232	130	1,512	1,800
	2300	Maximum	"	5.20	15.385		1,525	520	290	2,335	2,900
	2800	Power stations, fossil fuels, minimum	E-6	11	11.636		1,225	395	150	1,770	2,200
	2900	Maximum		5.70	22.456		1,850	760	289	2,899	3,675
	2950	Nuclear fuels, non-safety steel, minimum		7	18.286		1,225	615	236	2,076	2,675
	3000	Maximum		5.50	23.273		1,850	785	300	2,935	3,725
	3040	Safety steel, minimum		2.50	51.200		1,800	1,725	660	4,185	5,725
	3070	Maximum	↓	1.50	85.333		2,350	2,875	1,100	6,325	8,825
	3100	Roof trusses, minimum	E-5	13	6.154		1,725	207	116	2,048	2,375
	3200	Maximum		8.30	9.639		2,075	325	182	2,582	3,075
	3210	Schools, minimum		14.50	5.517		1,225	186	104	1,515	1,800
	3220	Maximum	↓	8.30	9.639		1,800	325	182	2,307	2,750
	3400	Welded construction, simple commercial bldgs., 1 to 2 stories	E-7	7.60	10.526		1,250	355	210	1,815	2,225
	3500	7 to 15 stories	E-9	8.30	15.422		1,450	520	225	2,195	2,750
	3700	Welded rigid frame, 1 story, minimum	E-7	15.80	5.063		1,275	171	101	1,547	1,800
	3800	Maximum	"	5.50	14.545		1,650	490	291	2,431	3,000
	3900	High strength steel mill spec extras: A242, A441,									
	3950	A529, A572 (42 ksi) and A992: same as A36 steel									
	4000	Add to A36 price for A572 (50, 60, 65 ksi)				Ton	25			25	27.50
	4100	A588 Weathering				"	60			60	66
	4200	Mill size extras for W-Shapes: 0 to 30 plf: no extra charge									
	4210	Member sizes 31 to 65 plf, add				Ton	30			30	33
	4220	Member sizes 66 to 100 plf, add					60			60	66
	4230	Member sizes 101 to 387 plf, add				↓	100			100	110
	4300	Column base plates, light, up to 150 lb	2 Sswk	2,000	.008	Lb.	.67	.27		.94	1.23
	4400	Heavy, over 150 lb	E-2	7,500	.007	"	.70	.25	.19	1.14	1.41
	4600	Castellated beams, light sections, to 50#/L.F., minimum		10.70	5.234	Ton	1,275	174	133	1,582	1,875
	4700	Maximum		7	8		1,400	265	203	1,868	2,225
	4900	Heavy sections, over 50# per L.F., minimum		11.70	4.786		1,350	159	121	1,630	1,875
	5000	Maximum	↓	7.80	7.179	↓	1,475	238	182	1,895	2,225
	5500	Steel domes - see R13128-310									
	5700	Steel estimating weights per S.F. - see R05120-220									

Figure 9.65

Item	Unit	Item	Unit
Aluminum	Lb.	Frames	S.F.
Area Walls	Ea.	Ladders	V.L.F.
Bumper Rails	L.F.	Lampposts	Ea.
Checkered Plate	Lb.	Lintels	Lb.
Castings Misc.	Lb.	Pipe Support Framing	Lb.
Chain Link Fabric	C.S.F.	Railing	L.F.
Corner Guards	L.F.	Solar Screens	S.F.
Crane Rail	Lb.	Stairs	Riser
Curb Edging	L.F.	Toilet Part. Supports	Stall
Decorative Covering	S.F.	Window Guards	S.F.
Door Frames	Ea.	Wire	L.F.
Fire Escapes	Lb.	Wire Rope	L.F.
Floor Grating	S.F.		

Expansion Control and Fasteners

This portion of Division 5 includes fasteners for wood construction (nails, timber connectors, lag screws) as well as those for metal. Expansion anchors, shear studs and expansion joint assemblies are also included. It is up to the estimator to determine how costs for these items are to be listed in the estimate. Wood fasteners may be included in Division 6; shear studs may be considered part of the structural steel or metal deck installation. As previously emphasized, it is not as important where items are included as long as all required items are included.

Sample Estimate: Division 5

The increased use of computers and the development of estimating software in the construction industry provides the estimator with tools to increase speed, improve documentation and to lessen the chance of mathematical error. In essence, estimates can be produced better and faster with the use of these tools.

Figure 9.66 is an example of a computer generated drawing that the estimator (contractor or engineer) might produce to aid in estimating the structural steel portion of the sample building project. Such a drawing would compile data from the specifications and from different plans, sections and details of the construction documents. The drawing in Figure 9.66 is of the framing for the second and third floors. Similar drawings might show column heights, base plates, roof and penthouse framing, including metal joists and bridging. These are not construction or shop drawings and are used only for takeoff. Figure 9.67 is the estimate sheet for the drawing in Figure 9.66. Note that all information, as provided in *Means Building Construction Cost Data*, is displayed for each item on the estimate sheet (in this case, burdened, or including overhead and profit).

For the sample project, the estimator is developing costs for comparison to and verification of subcontract bids. Depending on the estimating accuracy, an estimate developed as a crosscheck may be included in the project bid if no reasonable subcontract bids are submitted. This practice involves risk, but often no other option is available. A computer generated estimate as described above would be ideal for such a purpose.

Returning to pencil and paper, the quantity sheets for the structural steel, metal joists and deck are shown in Figures 9.68 and 9.69.

Quantities are calculated and subtotalled by the pound. The total weight is not converted to tons until it is entered on the estimate sheet. In this way, errors due

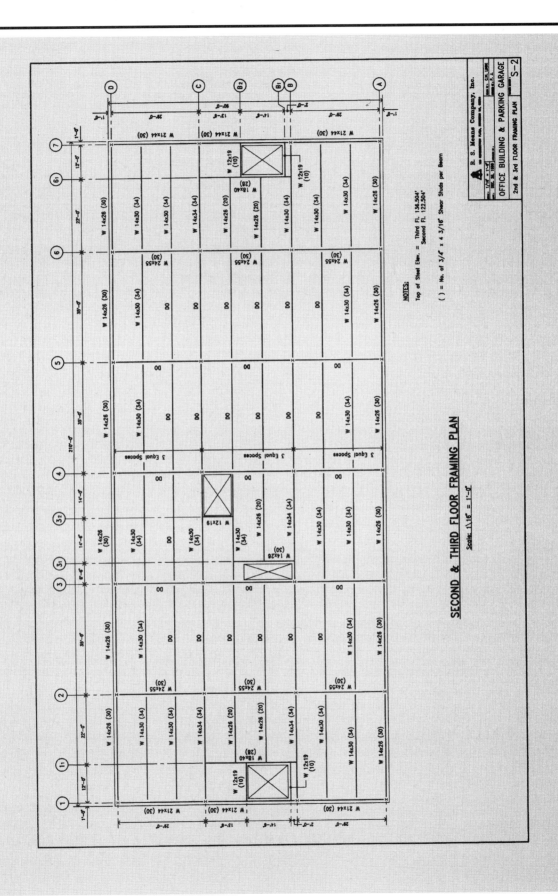

Figure 9.66

R.S. Means Incorporated
63 Smiths Lane
Kingston, MA 02364-0800
Unit Costlist

Division 5 Metals

Line Number	Description	Quantity	Unit	Labor Hours	Ext. Material	Ext. Labor	Ext. Equipment	Ext. Total	Ext. Total Incl O&P
059084400300	Welded shear connectors, 3/4" diameter, 4-3/16" long	1,450	Ea.	0.016	$565.00	$800.00	$755.00	$2,125.00	$2,875.00
051206401300	Strl st mbrs,com WF sizes,spans 10'to 45',incl bltd conns&erect,12x22	50	L.F.	0.064	$740.00	$106.00	$81.00	$925.00	$1,075.00
051206401900	Strl st mbrs,com WF sizes,spans 10'-45',incl bltd conns&erect,W 14x26	540	L.F.	0.057	$9,450.00	$1,025.00	$780.00	$11,200.00	$13,000.00
051206402100	Strl st mbrs,com WF sizes,spans 10' to 45',incl bltd conns&erect,14x30	1,410	L.F.	0.062	$28,200.00	$2,900.00	$2,225.00	$33,300.00	$38,800.00
051206404100	Strl st mbrs,com WF sizes,spans 10'-45',incl bltd conns&erect,W 21x44	156	L.F.	0.075	$4,600.00	$395.00	$222.00	$5,225.00	$6,000.00
051206404100	Strl st mbrs,com WF sizes,spans 10'-45',incl bltd conns&erect,W 21x44	395	L.F.	0.075	$11,700.00	$1,000.00	$560.00	$13,200.00	$15,200.00
Totals for Division 5 Metals					**$55,255.00**	**$6,226.00**	**$4,623.00**	**$65,975.00**	**$76,950.00**

Figure 9.67

QUANTITY SHEET

PROJECT: Office Building
LOCATION:
TAKE OFF BY:
ARCHITECT:
EXTENSIONS BY:

Division 5

ESTIMATE NO:
DATE:
CHECKED BY:

DESCRIPTION	NO.	DIMENSIONS L	W	Lbs.	UNIT	Lbs.	UNIT	Total Lbs.	UNIT	Acc.	UNIT
Columns											
W8 x 40	18	31.27'				22514	Lbs.				
W8 x 24	18	11.33'				4895					
W8 x 67	4	31.23'				8370					
W8 x 31	4	11.33'				1405					
W12 x 58	6	31.20'				10857					
W8 x 48	6	20.29'				5844		(53885) Lbs.			
PL 1 1/2" THICK	18	1.33'	1.33'	61.2		1949	Lbs.				
2" "	4	1.75'	1.75'	81.6		1000					
2 3/8" "	6	2'	2'	96.9		2326		(5275) Lbs.			
3/4" Anchor Bolts		24"								112 Ea.	
Beams - 2nd & 3rd Floors											
W14 x 26	4	34'				3536	Lbs.				
	8	35'				7280					
	4	25'				2600					
	1	27'				702					
	1	20'				520					
W14 x 30	8	34'				8160					
	30	35'				31500					
W14 x 34	5	34'				5780					
W12 x 19	5	9'				855					
W21 x 44	4	29'				5104					
	2	30'				2640					
W18 x 40	2	30'				2400					
W24 x 55	10	29'				15950					
	5	30'				8250					
Each Floor						95277	Lbs.				
Both Floors								(190554) Lbs.			
3/4" Shear Studs		4 3/16"								(2636) Ea.	

Figure 9.68

216

QUANTITY SHEET

PROJECT: Office Building
LOCATION:
TAKE OFF BY:

Division 5
ARCHITECT:
EXTENSIONS BY:

DESCRIPTION	NO.	L	W	Lbs.	UNIT	Lbs.	UNIT	Total Lbs.	UNIT	Acc.	UNIT
Beams - Roof											
W16 x 26	4	34'				3536	Lbs.				
	8	35'				7280	Lbs.				
W21 x 44	7	35'				10780	Lbs.				
W18 x 24 ,	1	26'				1040					
W21 x 49	4	29'				5684					
	2	30'				2940					
W21 x 62	10	29'				17980					
	3	30'				5580					
W27 x 94	2	30'				5640					
W12 x 19	1	9'				171	Lbs.	60631 Lbs		536 Ea.	
3/4" Shear Studs	4	3/16"									
Beams - Penthouse											
W21 x 49	2	30'				2940	Lbs.				
W21 x 62	1	30'				1860	Lbs.				
W16 x 26	4	35'				3640	Lbs.	8440 Lbs			
Joists - Roof & Penthouse											
2447 (R)	30	34'		11.5		11730	Lbs.				
(R)	52	35'		11.5		20930	Lbs.				
(P)	10	30'		11.5		3450	Lbs.	36110 Lbs			
Bottom Chord Ext.	32									32 Ea.	
Bridging (R)	36	88'		0.8		2534	Lbs.				
(R)	24	29'		0.8		557	Lbs.				
(P)	12	30'		0.8		288	Lbs.	3379 Lbs			
Deck (22 ga.)											
3" - 2nd & 3rd Floor	2	210'	90'		39900 SF						
Incl. Penthouse		70'	30'								
1 1/2" - R & P		210'	90'		18900 SF						
Edge Form	3	210'	90' + Opening							2160	LF

Figure 9.69

217

to rounding can be avoided. Quantities are entered and priced on the estimate sheets in Figures 9.70 to 9.72.

In Figure 9.70, the 10% addition for connections is included; it is based on the above discussion of allowances for structural steel. Base plate weights are added after the connections. Because this portion of the estimate is to be used for verification, a single price per ton for all structural steel is used (from Figure 9.65, Line 05120-680-0800). Note that the price per ton includes bolting. High strength bolts are specified for the project and require more labor than standard bolts. Standard bolts are usually used for connecting and are replaced with high strength bolts during the bolting up procedure. No deduction (from the $2,025 per ton) is used for deleting the standard bolting. The approximate quantity of high strength bolts may be determined from Figure 9.73, Reference Table R05090-510.

Material costs only are included for the anchor bolts and lintels. These items will be installed by different trades.

On Sheet 3 of the estimate (Figure 9.72) the subtotals for Division 5 are separated and listed according to the major subdivisions of metals work. These costs may be used as part of the historical cost development for estimating and comparison to future projects.

Division 6: Wood and Plastics

Wood frame construction is still dominant in the residential construction industry in the United States. However, its use for large scale commercial buildings has declined. There are many reasons for this trend; among them are design criteria, cost and, in some localities, building and fire code restrictions. Nevertheless, the use of wood framing for smaller suburban office buildings is still common.

Material prices for lumber fluctuate more and with greater frequency than any other building material. For this reason, when the material list is complete it is important to obtain current, local prices for the lumber. Installation costs depend on productivity. For Division 6, accurate cost records from past jobs can be most helpful. The estimate can be tailored to the productivity of specific crews as shown in Division 4 of the Sample Estimate.

Carpentry work can be broken down into the following categories: rough carpentry, finish carpentry and millwork, and laminated framing and decking. The rough carpentry materials can be sticks of lumber or sheets of plywood—job site fabricated and installed, or may consist of trusses, and truss joists, and panelized roof systems—prefabricated, delivered and erected by a specialty subcontractor.

Rough Carpentry

Lumber is usually estimated in board feet and purchased in 1,000 board foot quantities. A board foot is 1" × 12" × 12" (nominal) or 3/4" × 11-1/2" × 12" milled (actual). To determine board feet of a piece of framing, the nominal dimensions can be multiplied, and the result divided by 12. The final result represents the number of board feet per linear foot of that framing size.

Example: 2 × 10 joists

2 × 10 = 20

$$\frac{20}{12} = 1.67 \quad \frac{\text{Board feet}}{\text{Linear foot}}$$

CONSOLIDATED ESTIMATE

PROJECT: Office Building CLASSIFICATION: Division 5
LOCATION: ARCHITECT:
TAKE OFF BY: ABC   QUANTITIES BY: ABC   PRICES BY: As Shown   EXTENSIONS BY: DEF

ESTIMATE NO:
DATE: Jan-02
CHECKED BY: GHI

DESCRIPTION	SOURCE			QUANT	UNIT	MATERIAL COST	MATERIAL TOTAL	LABOR COST	LABOR TOTAL	EQUIPMENT COST	EQUIPMENT TOTAL	SUBCONTRACT COST	SUBCONTRACT TOTAL	TOTAL COST	TOTAL TOTAL
Division 5: Metals															
Structural Steel															
Columns				53,885	Lbs.										
2nd & 3rd Floor Beams				190,554											
Roof				60,631											
Penthouse				8440											
10% Connections				313,510	Lbs.										
				31,351											
Base PL				344,861	Lbs.										
				5,275											
				350,136	Lbs.										
Total	05120	680	0800	175	Ton							2025	354375		
High Strength Bolts 20 Ea. per Ton	05090	420	0300	3500	Ea.							5.80	20300		
Anchor Bolts (material Only)	03150	080	0500	112	Ea.							3.20	358		
Subtotals													$ 375,033		

Figure 9.70

CONSOLIDATED ESTIMATE

PROJECT: Office Building
LOCATION:
TAKE OFF BY: ABC
CLASSIFICATION: Division 5
ARCHITECT:
QUANTITIES BY: ABC
PRICES BY:
EXTENSIONS BY:
ESTIMATE NO:
DATE: Jan-02
CHECKED BY: GHI

DESCRIPTION	SOURCE			QUANT.	UNIT	MATERIAL			LABOR			EQUIPMENT			SUBCONTRACT		TOTAL	
						DEF	COST	TOTAL	DEF	COST	TOTAL	DEF	COST	TOTAL	COST	TOTAL	COST	TOTAL
Division 5: (Cont'd)																		
Metal Joists & Decks																		
Joists				36,110	Lbs.													
Bridging				3,379														
				39,489	Lbs.													
Total	05120	680	0800	19.75	Ton										2025	39994		
Bottom Chord Extensions	05210	600	6300	32	Ea.										24.50	784		
Composite Deck - 3" 22 gauge	05310	300	5200	40570	SF										1.44	58421		
Roof - 1-1/2" 22 gauge	05310	300	2400	18900	SF										1.12	21168		
Edge Form	05310	300	7100	2160	LF										3.84	8294		
Sheer Studs, 3/4" x 4-3/16"	05090	840	0300	3172	Ea.										1.99	6312		
Subtotals															$	134,973		

Figure 9.71

CONSOLIDATED ESTIMATE

PROJECT: Office Building
LOCATION:
CLASSIFICATION: Division 5
ARCHITECT:
ESTIMATE NO:
DATE: Jan-02
TAKE OFF BY: ABC
QUANTITIES BY: ABC
PRICES BY: As Shown
EXTENSIONS BY:
CHECKED BY: GHI

SOURCE	DESCRIPTION	QUANT.	UNIT	MATERIAL COST	DEF	MATERIAL TOTAL	LABOR COST	LABOR TOTAL	EQUIPMENT COST	DEF	EQUIPMENT TOTAL	SUBCONTRACT COST	SUBCONTRACT TOTAL	TOTAL COST	TOTAL
	Division 5: (Cont'd)														
	Miscellaneous Metals														
05517 700 0200	Stairs 3'-6" w/ Rails	188	Riser									233.00	43804		
05517 700 1500	Landings 10'-8" x 4'	320	SF									36.00	11520		
05520 700 0020	Exterior Aluminum Rails	150	LF									28.00	4200		
05120 480 2100	Lintels (Material Only)	34	Ea.									27.50	935		
	Subtotals												$ 60,459		
	Sheet 1 Subtotals												$ 375,033		
	Sheet 2 Subtotals												$ 134,973		
	Sheet 3 Subtotals												$ 60,459		
	Division 5 Totals												$ 570,465		

Figure 9.72

R05080-310   Coating Structural Steel

On field-welded jobs, the shop-applied primer coat is necessarily omitted. All painting must be done in the field and usually consists of red oxide rust inhibitive paint or an aluminum paint (see Division 09910-650 for paint material costs). The table below shows paint coverage and daily production for field painting.

See Division 05120-680 for hot-dipped galvanizing and for field-applied cold galvanizing and other paints and protective coatings.

See Division 05120-680 for steel surface preparation treatments such as wire brushing, pressure washing and sand blasting.

Type Construction	Surface Area per Ton	Coat	One Gallon Covers		In 8 Hrs. Person Covers		Average per Ton Spray	
			Brush	Spray	Brush	Spray	Gallons	Labor-hours
Light Structural	300 S.F. to 500 S.F.	1st	500 S.F.	455 S.F.	640 S.F.	2000 S.F.	0.9 gals.	1.6 L.H.
		2nd	450	410	800	2400	1.0	1.3
		3rd	450	410	960	3200	1.0	1.0
Medium	150 S.F. to 300 S.F.	All	400	365	1600	3200	0.6	0.6
Heavy Structural	50 S.F. to 150 S.F.	1st	400	365	1920	4000	0.2	0.2
		2nd	400	365	2000	4000	0.2	0.2
		3rd	400	365	2000	4000	0.2	0.2
Weighted Average	225 S.F.	All	400	365	1350	3000	0.6	0.6

R05090-510   High Strength Bolts

Common bolts (A307) are usually used in secondary connections (see Division 05090-150).

High strength bolts (A325 and A490) are usually specified for primary connections such as column splices, beam and girder connections to columns, column bracing, connections for supports of operating equipment or of other live loads which produce impact or reversal of stress, and in structures carrying cranes of over 5-ton capacity.

Allow 20 field bolts per ton of steel for a 6 story office building, apartment house or light industrial building. For 6 to 12 stories allow 25 bolts per ton, and above 12 stories, 30 bolts per ton. On power stations, 20 to 25 bolts per ton are needed.

R05090-520   Welded Structural Steel

Usual weight reductions with welded design run 10% to 20% compared with bolted or riveted connections. This amounts to about the same total cost compared with bolted structures since field welding is more expensive than bolts. For normal spans of 18' to 24' figure 6 to 7 connections per ton.

Trusses — For welded trusses add 4% to weight of main members for connections. Up to 15% less steel can be expected in a welded truss compared to one that is shop bolted. Cost of erection is the same whether shop bolted or welded.

General — Typical electrodes for structural steel welding are E6010, E6011, E60T and E70T. Typical buildings vary between 2# to 8# of weld rod per

ton of steel. Buildings utilizing continuous design require about three times as much welding as conventional welded structures. In estimating field erection by welding, it is best to use the average linear feet of weld per ton to arrive at the welding cost per ton. The type, size and position of the weld will have a direct bearing on the cost per linear foot. A typical field welder will deposit 1.8# to 2# of weld rod per hour manually. Using semiautomatic methods can increase production by as much as 50% to 75%.

Figure 9.73

The Quantity Sheet should indicate species, grade, and any type of wood preservative or fire retardant treatment specified or required by code.

Sills, Posts, and Girders used in subfloor framing should be taken off by length and quantity. The standard lengths available vary by species and dimensions. Cut-offs are often used for blocking. Careful selection of lengths will decrease the waste factor required.

Floor Joists, shown or specified by size and spacing, should be taken off by nominal length and the quantity required. Add for double joists under partitions, headers and cripple joists at openings, overhangs, laps at bearings, and blocking or bridging.

Ceiling Joists, similar to floor joists, carry roof loads and or/ceiling finishes. Soffits and suspended ceilings should be noted and taken off separately. Ledgers may be a part of the ceiling joist system. In a flat roof system, the rafters are called joists and are usually shown as a ceiling system.

Studs required are noted on the drawings by spacing, usually 16" O.C. or 24" O.C., with the stud size given. The linear feet of partitions with the same stud size, height and spacing, and divided by the spacing will give the estimator the approximate number of studs required. Additional studs for openings, corners, double top plates, sole plates, and intersecting partitions must be taken off separately. An allowance for waste should be included (or heights should be recorded as a standard length). A rule of thumb is to allow one stud for every linear foot of wall, for 16" O.C.

Number and Size of Openings are important. Even though there are no studs in these areas, the estimator must take off headers, subsills, king studs, trimmers, cripples, and knee studs. Where bracing and fire blocking are noted, indicate the type and quantity.

Roof Rafters vary with different types of roofs. A hip and valley, because of its complexity, has a greater material waste factor than most other roof types. Although straight gable, gambrel, and mansard roofs are not as complicated, care should be taken to ensure a good material takeoff. Roof pitches, overhangs, and soffit framing all affect the quantity of material and therefore, the costs. Rafters must be measured along the slope, not the horizontal. (See Figure 9.79.)

Roof Trusses are usually furnished and delivered to the job site by the truss fabricator. The high cost of job site labor and new gang nailing technology have created a small boom in truss manufacturing. Many architects' designs of wood frame and masonry bearing wall structures now include wood trusses of both the trussed rafter type and the flat chord type (also used for floors). Depending upon the size of truss, hoisting equipment may be needed for erection. The estimator should obtain prices and weights from the fabricator and should determine whether or not erection is included in the fabricator's cost. Architecturally exposed trusses are typically more expensive to fabricate and erect.

Tongue and Groove Roof Decks of various woods, solid planks, or laminated construction are nominally 2" to 4" thick and are often used with glued laminated beams or heavy timber framing. The square foot method is used to determine quantities and consideration is given to roof pitches and non-modular areas for the amount of waste involved. The materials are purchased by board foot measurement. The conversion from square foot to board foot must allow for net sizes as opposed to board measure. In this way, loss of coverage due to the available tongue and mill lengths can be taken into account.

Sheathing on walls can be plywood of different grades and thicknesses, wallboard, or solid boards nailed directly to the studs. Insulating sheets with air infiltration barriers are often used as sheathing in colder climates. Plywood can be applied with the grain vertical, horizontal, or rarely, diagonal to the studding. Solid boards are usually nailed diagonally, but can be applied horizontally when lateral forces are not present. For solid board sheathing, add 15% to 20% more material to the takeoff when using tongue and groove, as opposed to square edge sheathing. Wallboard can be installed either horizontally or vertically, depending upon wall height and fire code restrictions. When estimating quantities of plywood or wallboard sheathing, the estimator calculates the number of sheets required by measuring the square feet of area to be covered and then dividing by sheet size. Applying these materials diagonally or on non-modular areas will create waste. This waste factor must be included in the estimate. For diagonal application of boards, plywood, or wallboard, include an additional 10% to 15% material waste factor.

Subfloors can be CDX type plywood (with the thickness dependent on the load and span), solid boards laid diagonally or perpendicular to the joists, or tongue and groove planks. The quantity takeoff for subfloors is similar to sheathing (noted above).

Stressed Skin Plywood includes prefabricated roof panels, with or without bottom skin or tie rods, and folded plate roof panels with intermediate rafters. Stressed skin panels are typically custom prefabricated. Takeoff is by the square foot or panel.

Structural Joists are prefabricated "beams" with wood flanges and plywood or tubular steel webs. This type of joist is spaced in accordance with the load and requires bridging and blocking supplied by the fabricator. Quantity takeoff should include the following: type, number required, length, spacing, end-bearing conditions, number of rows, length of bridging, and blocking.

Grounds are normally 1" × 2" wood strips used for case work or plaster; the quantities are estimated in L.F.

Furring (1" × 2" or 3") wood strips are fastened to wood masonry or concrete walls so that wall coverings may be attached thereto. Furring may also be used on the underside of ceiling joists to fasten ceiling finishes. Quantities are estimated by L.F.

Lumber and Plywood Treatments can sometimes double the costs for material. The plans and specifications should be carefully checked for required treatments—against insects, fire or decay—as well as grade, species and drying specifications.

An alternative method to pricing rough carpentry by the piece or linear foot is to determine quantities (in board feet) based on square feet of surface area. Appendix B of this book contains charts and tables that may be used for this second method. Also included are quantities of nails required for each type and spacing of rough framing. A rule of thumb for this method can be used to determine linear feet of framing members (such as studs, joists) based on square feet of surface area (wall, floor, ceiling):

Spacing of Framing Members	Board Feet per Square Foot Surface
12" O.C.	1.2 B.F./S.F.
16" O.C.	1.0 B.F./S.F.
24" O.C.	0.8 B.F./S.F.

The requirements for rough carpentry, especially those for temporary construction, may not all be directly stated in the plans and specifications. These additional items may include blocking, temporary stairs, wood inserts for metal pan stairs, and railings, along with various other requirements for different trades. Temporary construction may also be included in Division 1 of the General Requirements.

Finish Carpentry and Millwork

Finish carpentry and millwork—wood rails, paneling, shelves, casements and cabinetry—are common in buildings that have no other wood.

Upon examination of the plans and specifications, the estimator must determine which items will be built on-site, and which will be fabricated off-site by a millwork subcontractor. Shop drawings are often required for architectural woodwork and are usually included in the subcontract price.

Window and Door Trim may be taken off and priced by the "set" or by the linear foot. Check for jamb extensions at exterior walls. The common use of pre-hung doors and windows makes it convenient to take off this trim with the doors and windows. Exterior trim, other than door and window trim, should be taken off with the siding, since the details and dimensions are interrelated.

Paneling is taken off by the type, finish, and square foot (converted to full sheets). Be sure to list any millwork that would show up on the details. Panel siding and associated trim are taken off by the square foot and linear foot, respectively. Be sure to provide an allowance for waste.

Decorative Beams and Columns that are non-structural should be estimated separately. Decorative trim may be used to wrap exposed structural elements. Particular attention should be paid to the joinery. Long, precise joints are difficult to construct in the field.

Cabinets, Counters, and Shelves are most often priced by the linear foot or by the unit. Job-fabricated, prefabricated, and subcontracted work should be estimated separately.

Stairs should be estimated by individual component unless accurate, complete system costs have been developed from previous projects. Typical components and units for estimating are shown in Figure 9.74.

A general rule for budgeting millwork is that total costs will be two to three times the cost of the materials. Millwork is often ordered and purchased directly by the owner; when installation is the responsibility of the contractor, costs for handling, storage, and protection should be included.

Laminated Construction

Laminated construction should be listed separately, as it is frequently supplied by a specialty subcontractor. Sometimes the beams are supplied and erected by one subcontractor, and the decking installed by the general contractor or another subcontractor. The takeoff units must be adapted to the system: square foot—floor, linear foot—members, or board foot—lumber. Since the members are factory fabricated, the plans and specifications must be submitted to a fabricator for takeoff and pricing.

Sample Estimate: Division 6

The estimate for Division 6—Wood and Plastics—of the sample project is minimal and straightforward. The estimate sheet is shown in Figure 9.75. The only item included for rough carpentry is fire retardant treated blocking.

		06430 Stairs & Railings	CREW	DAILY OUTPUT	LABOR-HOURS	UNIT	2002 BARE COSTS				TOTAL INCL O&P	
							MAT.	LABOR	EQUIP.	TOTAL		
505	0400	Decking, 1" x 4"	1 Carp	275	.029	S.F.	1.90	.87		2.77	3.45	**505**
	0500	2" x 4"		300	.027		1.69	.80		2.49	3.11	
	0600	2" x 6"		320	.025		1.69	.75		2.44	3.03	
	0650	5/4" x 6"		320	.025		1.68	.75		2.43	3.02	
	0700	Redwood decking, 1" x 4"		275	.029		4	.87		4.87	5.75	
	0800	2" x 6"		340	.024		11.40	.71		12.11	13.65	
	0900	5/4" x 6"	▼	320	.025	▼	7.15	.75		7.90	9	
620	0011	**STAIRS, PREFABRICATED**										**620**
	0100	Box stairs, prefabricated, 3'-0" wide										
	0110	Oak treads, no handrails, 2' high	2 Carp	5	3.200	Flight	216	96		312	385	
	0200	4' high		4	4		430	120		550	660	
	0300	6' high		3.50	4.571		620	137		757	895	
	0400	8' high		3	5.333		775	160		935	1,100	
	0600	With pine treads for carpet, 2' high		5	3.200		88.50	96		184.50	246	
	0700	4' high		4	4		164	120		284	365	
	0800	6' high		3.50	4.571		240	137		377	475	
	0900	8' high	▼	3	5.333		274	160		434	550	
	1100	For 4' wide stairs, add				▼	25%					
	1500	Prefabricated stair rail with balusters, 5 risers	2 Carp	15	1.067	Ea.	229	32		261	300	
	1700	Basement stairs, prefabricated, soft wood,										
	1710	open risers, 3' wide, 8' high	2 Carp	4	4	Flight	575	120		695	815	
	1900	Open stairs, prefabricated prefinished poplar, metal stringers,										
	1910	treads 3'-6" wide, no railings										
	2000	3' high	2 Carp	5	3.200	Flight	229	96		325	400	
	2100	4' high		4	4		485	120		605	720	
	2200	6' high		3.50	4.571		555	137		692	830	
	2300	8' high	▼	3	5.333	▼	730	160		890	1,050	
	2500	For prefab. 3 piece wood railings & balusters, add for										
	2600	3' high stairs	2 Carp	15	1.067	Ea.	31.50	32		63.50	84.50	
	2700	4' high stairs		14	1.143		51	34.50		85.50	110	
	2800	6' high stairs		13	1.231		63	37		100	127	
	2900	8' high stairs		12	1.333		96.50	40		136.50	169	
	3100	For 3'-6" x 3'-6" platform, add	▼	4	4	▼	72.50	120		192.50	267	
	3300	Curved stairways, 3'-3" wide, prefabricated, oak, unfinished,										
	3310	incl. curved balustrade system, open one side										
	3400	9' high	2 Carp	.70	22.857	Flight	6,375	685		7,060	8,100	
	3500	10' high		.70	22.857		7,200	685		7,885	9,000	
	3700	Open two sides, 9' high		.50	32		10,000	960		10,960	12,500	
	3800	10' high		.50	32		10,800	960		11,760	13,400	
	4000	Residential, wood, oak treads, prefabricated		1.50	10.667	▼	930	320		1,250	1,525	
	4200	Built in place	▼	.44	36.364	▼	1,325	1,100		2,425	3,150	
	4400	Spiral, oak, 4'-6" diameter, unfinished, prefabricated,										
	4500	incl. railing, 9' high	2 Carp	1.50	10.667	Flight	4,000	320		4,320	4,900	
630	0010	**STAIR PARTS** Balusters, turned, 30" high, pine, minimum	1 Carp	28	.286	Ea.	4.11	8.55		12.66	17.85	**630**
	0100	Maximum		26	.308		9	9.25		18.25	24.50	
	0300	30" high birch balusters, minimum		28	.286		6.30	8.55		14.85	20.50	
	0400	Maximum		26	.308		10.20	9.25		19.45	25.50	
	0600	42" high, pine balusters, minimum		27	.296		6	8.90		14.90	20.50	
	0700	Maximum		25	.320		13	9.60		22.60	29.50	
	0900	42" high birch balusters, minimum		27	.296		7.65	8.90		16.55	22.50	
	1000	Maximum		25	.320	▼	27	9.60		36.60	44.50	
	1050	Baluster, stock pine, 1-1/16" x 1-1/16"		240	.033	L.F.	1.98	1		2.98	3.74	
	1100	1-5/8" x 1-5/8"		220	.036	"	2.19	1.09		3.28	4.11	
	1200	Newels, 3-1/4" wide, starting, minimum		7	1.143	Ea.	33.50	34.50		68	90.50	
	1300	Maximum	▼	6	1.333	↓	126	40		166	202	

Figure 9.74

Blocking requirements may or may not be shown on the drawings or stated in the specifications, but in this case, previous experience dictates that an allowance be included. The only other carpentry for the project involves wall paneling at the elevator lobby of each floor, and vanities in the women's rest rooms. For custom quality finish work, it is recommended that the craftsman to perform the work be consulted. This person is able to call upon experience to best estimate the required time.

Division 7: Thermal and Moisture Protection

This division includes materials for sealing the outside of a building—for protection against moisture and air infiltration, as well as insulation and associated accessories. When reviewing the plans and specifications, the estimator should visualize the construction process, and thus determine all probable areas where these materials will be found in or on a building. The technique used for quantity takeoff depends on the specific materials and installation methods.

Waterproofing

- Dampproofing
- Vapor Barriers
- Caulking and Sealants
- Sheet and Membrane
- Integral Cement Coatings

A distinction should be made between dampproofing and waterproofing. Dampproofing is used to inhibit the migration of moisture or water vapor. In most cases, dampproofing will not stop the flow of water (even at minimal pressures). Waterproofing, on the other hand, consists of a continuous, impermeable membrane and is used to prevent or stop the flow of water.

Dampproofing usually consists of one or two bituminous coatings applied to foundation walls from about the finished grade line to the bottom of the footings. The areas involved are calculated from the total height of the dampproofing and the length of the wall. After separate areas are figured and added together to provide a total square foot area, a unit cost per square foot can be selected for the type of material, the number of coats, and the method of application specified for the building.

Waterproofing at or below grade with elastomeric sheets or membranes is estimated on the same basis as dampproofing, with two basic exceptions. First, the installed unit costs for the elastomeric sheets do not include bonding adhesive or splicing tape, which must be figured as an additional cost. Second, the membrane waterproofing under slabs must be estimated separately from the higher cost installation on walls. In all cases, unit costs are per square foot of covered surface.

For walls below grade, protection board is often specified to prevent damage to the barrier when the excavation is backfilled. Rigid foam insulation installed outside of the barrier may also serve a protective function. Metallic coating material may be applied to floors or walls, usually on the interior or dry side, after the masonry surface has been prepared (usually by chipping) for bonding to the new material. The unit cost per square foot for these materials depends on the thickness of the material, the position of the area to be covered and the preparation required. In many cases, these materials must be applied in locations where access is difficult and under the control of others. The estimator should make an allowance for delays caused by this problem.

CONSOLIDATED ESTIMATE

PROJECT: Office Building	CLASSIFICATION: Division 6		ESTIMATE NO:
LOCATION:	ARCHITECT:		DATE: Jan-02
TAKE OFF BY: ABC	QUANTITIES BY: ABC	PRICES BY: As Shown	EXTENSIONS BY: DEF	CHECKED BY: GHI

DESCRIPTION	SOURCE			QUANT	UNIT	MATERIAL		LABOR		EQUIPMENT		SUBCONTRACT		TOTAL
						COST	DEF TOTAL	COST	TOTAL	COST	DEF TOTAL	COST	TOTAL	COST	TOTAL
Division 6: Wood & Plastic														
Rough Carpentry														
Blocking	06110	100	2740	0.1	MBF	535.00	54	1275	128					
Fire Treatment	06073	400	0400	0.1	MBF	279.00	28							
Finish Carpentry														
Paneling@ Elevator Lobby	06250	500	2600	1800	SF	2.10	3780	1.20	2160					
Vanities	06410	400	8150	6	Ea.	295.00	1770	42.00	252					
Arch. Woodwork														
Vanity Top, with Backsplash	06055	740	1000	24	LF	26.00	624	8.00	192					
Cutouts	06055	740	1900	9	Ea.	2.63	24	7.50	68					
Division 6 Total							$ 6,279		$ 2,799					

Figure 9.75

Caulking and sealants are usually applied on the exterior of the building except for certain special conditions on the interior. In most cases, caulking and sealing is done to prevent water and/or air from entering a building. Caulking and sealing are usually specified at joints, expansion joints, control joints, door and window frames, and in places where dissimilar materials meet over the surface of the building exterior. To estimate the installed cost of this type of material, two things must be determined. First, the estimator must note (from the specifications) the kind of material to be used for each caulking or sealing job. Second, the dimensions of the joints to be caulked or sealed must be measured on the plans, with attention given to any requirements for backer rods. With this information, the estimator can select the applicable cost per linear foot and multiply it by the total length in feet. The result is an estimated cost for each kind of caulking or sealing on the job. Caulking and sealing may often be overlooked as incidental items. They may, in fact, represent a significant cost, depending upon the type of construction.

The specifications may require testing of the integrity of installed waterproofing in certain cases. If required, adequate time must be allowed and costs included in the estimate.

Insulation

- Batt or Roll
- Blown-in
- Board (Rigid and Semi-rigid)
- Cavity Masonry
- Perimeter Foundation
- Poured in Place
- Reflective
- Roof
- Sprayed

Insulation is primarily used to reduce heat transfer through the exterior enclosure of the building. The type and form of this insulation will vary according to its location in the structure and the size of the space it occupies. Major insulation types include mineral granules, fibers and foams, vegetable fibers and solids, and plastic foams. These materials may be required around foundations, on or inside walls, and under roofing. Many different details of the drawings must be examined in order to determine types, methods and quantities of insulation. The cost of insulation depends on the type of material, its form (loose, granular, batt or boards), its thickness in inches, the method of installation, and the total area in square feet.

It is becoming popular to specify insulation by "R" value only. The estimator may have the choice of materials, given a required "R" value and a certain cavity space (which may dictate insulation thickness). For example, the specifications may require an "R" value of 10, and only 2" of wall cavity is available for the thickness of the insulation. From Figure 9.76 and 9.77 it is seen that only fiberglass (Line 07220-700-0700) meets the design criteria. Note that if more cavity space were available, 3-1/2" non-rigid fiberglass (Line 07210-950-0420) would be a much less expensive alternative. The estimator may have to do some shopping to find the least expensive material for the specified "R" value and thickness. Installation costs may vary from one material to another. Also, wood blocking and/or nailers are often required to match the insulation thickness in some instances.

Working with the above data, the estimator can accurately select the installed cost per square foot and estimate the total cost. The estimate for insulation

			CREW	DAILY OUTPUT	LABOR-HOURS	UNIT	2002 BARE COSTS				TOTAL INCL O&P	
		07210	**Building Insulation**				MAT.	LABOR	EQUIP.	TOTAL		
950	1340	6" thick, R19	1 Carp	1,600	.005	S.F.	.39	.15		.54	.66	950
	1380	10" thick, R30	↓	1,350	.006	↓	.62	.18		.80	.96	
	1850	Friction fit wire insulation supports, 16" O.C.	↓	960	.008	Ea.	.05	.25		.30	.45	
	1900	For foil backing, add				S.F.	.04			.04	.04	
		07220	**Roof & Deck Insulation**									
700	0010	**ROOF DECK INSULATION**										700
	0020	Fiberboard low density, 1/2" thick R1.39	1 Rofc	1,000	.008	S.F.	.19	.21		.40	.57	
	0030	1" thick R2.78		800	.010		.34	.27		.61	.82	
	0080	1 1/2" thick R4.17		800	.010		.50	.27		.77	1	
	0100	2" thick R5.56		800	.010		.68	.27		.95	1.20	
	0110	Fiberboard high density, 1/2" thick R1.3		1,000	.008		.20	.21		.41	.58	
	0120	1" thick R2.5		800	.010		.36	.27		.63	.85	
	0130	1-1/2" thick R3.8		800	.010		.59	.27		.86	1.10	
	0200	Fiberglass, 3/4" thick R2.78		1,000	.008		.46	.21		.67	.87	
	0400	15/16" thick R3.70		1,000	.008		.61	.21		.82	1.03	
	0460	1-1/16" thick R4.17		1,000	.008		.76	.21		.97	1.20	
	0600	1-5/16" thick R5.26		1,000	.008		1.05	.21		1.26	1.52	
	0650	2-1/16" thick R8.33		800	.010		1.12	.27		1.39	1.68	
	0700	2-7/16" thick R10		800	.010		1.28	.27		1.55	1.86	
	1500	Foamglass, 1-1/2" thick R4.5		800	.010		1.46	.27		1.73	2.06	
	1530	3" thick R9		700	.011	↓	3.12	.30		3.42	3.95	
	1600	Tapered for drainage		600	.013	B.F.	.94	.35		1.29	1.63	
	1650	Perlite, 1/2" thick R1.32		1,050	.008	S.F.	.27	.20		.47	.64	
	1655	3/4" thick R2.08		800	.010		.33	.27		.60	.81	
	1660	1" thick R2.78		800	.010		.29	.27		.56	.77	
	1670	1-1/2" thick R4.17		800	.010		.39	.27		.66	.88	
	1680	2" thick R5.56		700	.011		.57	.30		.87	1.15	
	1685	2-1/2" thick R6.67		700	.011	↓	.77	.30		1.07	1.37	
	1690	Tapered for drainage		800	.010	B.F.	.59	.27		.86	1.10	
	1700	Polyisocyanurate, 2#/CF density, 3/4" thick, R5.1		1,500	.005	S.F.	.31	.14		.45	.58	
	1705	1" thick R7.14		1,400	.006		.33	.15		.48	.62	
	1715	1-1/2" thick R10.87		1,250	.006		.35	.17		.52	.68	
	1725	2" thick R14.29		1,100	.007		.45	.19		.64	.83	
	1735	2-1/2" thick R16.67		1,050	.008		.50	.20		.70	.89	
	1745	3" thick R21.74		1,000	.008		.71	.21		.92	1.14	
	1755	3-1/2" thick R25		1,000	.008	↓	.74	.21		.95	1.17	
	1765	Tapered for drainage	↓	1,400	.006	B.F.	.38	.15		.53	.68	
	1900	Extruded Polystyrene										
	1910	15 PSI compressive strength, 1" thick, R5	1 Rofc	1,500	.005	S.F.	.23	.14		.37	.49	
	1920	2" thick, R10		1,250	.006		.36	.17		.53	.69	
	1930	3" thick R15		1,000	.008		.77	.21		.98	1.21	
	1932	4" thick R20		1,000	.008	↓	1.07	.21		1.28	1.54	
	1934	Tapered for drainage		1,500	.005	B.F.	.35	.14		.49	.63	
	1940	25 PSI compressive strength, 1" thick R5		1,500	.005	S.F.	.39	.14		.53	.67	
	1942	2" thick R10		1,250	.006		.76	.17		.93	1.13	
	1944	3" thick R15		1,000	.008		1.15	.21		1.36	1.63	
	1946	4" thick R20		1,000	.008	↓	1.11	.21		1.32	1.58	
	1948	Tapered for drainage		1,500	.005	B.F.	.40	.14		.54	.68	
	1950	40 psi compressive strength, 1" thick R5		1,500	.005	S.F.	.35	.14		.49	.63	
	1952	2" thick R10		1,250	.006		.69	.17		.86	1.05	
	1954	3" thick R15		1,000	.008		1.01	.21		1.22	1.47	
	1956	4" thick R20		1,000	.008	↓	1.35	.21		1.56	1.85	
	1958	Tapered for drainage		1,400	.006	B.F.	.50	.15		.65	.81	
	1960	60 PSI compressive strength, 1" thick R5		1,450	.006	S.F.	.42	.15		.57	.71	
	1962	2" thick R10	↓	1,200	.007	↓	.75	.18		.93	1.13	

Figure 9.76

07210	Building Insulation	CREW	DAILY OUTPUT	LABOR-HOURS	UNIT	2002 BARE COSTS				TOTAL INCL O&P	
						MAT.	LABOR	EQUIP.	TOTAL		
900 0721	2-1/2" thick, R10.9	1 Carp	800	.010	S.F.	1.84	.30		2.14	2.49	**900**
0741	3" thick, R13		730	.011		2.20	.33		2.53	2.93	
0821	Foil faced, 1" thick, R4.3		1,000	.008		1.09	.24		1.33	1.57	
0840	1-1/2" thick, R6.5		890	.009		1.57	.27		1.84	2.15	
0850	2" thick, R8.7		800	.010		2.05	.30		2.35	2.73	
0880	2-1/2" thick, R10.9		800	.010		2.46	.30		2.76	3.18	
0900	3" thick, R13		730	.011		2.94	.33		3.27	3.74	
1500	Foamglass, 1-1/2" thick, R4.5		800	.010		1.60	.30		1.90	2.23	
1550	3" thick, R9	▼	730	.011	▼	2.79	.33		3.12	3.58	
1600	Isocyanurate, 4' x 8' sheet, foil faced, both sides										
1610	1/2" thick, R3.9	1 Carp	800	.010	S.F.	.29	.30		.59	.79	
1620	5/8" thick, R4.5		800	.010		.30	.30		.60	.80	
1630	3/4" thick, R5.4		800	.010		.31	.30		.61	.81	
1640	1" thick, R7.2		800	.010		.34	.30		.64	.84	
1650	1-1/2" thick, R10.8		730	.011		.38	.33		.71	.93	
1660	2" thick, R14.4 **CN**		730	.011		.47	.33		.80	1.03	
1670	3" thick, R21.6		730	.011		1.11	.33		1.44	1.73	
1680	4" thick, R28.8		730	.011		1.37	.33		1.70	2.02	
1700	Perlite, 1" thick, R2.77		800	.010		.26	.30		.56	.76	
1750	2" thick, R5.55		730	.011		.50	.33		.83	1.06	
1900	Extruded polystyrene, 25 PSI compressive strength, 1" thick, R5		800	.010		.34	.30		.64	.84	
1940	2" thick R10		730	.011		.67	.33		1	1.25	
1960	3" thick, R15		730	.011		.96	.33		1.29	1.57	
2100	Expanded polystyrene, 1" thick, R3.85		800	.010		.14	.30		.44	.62	
2120	2" thick, R7.69		730	.011		.38	.33		.71	.93	
2140	3" thick, R11.49	▼	730	.011	▼	.52	.33		.85	1.08	
950 0010	**WALL OR CEILING INSUL., NON-RIGID**										**950**
0040	Fiberglass, kraft faced, batts or blankets										
0060	3-1/2" thick, R11, 11" wide	1 Carp	1,150	.007	S.F.	.23	.21		.44	.58	
0080	15" wide **CN**		1,600	.005		.23	.15		.38	.48	
0100	23" wide		1,600	.005		.23	.15		.38	.48	
0140	6" thick, R19, 11" wide		1,000	.008		.33	.24		.57	.73	
0160	15" wide		1,350	.006		.33	.18		.51	.64	
0180	23" wide		1,600	.005		.33	.15		.48	.59	
0200	9" thick, R30, 15" wide		1,150	.007		.60	.21		.81	.99	
0220	23" wide		1,350	.006		.60	.18		.78	.94	
0240	12" thick, R38, 15" wide		1,000	.008		.76	.24		1	1.21	
0260	23" wide	▼	1,350	.006	▼	.76	.18		.94	1.12	
0400	Fiberglass, foil faced, batts or blankets										
0420	3-1/2" thick, R11, 15" wide	1 Carp	1,600	.005	S.F.	.34	.15		.49	.60	
0440	23" wide		1,600	.005		.34	.15		.49	.60	
0460	6" thick, R19, 15" wide		1,350	.006		.41	.18		.59	.73	
0480	23" wide		1,600	.005		.41	.15		.56	.68	
0500	9" thick, R30, 15" wide		1,150	.007		.71	.21		.92	1.11	
0550	23" wide	▼	1,350	.006	▼	.71	.18		.89	1.06	
0800	Fiberglass, unfaced, batts or blankets										
0820	3-1/2" thick, R11, 15" wide	1 Carp	1,350	.006	S.F.	.21	.18		.39	.51	
0830	23" wide		1,600	.005		.21	.15		.36	.46	
0860	6" thick, R19, 15" wide		1,150	.007		.34	.21		.55	.70	
0880	23" wide		1,350	.006		.34	.18		.52	.65	
0900	9" thick, R30, 15" wide		1,000	.008		.60	.24		.84	1.03	
0920	23" wide		1,150	.007		.60	.21		.81	.99	
0940	12" thick, R38, 15" wide		1,000	.008		.76	.24		1	1.21	
0960	23" wide	▼	1,150	.007	▼	.76	.21		.97	1.17	
1300	Mineral fiber batts, kraft faced										
1320	3-1/2" thick, R12	1 Carp	1,600	.005	S.F.	.26	.15		.41	.52	

Figure 9.77

231

should also include associated costs, such as cutting and patching for difficult installation, or requirements for air vents and other accessories.

Insulation is not only used for controlling heat transfer. It is also specified for use in internal walls and ceilings for controlling *sound* transfer. Although the noise reduction coefficient of batt insulation is not as great as specialized sound attenuation blankets, the costs are considerably less.

Shingles

Most residences and many smaller types of commercial buildings have sloping roofs covered with some form of shingle or watershed material. The materials used in shingles vary from the more common granular-covered asphalt and fiberglass units to wood, metal, clay, concrete, or slate.

The first step in estimating the cost of a shingle roof is to determine the material specified, shingle size and weight, and installation method. With this information, the estimator can select the accurate installed cost of the roofing material.

In a sloping roof deck, the ridge and eave lengths, as well as the ridge to eaves dimension, must be known or measured before the actual roof area can be calculated. When the plan dimensions of the roof are known and the sloping dimensions are not known, the actual roof area can still be estimated, providing the slope of the roof is known. Figure 9.78 is a table of multipliers that can be used for this purpose. The roof slope is given in both the inches of rise per foot of horizontal run, and in the degree of slope, which allows direct conversion of the horizontal plan dimension into the dimension on the slope.

After the roof area has been estimated in square feet, it must be divided by 100 to convert it into roofing squares (the conventional "unit" for roofing—one square equals 100 square feet). To determine the quantity of shingles required for hips or ridges, add one square for each 100 linear feet of hips and/or ridges.

When the total squares of roofing have been calculated, the estimator should make an allowance for waste based on the design of the roof. A minimum allowance of 3% to 5% is needed if the roof has two straight sides with two gable ends and no breaks. At the other extreme any roof with several valleys, hips, and ridges may need a waste allowance of 15% or more to cover the excess cutting required.

Accessories that are part of a shingle roof include drip edges, flashings at chimneys, dormers, skylights, vents, valleys, and walls. These are necessary to complete the roof and should be included in the estimate for the shingles.

Roofing and Siding

In addition to shingles, many types of roofing and siding are used on commercial and industrial buildings. These are made of several kinds of material and come in many forms for both roofing and siding—panels, sheets, membranes, and boards.

The materials used in roofing and siding panels include: aluminum, mineral fibercement epoxy, fibrous glass, steel, vinyl, many types of synthetic sheets and membranes, coal tar, asphalt felt, tar felt, and asphalt asbestos felt. Most of the latter materials are used in job-fabricated, built-up roofs, and as backing for other materials, such as shingles.

The basic data required for estimating either roofing or siding includes the specification of the material the supporting structure, the method of installation, and the area to be covered. When selecting the current unit price for these materials, the estimator must remember that basic installed unit costs are per square foot for siding and per square for roofing. The major exceptions to this general rule are prefabricated roofing panels and single-ply roofing, which are priced per square foot.

Single-ply Roofs

- Chlorinated Polyethylene (CPE)
- Chlorosulfonated Polyethylene
- Ethylene Propylene Diene Monomer (EPDM)
- Polychloroprene (Neoprene)
- Polyisobutylene (PIB)
- Polyvinyl Chloride (PVC)
- Modified Bitumen

Since the early 1970s, the use of single-ply roofing membranes in the construction industry has been on the rise. Market surveys have recently shown that of all the single-ply systems being installed, about one in three is on new construction. Materially, these roofs are more expensive than other, more conventional roofs; however, labor costs are much lower because of a faster installation. Re-roofing represents the largest market for single-ply roof systems today. Single-ply roof systems are normally installed in one of the following ways:

Factors for Converting Inclined to Horizontal					
Roof Slope	Approx. Angle	Factor	Roof Slope	Approx. Angle	Factor
Flat	0	1.000	12 in 12	45.0	1.414
1 in 12	4.8	1.003	13 in 12	47.3	1.474
2 in 12	9.5	1.014	14 in 12	49.4	1.537
3 in 12	14.0	1.031	15 in 12	51.3	1.601
4 in 12	18.4	1.054	16 in 12	53.1	1.667
5 in 12	22.6	1.083	17 in 12	54.8	1.734
6 in 12	26.6	1.118	18 in 12	56.3	1.803
7 in 12	30.3	1.158	19 in 12	57.7	1.873
8 in 12	33.7	1.202	20 in 12	59.0	1.943
9 in 12	36.9	1.250	21 in 12	60.3	2.015
10 in 12	39.8	1.302	22 in 12	61.4	2.088
11 in 12	42.5	1.357	23 in 12	62.4	2.162

Example:

[20' (1.302) 90'] 2 = 4,687.2 S.F. = 46.9 Sq.

OR

[40' (1.302) 90'] = 4,687.2 S.F. = 46.9 Sq.   40'

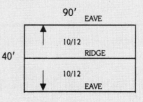

Figure 9.78

Loose-laid and ballasted: Generally this is the easiest type of single-ply roof to install. Some special consideration must be given, however, when flashing is attached to the roof. The membrane is typically fused together at the seams stretched out flat and ballasted with stone (1-1/2" @ 12 PSF) to prevent wind blow-off. This extra load must be considered during design stages. It is particularly important if re-roofing over an existing built-up roof that already weighs 10-15 PSF. A slip-sheet or vapor barrier is sometimes required to separate the new roof from the old.

Partially-adhered: This method of installation uses a series of bar or point attachments which adhere the membrane to a substrate. The membrane manufacturer typically specifies the method to be used based on the material and substrate. Partially-adhered systems do not use ballast material. A slip-sheet may be required.

Fully-adhered: This is generally the most time-consuming of the single-plies to install, because these roofs employ a contact cement, cold adhesive, or hot bitumen to adhere the membrane uniformly to the substrate. Only manufacturer-approved insulation board or substrate should be used to receive the membrane. No ballast is required.

The materials available can be broken down into three categories:

- Thermo-Setting: EDPM, Neoprene, and PIB
- Thermo-Plastic: Hypalon, PVC and CPE
- Composites: Modified Bitumen

Each has its own requirements and performance characteristics. Most are available in all three installation methods. See Figure 9.79.

Single-ply roof systems are available from many sources. Most if not all manufacturers, however, sell their materials only to franchised installers. As a result, there may be only one source for a price in any given area. Read the specifications carefully. Estimate the system required, exactly as specified; substitutes are usually not allowed.

Sheet Metal

- Copper and Stainless Steel
- Gutters and Downspouts
- Edge Cleats and Gravel Stops
- Flashings
- Trim
- Miscellaneous

Sheet metal work included in this division is limited to that used on roofs or sidewalls of buildings, usually on the exterior exposed to the weather. Many of the items covered are wholly or partially prefabricated with labor added for installation. Several are materials that require labor added for on-site fabrication; this cost must be estimated separately.

Pricing shop-made items such as downspouts, drip edges, expansion joints, gravel stops, gutters, reglets, and termite shields requires that the estimator determine the material, size, and shape of the fabricated section, and the linear feet of the item. From this data, an accurate unit can be selected and multiplied by the linear footage in order to obtain a total cost.

The cost of items like copper roofing and metal flashing is estimated in a similar manner, except that unit costs are per square foot. Some roofing systems, particularly single-ply, require flashing materials that are unique to that roofing system.

Roofing materials like monel, stainless steel, and zinc copper alloy are also estimated by the same method, except that the unit costs are per square (100 square feet). Prefabricated items like strainers and louvers are priced on a cost-per-unit basis. Adhesives are priced by the gallon. The installed cost of roofing adhesives depends on the cost per gallon and the coverage per gallon. With trowel grade adhesive, the coverage will vary from a light coating at 25 S.F. per gallon to a heavy coating at 10 S.F. per gallon. With most flashing work, the asphalt adhesive will cover an average of 15 S.F. for each layer or course. In many specifications the coverage of special materials like adhesives will be stated and should be used as the basis for the estimate.

Roof Accessories

- Hatches
- Skylights
- Vents
- Snow Guards

Single-Ply Roofing Membrane Installation Guide															
Generic Materials (Classification)	Compatible Substrates						Attachment Method				Sealing Method				
	Slip-Sheet Req'd.	Concrete	Exist. Asphalt Memb.	Insulation Board	Plywood	Spray Urethane Foam	Adhesive	Fully Adhered	Loose Laid/Ballast	Partially-Adhered	Adhesive	Hot Air Gun	Self-Sealing	Solvent	Torch Heating
Thermo Setting															
EPDM (Ethylene, propylene diene monomer)	X	X	X	X	X	X	X	X		X	X	X		X	X
Neoprene (synthetic rubber)	X	X		X	X		X	X	X		X				
PIB (Polyisobutylene)	X	X	X	X	X	X	X		X		X	X		X	
Thermo Plastic															
CSPE (Chlorosulfenated polyethyene)	X	X		X	X	X	X	X	X	X	X	X			
CPE (Chlorinated polyethylene)	X	X		X	X			X	X	X	X				
PVC (Ployvinyl chloride)	X	X		X	X	X			X	X	X			X	
Composites															
Glass reinforced EPDM/neoprene	X	X		X	X	X			X		X				
Modified bitumen/polyester	X		X	X	X			X			X	X			X
Modified bitumen/polyethylene & aluminum	X	X		X	X		X	X	X		X				X
Modified bitumen/polyethylene sheet	X	X		X	X							X			X
Modified CPE				X	X			X			X				
Non-woven glass reinforced PVC							X	X	X		X				
Nylon reinforced PVC		X		X	X				X			X		X	
Nylon reinforced/butyl or neoprene	X							X				X		X	
Polyester reinforce CPE	X	X	X	X	X	X			X	X	X	X			
Polyester reinforced PVC	X	X		X	X	X			X	X	X	X		X	
Rubber asphalt/plastic sheet	X	X	X	X	X			X					X		

Figure 9.79

Roof accessories must be considered as part of the complete weatherproofing system. Standard size accessories, such as ceiling, roof, and smoke vents or hatches, and snow guards, are priced per installed unit. Accessories that must be fabricated to meet project specifications may be priced per square foot, per linear foot or per unit. Skylight costs, for example, are listed by the square foot with unit costs decreasing in steps as the nominal size of individual units increases.

Skyroofs are priced on the same basis, but due to the many variations in the shape and construction of these units, costs are per square foot of surface area. These costs will vary with the size and type of unit, and in many cases, maximum and minimum costs give the estimator a range of prices for various design variations. Because there are many types and styles, the estimator must determine the exact specifications for the skyroof being priced. The accuracy of the total cost figure will depend entirely on the selection of the proper unit cost and calculation of the skyroof area. Skyroofs are becoming widely used in the industry and the work is growing more and more specialized. Often a particular manufacturer is specified. Specialty installing subcontractors are often factory authorized and required to perform the installation to maintain warranties and waterproof integrity.

Accessories such as roof drains, plumbing vents, and duct penetrations are usually installed by other appropriate subcontractors. However, costs for flashing and seating these items are often included by the roofing subcontractor.

When estimating Division 7, associated costs must be included for items that may not be directly stated in the specifications. Placement of materials, for example, may require the use of a crane or conveyors. Pitch pockets, sleepers, pads, and walkways may be required for rooftop equipment. Blocking and cant strips and items associated with different trades must also be coordinated. Once again, the estimator must visualize the construction process.

Sample Estimate: Division 7

The thermal and moisture protection portion of the sample building project is to be done partially by crews of the general contractor and partially by a subcontractor. The estimate sheets are shown in Figures 9.80 and 9.81. It is important to verify that the "units" used for takeoff are the same as those used for pricing. Note in Figure 9.82, the polyethylene vapor barrier is priced by the square (sq.). It could be an easy mistake to enter 20,800 (18,900 square feet of floor area plus 10% overlap) on the estimate sheet (Figure 9.80). Even if one person performs the takeoff and another the pricing, such an error could possibly occur in haste. Ideally time should be allowed for someone to always cross-check the work and calculations of another.

Smoke hatches are specified to be installed at the tops of the stair towers. Costs for smoke hatches in *Means Building Construction Cost Data* are given only as percentages to be added to the *bare costs* of roof hatches. See Figures 9.83 and 9.84. Calculations to add the percentages are shown in Figure 9.85. The bare costs are increased before overhead and profit are added. 10% is added to the bare material cost for handling. In Figure 9.83, the installing crew for the roof hatches is G3 (see Figure 9.86). The crews consists of sheet metal workers and laborers. The percentage added to the labor costs, for overhead and profit, is the average of the percentages for the two trades from Figure 9.87—54.2% and 55.6%.

The development and application of these percentages for overhead and profit are discussed in Chapter 4 and Chapter 7 of this book.

CONSOLIDATED ESTIMATE

PROJECT: Office Building
LOCATION:
TAKE OFF BY: ABC

CLASSIFICATION: Division 7
ARCHITECT:
QUANTITIES BY: ABC   PRICES BY: As Shown   EXTENSIONS BY:

ESTIMATE NO:
DATE: Jan-02
CHECKED BY: GHI

SOURCE			DESCRIPTION	QUANT	UNIT	MATERIAL DEF COST	MATERIAL TOTAL	LABOR DEF COST	LABOR TOTAL	EQUIPMENT DEF COST	EQUIPMENT TOTAL	SUBCONTRACT COST	SUBCONTRACT TOTAL	TOTAL COST	TOTAL
Division 7: Moisture & Thermal Protection															
			Waterproofing												
07110	100	0700	Asphalt Coating	3000	SF	0.29	870	0.53	1590						
07260	100	0900	Vapor Barrier - 6 mil poly	208	Sq.	2.93	609	6.50	1352						
			Insulation												
07210	950	0420	3-1/2" Fiberglass Exterior	17600	SF	0.34	5984	0.15	2640						
07210	950	0820	3-1/2" Fiberglass Interior	7620	Sf	0.21	1600	0.18	1372						
07220	700	0700	Roof Deck (Incl. Penthouse)	18900	SF			SUBCONTRACTED				1.86	35154		
			Roofing												
07510	300	0500	4 - ply Built-up	189	Sq.							199.00	37611		
07510	400	0010	4 x 4 Cant	800	LF							2.47	1976		
			Subtotals			$	9,064	$	6,954			$	74,741		

Figure 9.80

237

CONSOLIDATED ESTIMATE

PROJECT: Office Building
LOCATION:
TAKE OFF BY: ABC
QUANTITIES BY: ABC

CLASSIFICATION: Division 7
ARCHITECT:
PRICES BY: As Shown
EXTENSIONS BY:

ESTIMATE NO:
DATE: Jan-02
CHECKED BY: GHI

DESCRIPTION	SOURCE			QUANT	UNIT	MATERIAL DEF		LABOR DEF		EQUIPMENT DEF		SUBCONTRACT		TOTAL	
						COST	TOTAL	COST	TOTAL	COST	TOTAL	COST	TOTAL	COST	TOTAL
Division 7: (Cont'd)															
Sheet Metals															
Gravel Stop	07710	550	0350	800	LF			SUBCONTRACTED				8.00	6400		
Aluminum Flashing	07650	600	0100	200	SF							3.64	728		
Accessories															
Smoke Hatches (Stairs)	07720	700	1200	3	Ea.							1500	4500		
Smoke Vent (Elevator)	07720	860	0200	1	Ea.							1150	1150		
Subtotals													$ 12,778		
Sheet 1 Subtotals							$ 9,064		$ 6,954				$ 74,741		
Sheet 2 Subtotals													$ 12,778		
Division 7 Totals							$ 9,064		$ 6,954				$ 87,519		

Figure 9.81

07200 | Thermal Protection

07240 | Ext. Insulation/Finish Systems

			CREW	DAILY OUTPUT	LABOR-HOURS	UNIT	MAT.	LABOR	EQUIP.	TOTAL	TOTAL INCL O&P	
							2002 BARE COSTS					
100	0430	Crown moulding 16" high x 12" wide	1 Plas	150	.053	L.F.	5.65	1.50		7.15	8.55	**100**
	0440	For higher than one story, add						25%				

07260 | Vapor Retarders

			CREW	DAILY OUTPUT	LABOR-HOURS	UNIT	MAT.	LABOR	EQUIP.	TOTAL	TOTAL INCL O&P	
100	0010	**BUILDING PAPER** Aluminum and kraft laminated, foil 1 side	1 Carp	37	.216	Sq.	3.64	6.50		10.14	14.10	**100**
	0100	Foil 2 sides		37	.216		5.80	6.50		12.30	16.50	
	0300	Asphalt, two ply, 30#, for subfloors		19	.421		9.85	12.65		22.50	30.50	
	0400	Asphalt felt sheathing paper, 15#		37	.216		2.13	6.50		8.63	12.45	
	0450	Housewrap, exterior, spun bonded polypropylene										
	0470	Small roll	1 Carp	3,800	.002	S.F.	.16	.06		.22	.27	
	0480	Large roll	"	4,000	.002	"	.10	.06		.16	.20	
	0500	Material only, 3' x 111.1' roll				Ea.	52.50			52.50	58	
	0520	9' x 111.1' roll				"	96			96	105	
	0600	Polyethylene vapor barrier, standard, .002" thick	1 Carp	37	.216	Sq.	1.11	6.50		7.61	11.30	
	0700	.004" thick		37	.216		2.19	6.50		8.69	12.50	
	0900	.006" thick		37	.216		2.93	6.50		9.43	13.30	
	1200	.010" thick		37	.216		5.75	6.50		12.25	16.40	
	1300	Clear reinforced, fire retardant, .008" thick		37	.216		8.55	6.50		15.05	19.55	
	1350	Cross laminated type, .003" thick		37	.216		6.65	6.50		13.15	17.40	
	1400	.004" thick		37	.216		7.40	6.50		13.90	18.25	
	1500	Red rosin paper, 5 sq rolls, 4 lb per square		37	.216		1.58	6.50		8.08	11.85	
	1600	5 lbs. per square		37	.216		2.04	6.50		8.54	12.35	
	1800	Reinf. waterproof, .002" polyethylene backing, 1 side		37	.216		5	6.50		11.50	15.60	
	1900	2 sides		37	.216		6.65	6.50		13.15	17.40	
	2100	Roof deck vapor barrier, class 1 metal decks	1 Rofc	37	.216		8.40	5.75		14.15	19	
	2200	For all other decks	"	37	.216		6.25	5.75		12	16.70	
	2400	Waterproofed kraft with sisal or fiberglass fibers, minimum	1 Carp	37	.216		5.40	6.50		11.90	16.05	
	2500	Maximum	"	37	.216		13.50	6.50		20	25	

07300 | Shingles, Roof Tiles & Roof Coverings

07310 | Shingles

			CREW	DAILY OUTPUT	LABOR-HOURS	UNIT	MAT.	LABOR	EQUIP.	TOTAL	TOTAL INCL O&P	
							2002 BARE COSTS					
050	0010	**ALUMINUM** Shingles, mill finish, .019" thick	1 Carp	5	1.600	Sq.	150	48		198	240	**050**
	0100	.020" thick	"	5	1.600		149	48		197	238	
	0300	For colors, add					15.15			15.15	16.65	
	0600	Ridge cap, .024" thick	1 Carp	170	.047	L.F.	1.66	1.41		3.07	4.03	
	0700	End wall flashing, .024" thick		170	.047		1.29	1.41		2.70	3.62	
	0900	Valley section, .024" thick		170	.047		2.08	1.41		3.49	4.49	
	1000	Starter strip, .024" thick		400	.020		1.12	.60		1.72	2.16	
	1200	Side wall flashing, .024" thick		170	.047		1.28	1.41		2.69	3.61	
	1500	Gable flashing, .024" thick		400	.020		1.20	.60		1.80	2.25	
100	0010	**ASPHALT SHINGLES**										**100**
	0100	Standard strip shingles										
	0150	Inorganic, class A, 210-235 lb/sq **CN**	1 Rofc	5.50	1.455	Sq.	28.50	38.50		67	97	
	0155	Pneumatic nailed		7	1.143		28.50	30.50		59	82.50	
	0200	Organic, class C, 235-240 lb/sq		5	1.600		37.50	42.50		80	114	
	0205	Pneumatic nailed		6.25	1.280		37.50	34		71.50	99	
	0250	Standard, laminated multi-layered shingles										
	0300	Class A, 240-260 lb/sq	1 Rofc	4.50	1.778	Sq.	39.50	47.50		87	124	

Figure 9.82

239

07710 | Manufactured Roof Specialties

		Description	CREW	DAILY OUTPUT	LABOR-HOURS	UNIT	MAT.	LABOR	EQUIP.	TOTAL	TOTAL INCL O&P	
800	2400	Large, minimum	1 Rofc	9	.889	Ea.	112	23.50		135.50	163	800
	2450	Maximum	↓	3	2.667	↓	117	71		188	250	
	2500	Roof to wall joint with extruded aluminum cover	1 Shee	115	.070	L.F.	20.50	2.44		22.94	26.50	
	2600	See also divisions 03150 & 05810										
	2700	Wall joint, closed cell foam on PVC cover, 9" wide	1 Rofc	125	.064	L.F.	3.16	1.70		4.86	6.40	
	2800	12" wide	"	115	.070	"	3.57	1.85		5.42	7.10	

07720 | Roof Accessories

		Description	CREW	DAILY OUTPUT	LABOR-HOURS	UNIT	MAT.	LABOR	EQUIP.	TOTAL	TOTAL INCL O&P	
480	0010	**PITCH POCKETS**										480
	0100	Adjustable, 4" to 7", welded corners, 4" deep	1 Rofc	48	.167	Ea.	10.50	4.43		14.93	19.10	
	0200	Side extenders, 6"	"	240	.033	"	1.70	.89		2.59	3.38	
500	0010	**ROOF VENTS** Mushroom for built-up roofs, aluminum	1 Rofc	30	.267	Ea.	24	7.10		31.10	38.50	500
	0100	PVC, 6" high	"	30	.267	"	27.50	7.10		34.60	42.50	
550	0010	**RIDGE VENT**										550
	0100	Aluminum strips, mill finish	1 Rofc	160	.050	L.F.	1.15	1.33		2.48	3.53	
	0150	Painted finish		160	.050	"	2.12	1.33		3.45	4.60	
	0200	Connectors		48	.167	Ea.	1.88	4.43		6.31	9.60	
	0300	End caps		48	.167	"	.79	4.43		5.22	8.40	
	0400	Galvanized strips, with damper and bird screen		160	.050	L.F.	2.07	1.33		3.40	4.54	
	0430	Molded polyethylene, shingles not included		160	.050	"	2.55	1.33		3.88	5.05	
	0440	End plugs		48	.167	Ea.	.79	4.43		5.22	8.40	
	0450	Flexible roll, shingles not included	↓	160	.050	L.F.	1.99	1.33		3.32	4.45	
700	0010	**ROOF HATCHES** With curb, 1" fiberglass insulation, 2'-6" x 3'-0"										700
	0500	Aluminum curb and cover	G-3	10	3.200	Ea.	410	93.50		503.50	595	
	0520	Galvanized steel curb and aluminum cover CN		10	3.200		340	93.50		433.50	520	
	0540	Galvanized steel curb and cover		10	3.200		370	93.50		463.50	550	
	0600	2'-6" x 4'-6", aluminum curb and cover		9	3.556		580	104		684	800	
	0800	Galvanized steel curb and aluminum cover		9	3.556		485	104		589	695	
	0900	Galvanized steel curb and cover		9	3.556		575	104		679	795	
	1200	2'-6" x 8'-0", aluminum curb and cover		6.60	4.848		1,150	142		1,292	1,500	
	1400	Galvanized steel curb and aluminum cover		6.60	4.848		935	142		1,077	1,250	
	1500	Galvanized steel curb and cover	↓	6.60	4.848		945	142		1,087	1,275	
	1800	For plexiglass panels, 2'-6" x 3'-0", add to above				↓	345			345	380	
800	0010	**WALKWAY** For built-up roofs, asphalt impregnated, 3' x 6' x 1/2" thk	1 Rofc	400	.020	S.F.	.99	.53		1.52	2	800
	0100	3' x 3' x 3/4" thick	"	400	.020		1.50	.53		2.03	2.56	
	0300	Concrete patio blocks, 2" thick, natural	1 Clab	115	.070		1.39	1.63		3.02	4.07	
	0400	Colors	"	115	.070	↓	1.75	1.63		3.38	4.47	
850	0010	**SMOKE HATCHES** Unlabeled, not including hand winch operator										850
	0200	For 3'-0" long, add to roof hatches from division 07720-700				Ea.	25%	5%				
	0300	For 8'-0" long, add to roof hatches from division 07720-700				"	10%	5%				
860	0010	**SMOKE VENT**, insulated, 4' x 4'										860
	0100	Aluminum cover and frame	G-3	13	2.462	Ea.	1,050	72		1,122	1,250	
	0200	Galvanized steel cover and frame		13	2.462		950	72		1,022	1,150	
	0300	4' x 8' aluminum cover and frame		8	4		1,425	117		1,542	1,750	
	0400	Galvanized steel cover and frame	↓	8	4	↓	1,250	117		1,367	1,550	
870	0010	**VENTS, ONE-WAY** For insul. decks, 1 per M.S.F., plastic, min.	1 Rofc	40	.200	Ea.	12.50	5.30		17.80	23	870
	0100	Maximum		20	.400		28.50	10.65		39.15	49.50	
	0300	Aluminum	↓	30	.267		12.50	7.10		19.60	26	
	0800	Polystyrene baffles, 12" wide for 16" O.C. rafter spacing	1 Carp	90	.089		.45	2.67		3.12	4.65	
	0900	For 24" O.C. rafter spacing	"	110	.073	↓	1.05	2.18		3.23	4.56	

Figure 9.83

07812	Cementitious Fireproofing	CREW	DAILY OUTPUT	LABOR-HOURS	UNIT	2002 BARE COSTS				TOTAL INCL O&P
						MAT.	LABOR	EQUIP.	TOTAL	
600	0010	**SPRAYED** Mineral fiber or cementitious for fireproofing,									600
	0050	not incl tamping or canvas protection									
	0100	1" thick, on flat plate steel	G-2	3,000	.008	S.F.	.40	.20	.04	.64	.79
	0200	Flat decking		2,400	.010		.40	.25	.04	.69	.88
	0400	Beams		1,500	.016		.40	.40	.07	.87	1.14
	0500	Corrugated or fluted decks		1,250	.019		.61	.48	.08	1.17	1.50
	0700	Columns, 1-1/8" thick		1,100	.022		.45	.55	.10	1.10	1.45
	0800	2-3/16" thick	▼	700	.034	▼	.88	.86	.15	1.89	2.46
	0850	For tamping, add						10%			
	0900	For canvas protection, add	G-2	5,000	.005	S.F.	.06	.12	.02	.20	.28
	1000	Acoustical sprayed, 1" thick, finished, straight work, minimum		520	.046		.44	1.16	.20	1.80	2.48
	1100	Maximum		200	.120		.47	3.01	.52	4	5.75
	1300	Difficult access, minimum		225	.107		.47	2.68	.47	3.62	5.15
	1400	Maximum	▼	130	.185	▼	.51	4.63	.81	5.95	8.60
	1500	Intumescent epoxy fireproofing on wire mesh, 3/16" thick									
	1550	1 hour rating, exterior use	G-2	136	.176	S.F.	5.40	4.43	.77	10.60	13.60
	1600	Magnesium oxychloride, 35# to 40# density, 1/4" thick		3,000	.008		1.14	.20	.04	1.38	1.60
	1650	1/2" thick		2,000	.012		2.28	.30	.05	2.63	3.03
	1700	60# to 70# density, 1/4" thick		3,000	.008		1.50	.20	.04	1.74	2
	1750	1/2" thick		2,000	.012		3.03	.30	.05	3.38	3.85
	2000	Vermiculite cement, troweled or sprayed, 1/4" thick		3,000	.008		1.03	.20	.04	1.27	1.48
	2050	1/2" thick	▼	2,000	.012	▼	2.04	.30	.05	2.39	2.76

07840	Firestopping									
100	0010	**FIRESTOPPING** R07800 -030									100
	0100	Metallic piping, non insulated									
	0110	Through walls, 2" diameter	1 Carp	16	.500	Ea.	9.60	15		24.60	34
	0120	4" diameter		14	.571		14.65	17.15		31.80	42.50
	0130	6" diameter		12	.667		19.70	20		39.70	52.50
	0140	12" diameter		10	.800		35	24		59	76
	0150	Through floors, 2" diameter		32	.250		5.80	7.50		13.30	18.10
	0160	4" diameter		28	.286		8.35	8.55		16.90	22.50
	0170	6" diameter		24	.333		11	10		21	27.50
	0180	12" diameter	▼	20	.400	▼	18.50	12		30.50	39
	0190	Metallic piping, insulated									
	0200	Through walls, 2" diameter	1 Carp	16	.500	Ea.	13.60	15		28.60	38.50
	0210	4" diameter		14	.571		18.65	17.15		35.80	47
	0220	6" diameter		12	.667		23.50	20		43.50	57
	0230	12" diameter		10	.800		39	24		63	80
	0240	Through floors, 2" diameter		32	.250		9.80	7.50		17.30	22.50
	0250	4" diameter		28	.286		12.35	8.55		20.90	27
	0260	6" diameter		24	.333		15	10		25	32
	0270	12" diameter	▼	20	.400	▼	18.50	12		30.50	39
	0280	Non metallic piping, non insulated									
	0290	Through walls, 2" diameter	1 Carp	12	.667	Ea.	39.50	20		59.50	74.50
	0300	4" diameter		10	.800		49.50	24		73.50	92
	0310	6" diameter		8	1		69	30		99	123
	0330	Through floors, 2" diameter		16	.500		31	15		46	57.50
	0340	4" diameter		6	1.333		38.50	40		78.50	105
	0350	6" diameter	▼	6	1.333	▼	46	40		86	113
	0370	Ductwork, insulated & non insulated, round									
	0380	Through walls, 6" diameter	1 Carp	12	.667	Ea.	20	20		40	53
	0390	12" diameter		10	.800		40	24		64	81.50
	0400	18" diameter		8	1		65	30		95	118
	0410	Through floors, 6" diameter		16	.500		11	15		26	35.50
	0420	12" diameter	▼	14	.571	▼	20	17.15		37.15	48.50

Figure 9.84

Any one door assembly (door, frame, hardware) can be one of hundreds of combinations of many variable features:

Door:	Frame:	Hardware:
Size	Size	Lockset
Thickness	Throat	Passage set
Wood-type	Wood-type	Panic bar
Metal-gauge	Metal-gauge	Closer
Laminate	Casing	Hinges
Hollow-core type	Stops	Stops
Solid-core material	Fire rating	Bolts
Fire rating	Knock down	Finish
Finish	Welded	Plates

Most architectural plans and specifications include door, window, and hardware schedules to tabulate these combinations. The estimator should use these schedules and details in conjunction with the plans to avoid duplication or omission of units when determining the quantities. The schedules should identify the location, size, and type of each unit. Schedules should also include information regarding the frame, fire rating, hardware, and special notes. If no such schedules are included, the estimator should prepare them in order to provide an accurate quantity takeoff. Figure 9.88 is an example of a form that may be used to prepare such a schedule. Most suppliers will prepare separate schedules for approval by the architect or owner.

Metal Doors and Frames

A proper door schedule on the architectural drawings identifies each opening in detail. Define each opening in accordance with the items in the schedule and any other pertinent data. Installation information should be carefully reviewed in the specifications.

For the quantity survey, combine all similar doors and frames, checking each off as you go to ensure none has been left out. An easy and obvious check is to count the total number of openings, making certain that two doors and only one frame have been included where double doors are used. Important details to check for both door and frame are:

- Material
- Gauge
- Size
- Core Material
- Fire Rating Label
- Finish
- Style

Line Number	Bare Costs		Total Including O & P
	Material	Labor	
07720-700-1200	$1,150.00	$142.00	
07720-850-0300	(10%) 115.00	(5%) 7.10	
	1,265.00	149.10	
Overhead & Profit	(10%) 126.50	(54.8%) 81.71	
	$1,391.50	$230.81	$1622.31

Figure 9.85

Crew No.	Bare Costs		Incl. Subs O & P		Cost Per Labor-Hour	

Crew E-17

	Hr.	Daily	Hr.	Daily	Bare Costs	Incl. O&P
1 Structural Steel Foreman	$36.25	$290.00	$65.35	$522.80	$35.25	$63.55
1 Structural Steel Worker	34.25	274.00	61.75	494.00		
1 Power Tool		3.40		3.75	.21	.23
16 L.H., Daily Totals		$567.40		$1020.55	$35.46	$63.78

Crew E-18

	Hr.	Daily	Hr.	Daily	Bare Costs	Incl. O&P
1 Structural Steel Foreman	$36.25	$290.00	$65.35	$522.80	$34.04	$59.55
3 Structural Steel Workers	34.25	822.00	61.75	1482.00		
1 Equipment Operator (med.)	31.20	249.60	47.15	377.20		
1 Crane, 20 Ton		588.55		647.40	14.71	16.19
40 L.H., Daily Totals		$1950.15		$3029.40	$48.75	$75.74

Crew E-19

	Hr.	Daily	Hr.	Daily	Bare Costs	Incl. O&P
1 Structural Steel Worker	$34.25	$274.00	$61.75	$494.00	$33.43	$57.38
1 Structural Steel Foreman	36.25	290.00	65.35	522.80		
1 Equip. Oper. (light)	29.80	238.40	45.05	360.40		
1 Power Tool		3.40		3.75		
1 Crane, 20 Ton		588.55		647.40	24.66	27.13
24 L.H., Daily Totals		$1394.35		$2028.35	$58.09	$84.51

Crew E-20

	Hr.	Daily	Hr.	Daily	Bare Costs	Incl. O&P
1 Structural Steel Foreman	$36.25	$290.00	$65.35	$522.80	$33.31	$57.91
5 Structural Steel Workers	34.25	1370.00	61.75	2470.00		
1 Equip. Oper. (crane)	32.35	258.80	48.90	391.20		
1 Oiler	26.65	213.20	40.25	322.00		
1 Power Tool		3.40		3.75		
1 Crane, 40 Ton		847.60		932.35	13.30	14.63
64 L.H., Daily Totals		$2983.00		$4642.10	$46.61	$72.54

Crew E-22

	Hr.	Daily	Hr.	Daily	Bare Costs	Incl. O&P
1 Skilled Worker Foreman	$32.95	$263.60	$51.50	$412.00	$31.62	$49.40
2 Skilled Worker	30.95	495.20	48.35	773.60		
24 L.H., Daily Totals		$758.80		$1185.60	$31.62	$49.40

Crew E-24

	Hr.	Daily	Hr.	Daily	Bare Costs	Incl. O&P
3 Structural Steel Worker	$34.25	$822.00	$61.75	$1482.00	$33.49	$58.10
1 Equipment Operator (medium)	31.20	249.60	47.15	377.20		
1-25 Ton Crane		724.80		797.30	22.65	24.92
32 L.H., Daily Totals		$1796.40		$2656.50	$56.14	$83.02

Crew E-25

	Hr.	Daily	Hr.	Daily	Bare Costs	Incl. O&P
1 Welder Foreman	$36.25	$290.00	$65.35	$522.80	$36.25	$65.35
1 Cutting Torch		28.00		30.80		
1 Gases		34.00		37.40	7.75	8.53
8 L.H., Daily Totals		$352.00		$591.00	$44.00	$73.88

Crew F-3

	Hr.	Daily	Hr.	Daily	Bare Costs	Incl. O&P
4 Carpenters	$30.00	$960.00	$46.70	$1494.40	$30.47	$47.14
1 Equip. Oper. (crane)	32.35	258.80	48.90	391.20		
1 Hyd. Crane, 12 Ton		607.60		668.35	15.19	16.71
40 L.H., Daily Totals		$1826.40		$2553.95	$45.66	$63.85

Crew F-4

	Hr.	Daily	Hr.	Daily	Bare Costs	Incl. O&P
4 Carpenters	$30.00	$960.00	$46.70	$1494.40	$29.83	$45.99
1 Equip. Oper. (crane)	32.35	258.80	48.90	391.20		
1 Equip. Oper. Oiler	26.65	213.20	40.25	322.00		
1 Hyd. Crane, 55 Ton		972.40		1069.65	20.26	22.28
48 L.H., Daily Totals		$2404.40		$3277.25	$50.09	$68.27

Crew F-5

	Hr.	Daily	Hr.	Daily	Bare Costs	Incl. O&P
1 Carpenter Foreman	$32.00	$256.00	$49.80	$398.40	$30.50	$47.48
3 Carpenters	30.00	720.00	46.70	1120.80		
32 L.H., Daily Totals		$976.00		$1519.20	$30.50	$47.48

Crew F-6

	Hr.	Daily	Hr.	Daily	Bare Costs	Incl. O&P
2 Carpenters	$30.00	$480.00	$46.70	$747.20	$27.85	$43.06
2 Building Laborers	23.45	375.20	36.50	584.00		
1 Equip. Oper. (crane)	32.35	258.80	48.90	391.20		
1 Hyd. Crane, 12 Ton		607.60		668.35	15.19	16.71
40 L.H., Daily Totals		$1721.60		$2390.75	$43.04	$59.77

Crew F-7

	Hr.	Daily	Hr.	Daily	Bare Costs	Incl. O&P
2 Carpenters	$30.00	$480.00	$46.70	$747.20	$26.73	$41.60
2 Building Laborers	23.45	375.20	36.50	584.00		
32 L.H., Daily Totals		$855.20		$1331.20	$26.73	$41.60

Crew G-1

	Hr.	Daily	Hr.	Daily	Bare Costs	Incl. O&P
1 Roofer Foreman	$28.60	$228.80	$48.65	$389.20	$24.94	$42.44
4 Roofers, Composition	26.60	851.20	45.25	1448.00		
2 Roofer Helpers	19.80	316.80	33.70	539.20		
1 Application Equipment		159.60		175.55		
1 Tar Kettle/Pot		61.75		67.95		
1 Crew Truck		81.40		89.55	5.41	5.95
56 L.H., Daily Totals		$1699.55		$2709.45	$30.35	$48.39

Crew G-2

	Hr.	Daily	Hr.	Daily	Bare Costs	Incl. O&P
1 Plasterer	$28.10	$224.80	$43.10	$344.80	$25.08	$38.65
1 Plasterer Helper	23.70	189.60	36.35	290.80		
1 Building Laborer	23.45	187.60	36.50	292.00		
1 Grouting Equipment		104.95		115.45	4.37	4.81
24 L.H., Daily Totals		$706.95		$1043.05	$29.45	$43.46

Crew G-3

	Hr.	Daily	Hr.	Daily	Bare Costs	Incl. O&P
2 Sheet Metal Workers	$35.10	$561.60	$54.10	$865.60	$29.27	$45.30
2 Building Laborers	23.45	375.20	36.50	584.00		
32 L.H., Daily Totals		$936.80		$1449.60	$29.27	$45.30

Crew G-4

	Hr.	Daily	Hr.	Daily	Bare Costs	Incl. O&P
1 Labor Foreman (outside)	$25.45	$203.60	$39.60	$316.80	$24.12	$37.53
2 Building Laborers	23.45	375.20	36.50	584.00		
1 Light Truck, 1.5 Ton		155.60		171.15		
1 Air Compr., 160 C.F.M.		97.20		106.90	10.53	11.59
24 L.H., Daily Totals		$831.60		$1178.85	$34.65	$49.12

Crew G-5

	Hr.	Daily	Hr.	Daily	Bare Costs	Incl. O&P
1 Roofer Foreman	$28.60	$228.80	$48.65	$389.20	$24.28	$41.31
2 Roofers, Composition	26.60	425.60	45.25	724.00		
2 Roofer Helpers	19.80	316.80	33.70	539.20		
1 Application Equipment		159.60		175.55	3.99	4.39
40 L.H., Daily Totals		$1130.80		$1827.95	$28.27	$45.70

Crew G-6A

	Hr.	Daily	Hr.	Daily	Bare Costs	Incl. O&P
2 Roofers Composition	$26.60	$425.60	$45.25	$724.00	$26.60	$45.25
1 Small Compressor		19.15		21.05		
2 Pneumatic Nailers		40.90		45.00	3.75	4.13
16 L.H., Daily Totals		$485.65		$790.05	$30.35	$49.38

Figure 9.86

Installing Contractor's Overhead & Profit

Below are the **average** installing contractor's percentage mark-ups applied to base labor rates to arrive at typical billing rates.

Column A: Labor rates are based on union wages averaged for 30 major U.S. cities. Base rates including fringe benefits are listed hourly and daily. These figures are the sum of the wage rate and employer-paid fringe benefits such as vacation pay, employer-paid health and welfare costs, pension costs, plus appropriate training and industry advancement funds costs.

Column B: Workers' Compensation rates are the national average of state rates established for each trade.

Column C: Column C lists average fixed overhead figures for all trades. Included are Federal and State Unemployment costs set at 7.0%; Social Security Taxes (FICA) set at 7.65%; Builder's Risk Insurance costs set at 0.34%; and Public Liability costs set at 1.55%. All the percentages except those for Social Security Taxes vary from state to state as well as from company to company.

Columns D and E: Percentages in Columns D and E are based on the presumption that the installing contractor has annual billing of $1,500,000 and up. Overhead percentages may increase with smaller annual billing. The overhead percentages for any given contractor may vary greatly and depend on a number of factors, such as the contractor's annual volume, engineering and logistical support costs, and staff requirements. The figures for overhead and profit will also vary depending on the type of job, the job location, and the prevailing economic conditions. All factors should be examined very carefully for each job.

Column F: Column F lists the total of Columns B, C, D, and E.

Column G: Column G is Column A (hourly base labor rate) multiplied by the percentage in Column F (O&P percentage).

Column H: Column H is the total of Column A (hourly base labor rate) plus Column G (Total O&P).

Column I: Column I is Column H multiplied by eight hours.

		A Base Rate Incl. Fringes		B Workers' Comp. Ins.	C Average Fixed Over-head	D Over-head	E Profit	F Total Overhead & Profit	G	H Rate with O & P	I
Abbr.	Trade	Hourly	Daily					%	Amount	Hourly	Daily
Skwk	Skilled Workers Average (35 trades)	$30.95	$247.60	16.8%	16.5%	13.0%	10%	56.3%	$17.40	$48.35	$386.80
	Helpers Average (5 trades)	22.75	182.00	18.5		11.0		56.0	12.75	35.50	284.00
	Foreman Average, Inside ($.50 over trade)	31.45	251.60	16.8		13.0		56.3	17.70	49.15	393.20
	Foreman Average, Outside ($2.00 over trade)	32.95	263.60	16.8		13.0		56.3	18.55	51.50	412.00
Clab	Common Building Laborers	23.45	187.60	18.1		11.0		55.6	13.05	36.50	292.00
Asbe	Asbestos/Insulation Workers/Pipe Coverers	33.45	267.60	16.2		16.0		58.7	19.65	53.10	424.80
Boil	Boilermakers	36.25	290.00	14.7		16.0		57.2	20.75	57.00	456.00
Bric	Bricklayers	30.50	244.00	16.0		11.0		53.5	16.30	46.80	374.40
Brhe	Bricklayer Helpers	23.50	188.00	16.0		11.0		53.5	12.55	36.05	288.40
Carp	Carpenters	30.00	240.00	18.1		11.0		55.6	16.70	46.70	373.60
Cefi	Cement Finishers	28.70	229.60	10.6		11.0		48.1	13.80	42.50	340.00
Elec	Electricians	35.45	283.60	6.7		16.0		49.2	17.45	52.90	423.20
Elev	Elevator Constructors	37.10	296.80	7.7		16.0		50.2	18.60	55.70	445.60
Eqhv	Equipment Operators, Crane or Shovel	32.35	258.80	10.6		14.0		51.1	16.55	48.90	391.20
Eqmd	Equipment Operators, Medium Equipment	31.20	249.60	10.6		14.0		51.1	15.95	47.15	377.20
Eqlt	Equipment Operators, Light Equipment	29.80	238.40	10.6		14.0		51.1	15.25	45.05	360.40
Eqol	Equipment Operators, Oilers	26.65	213.20	10.6		14.0		51.1	13.60	40.25	322.00
Eqmm	Equipment Operators, Master Mechanics	32.80	262.40	10.6		14.0		51.1	16.75	49.55	396.40
Glaz	Glaziers	30.00	240.00	13.8		11.0		51.3	15.40	45.40	363.20
Lath	Lathers	28.75	230.00	11.1		11.0		48.6	13.95	42.70	341.60
Marb	Marble Setters	30.10	240.80	16.0		11.0		53.5	16.10	46.20	369.60
Mill	Millwrights	31.75	254.00	10.6		11.0		48.1	15.25	47.00	376.00
Mstz	Mosaic & Terrazzo Workers	29.25	234.00	9.8		11.0		47.3	13.85	43.10	344.80
Pord	Painters, Ordinary	27.15	217.20	13.8		11.0		51.3	13.95	41.10	328.80
Psst	Painters, Structural Steel	27.90	223.20	48.4		11.0		85.9	23.95	51.85	414.80
Pape	Paper Hangers	27.10	216.80	13.8		11.0		51.3	13.90	41.00	328.00
Pile	Pile Drivers	29.80	238.40	24.9		16.0		67.4	20.10	49.90	399.20
Plas	Plasterers	28.10	224.80	15.8		11.0		53.3	15.00	43.10	344.80
Plah	Plasterer Helpers	23.70	189.60	15.8		11.0		53.3	12.65	36.35	290.80
Plum	Plumbers	35.95	287.60	8.3		16.0		50.8	18.25	54.20	433.60
Rodm	Rodmen (Reinforcing)	34.25	274.00	28.3		14.0		68.8	23.55	57.80	462.40
Rofc	Roofers, Composition	26.60	212.80	32.6		11.0		70.1	18.65	45.25	362.00
Rots	Roofers, Tile & Slate	26.75	214.00	32.6		11.0		70.1	18.75	45.50	364.00
Rohe	Roofers, Helpers (Composition)	19.80	158.40	32.6		11.0		70.1	13.90	33.70	269.60
Shee	Sheet Metal Workers	35.10	280.80	11.7		16.0		54.2	19.00	54.10	432.80
Spri	Sprinkler Installers	36.20	289.60	8.7		16.0		51.2	18.55	54.75	438.00
Stpi	Steamfitters or Pipefitters	36.20	289.60	8.3		16.0		50.8	18.40	54.60	436.80
Ston	Stone Masons	30.65	245.20	16.0		11.0		53.5	16.40	47.05	376.40
Sswk	Structural Steel Workers	34.25	274.00	39.8		14.0		80.3	27.50	61.75	494.00
Tilf	Tile Layers	29.15	233.20	9.8		11.0		47.3	13.80	42.95	343.60
Tilh	Tile Layers Helpers	23.35	186.80	9.8		11.0		47.3	11.05	34.40	275.20
Trlt	Truck Drivers, Light	24.30	194.40	14.9		11.0		52.4	12.75	37.05	296.40
Trhv	Truck Drivers, Heavy	25.00	200.00	14.9		11.0		52.4	13.10	38.10	304.80
Sswl	Welders, Structural Steel	34.25	274.00	39.8		14.0		80.3	27.50	61.75	494.00
Wrck	*Wrecking	23.45	187.60	41.2		11.0		78.7	18.45	41.90	335.20

*Not included in averages

Figure 9.87

DOOR AND FRAME SCHEDULE

PROJECT

LOCATION

ARCHITECT

OWNER

PAGE

DATE

BY

OF

| DOOR NO. | SIZE | | | DOOR | | | | | FRAME | | | | | FIRE RATING | | HARDWARE | | REMARKS |
	W	H	T	MAT.	TYPE	GLASS	LOUVER	MAT.	TYPE	JAMB	HEAD	SILL	LAB	CON	SET NO.	KEYSIDE ROOM NO.	

Figure 9.88

245

Wood and Plastic Doors

The quantity survey for wood and plastic laminated doors is identical to that of metal doors. Where local work rules permit, pre-hung doors and windows are becoming prevalent in the industry. For these, locksets and interior casings are usually extra. As these may be standard for a number of doors in any particular building, they need only be counted. Remember that exterior pre-hung doors need casings on the interior.

Leave a space in the tabulation on the Quantity Sheet for casings, stops, grounds, and hardware. This can be done either on the same sheet or on separate sheets.

Special Doors

There are many types of specialty doors that may be included—for example, sliding glass doors, overhead garage doors and bulkhead doors. These items should be taken off individually. The estimator should thoroughly examine the plans and specifications to be sure to include all hardware, operating mechanisms, fire ratings, finishes, and any special installation requirements.

Fire Doors

The estimator must pay particular attention to fire doors when performing the quantity takeoff. It is important to determine the exact type of door required. Figure 9.89 is a table describing various types of fire doors. Please note that a "B" label door can be one of four types. If the plans or door schedule do not specify exactly which temperature rise is required, the estimator should consult the architect or local building inspector. Many building and fire codes also require that frames and hardware at fire doors be fire-rated and labelled as such. When determining quantities, the estimator must also include any glass (usually wired) or special inserts to be installed in fire doors (or in any doors).

Entrances and Storefronts

Entrances and storefronts are almost all special designs and combinations of unit items to fit a unique situation. The estimator should submit the plans and specifications to a specialty installer for takeoff and pricing.

The general procedure for the installer's takeoff is:

For stationary units:

- Determine height and width of each like unit.
- Determine linear feet of intermediate, horizontal, and vertical members, rounded to next higher foot.
- Determine number of joints,

For entrance units:

- Determine number of joints.
- Determine special frame hardware per unit.
- Determine special door hardware per unit.
- Determine thresholds and closers.

Windows

As with doors, a window schedule should be included in the architectural drawings. Items that merit special attention are:

- Material
- Gauge/Thickness
- Screens

	Fire Door			
Classification	Time Rating (as shown on Label)		Temperature Rise (as shown on Label)	Maximum Glass Area
3 Hour fire doors (A) are for use in openings in walls separating buildings or dividing a single building into the areas.	3 Hr.	(A)	30 min. 250°F Max	None
	3 Hr.	(A)	30 Min. 450°F Max	
	3 Hr.	(A)	30 Min. 650°F Max	
	3 Hr.	(A)	*	
1-1/2 Hour fire doors (B) and (D) are for use in openings in 2 Hour enclosures of vertical communication through buildings (stairs, elevators, etc.) or in exterior walls which are subject to severe fire exposure from outside of the building. 1 Hour fire doors (B) are for use in openings in 1 Hour enclosures of vertical communication through buildings (stairs, elevators, etc.)	1-1/2 Hr.	(B)	30 Min. 250°F Max	100 square inches per door
	1-1/2 Hr.	(B)	30 Min. 450°F Max	
	1-1/2 Hr.	(B)	30 Min. 650°F Max	
	1-1/2 Hr.	(B)	*	
	1 Hr.		30 Min. 250°F Max	
	1-1/2 Hr.	(D)	30 Min. 250°F Max	None
	1-1/2 Hr.	(D)	30 Min. 450°F Max	
	1-1/2 Hr.	(D)	30 Min. 650°F Max	
	1-1/2 Hr.	(D)	*	
3/4 Hour fire doors (C) and (E) are for use in openings in corridor and room partitions or in exterior walls which are subject to moderate fire exposure from outside of the building.	3/4 Hr.	(C)	**	1296 square
	3/4 Hr.	(E)	**	720 square inches per light
1/2 Hour fire doors and 1/3 Hour fire doors are for use where smoke control is a primary consideration and are for the protection of openings in partitions between a habitable room and a corridor when the wall has a fire-resistance rating of not more than one hour.	1/2 Hr.		**	No limit
	1/3 Hr.		**	

*The labels do not record any temperature rise limits. This means that the temperature rise on the unexposed face of the door at the end of 30 minutes of test is in excess of 650°F.
**Temperature rise is not recorded.

Figure 9.89

247

- Glazing (type of glass and setting specifications)
- Trim
- Hardware
- Special installation requirements

When using pre-hung units, be sure to add interior trim and stools.

Finish Hardware and Specialties

The estimator should list the hardware separately or on the door and window schedule. Remember that most pre-hung doors and windows do not include locksets. Some casement, awning, and jalousie windows include cranks and locks.

Be sure to check the specifications for:

- Base Metal
- Finish
- Service (Heavy, Light, Medium)
- Any other special detail

Also check the specifications and code for both handicap and exit requirements. Metal thresholds and astragals are also included in this division. Weatherstripping may or may not be included with pre-hung doors and windows.

Glass and Glazing

Glazing quantities are a function of the material, method, and length to be glazed. Therefore, quantities are measured in united inches or united feet (length + width). Many installers, however, figure all glass and glazing by the square foot. Be sure to read the specifications carefully, as there are many different grades, thicknesses, and other variables in glass.

The types of glass include tempered, plate, safety, insulated, tinted, and various combinations of the above.

Sample Estimate: Division 8

There are two basic ways that doors may be shown on a schedule. The first is to list each type of door (usually accompanied by elevations) that have common characteristics. An example of this method is shown for the sample project in Figure 9.90. Quantities of each type are not shown. The second method is to list each door individually by door number. An example of this method (for another project) is shown in Figure 9.91. In both cases, the information on the schedules provides parameters for takeoff. The specifications will provide much more detailed information that will be required before pricing.

The estimate sheets for Division 8 are shown in Figures 9.92 to 9.94. Note on Sheet 2 of the estimate (Figure 9.93) that costs for hinges are for material only. Labor costs for hinges are usually included in the installation costs for doors. To a carpenter, hanging a door and installing the hinges is one operation. The costs in *Means Building Construction Cost Data* are listed accordingly.

In Figure 9.95 a percentage is to be added to the aluminum entrances for black anodized finish. This percentage is applied to bare material cost. Subsequently, 10% for handling is added:

Bare Material Cost: $650/pr.

08411-140-0400

For black finish: add 36%	$234.00

08411-140-1500

O & P (material handling): add 10%.	23.40
Total incl. O & P for black finish	$257.40

Window/curtain walls are almost always custom fabricated for each project. Relatively few manufacturers supply and install these highly specialized systems, so architects will often design curtain walls based on the specifications of a particular system. These reasons dictate that a firm subcontractor quotation is required for estimating curtain walls. For budget purposes, historical costs or those in *Means Building Construction Cost Data* may be used. The telephone quotation (which would be followed by a detailed written quotation) for the sample project is shown in Figure 9.96.

Division 9: Finishes

Some buildings today are built "on spec," or speculatively, before they are partially or fully tenanted. In this case, interior work (primarily finishes, and electrical and mechanical distribution) is not usually completed until a tenant is secured. "Interior contractors" are becoming more prevalent in the industry. This type of firm may perform or subcontract all work in Division 9, and may

Door and Frame Schedule								
	Door			Frame			Hardware Set	Remarks
	Size	Type	Rating	Type	Throat	Rating		
A	6° x 7°	Alum.	—	Alum.	—	—	—	Double Door w/2° x 6° Transom
B	3° x 7°	H.M. 18 ga.	—	H.M. 16 ga.	8″	—	4-1	Insulated
C	3° x 7°	H.M. 18 ga.	"B" 1-1/2 hr.	H.M. 16 ga.	8″	"B" 1-1/2 hr.	4-1	10″ x 10″ Lites
D	3° x 7°	Oak Veneer	"B" 1-1/2 hr.	H.M. 16 ga.	4-3/4″	"B" 1-1/2 hr.	4-2	10″ x 10″ Lites
E	3° x 7°	Oak Veneer	"B" 1 hr.	H.M. 16 ga.	4-3/4″	"B" 1 hr.	4-3	
F	3° x 7°	Oak Veneer	"B" 1 hr.	H.M. 16 ga.	4-3/4″	"B" 1 hr.	4-4	
G	3° x 7°	H.M. 18 ga.	"B" 1 hr.	H.M. 16 ga.	8″	"B" 1 hr.	4-5	
H	6° x 7°	H.M. 18 ga.	"B" 1 hr.	H.M. 16 ga.	8″	"B" 1 hr.	4-6	Double Door

Figure 9.90

								Hardware	
Door	Size	Type	Rating	Frame	Depth	Rating		Set	Remarks
B01	3° x 6⁸	Flush Steel 18 ga.	"B" 1-1/2 Hr.	Existing	—	—		HW-1	10" x 10" Vision Lite Shop-Primed
B02	3° x 6⁸	Flush Steel 18 ga.	"B" 1 Hr.	HMKD 16 ga.	4-7/8"	"B" 1 Hr.		HW-2	Shop-Primed
B03	3° x 6⁸	Flush Steel 18 ga.	"B" 1 Hr.	H.M. Welded 16 ga.	8"	"B" 1 Hr.		HW-3	w/Masonry Anchors Shop-Primed
B04	3° x 6⁸	Flush Steel 18 ga.	"B" 1-1/2 Hr.	H.M. Welded 16 ga.	8"	"B" 1-1/2 Hr.		HW-3	w/Masonry Anchors Shop-Primed
B05	3° x 6⁸	Flush Steel 18 ga.	"B" 1-1/2 Hr.	H.M. Welded 16 ga.	8"	"B" 1-1/2 Hr.		HW-3	w/Masonry Anchors Shop-Primed
B06	3° x 6⁸	Flush Steel 18 ga.	"B" 1 Hr.	HMKD 16 ga.	4-7/8"	"B" 1 Hr.		HW-2	Shop-Primed
101	3° x 6⁸	Flush Steel 18 ga.	—	HMKD 16 ga.	4-7/8"	—		HW-4	Transom Frame Above w/Masonry Anchors
102	3° x 6⁸	Flush Steel 18 ga.	"B" 1-1/2 Hr.	Existing	—	—		HW-1	Shop-Primed
103	3° x 6⁸	Flush Oak Face	"B" 1 Hr.	HMKD 16 ga.	4-7/8"	"B" 1 Hr.		HW-5	
104	2° x 6⁸	Flush Oak Face SC	—	HMKD 16 ga.	4-5/8"	—		HW-6	
105	3° x 6⁸	Flush Oak Face	"B" 1 Hr.	HMKD 16 ga.	4-7/8"	"B" 1 Hr.		HW-5	

Door Schedule

Figure 9.91

CONSOLIDATED ESTIMATE

PROJECT: Office Building   CLASSIFICATION: Division 8   ESTIMATE NO:
LOCATION: ARCHITECT: DATE: Jan-02
TAKE OFF BY: ABC   QUANTITIES BY: ABC   PRICES BY: ABC   CHECKED BY: GHI

EXTENSIONS BY:

DESCRIPTION	SOURCE	QUANT	UNIT	MATERIAL COST	MATERIAL DEF TOTAL	LABOR COST	LABOR DEF TOTAL	EQUIPMENT COST	EQUIPMENT TOTAL	SUBCONTRACT COST	SUBCONTRACT TOTAL	TOTAL COST	TOTAL TOTAL
Division 8: Doors & Windows													
H.M.													
Frames 16 gauge 3' x 7'													
8-3/4" deep	08110 820 4400	2	Ea.	101.00	202	32.00	64						
8-3/4" deep, "B" label	08110 820 6200	10	Ea.	108.00	1080	32.00	320						
5-3/4" deep, "B" label	08110 820 5400	21	Ea.	87.00	1827	32.00	672						
6' x 7'													
8-3/4" deep, "B" label	08110 820 6240	1	Ea.	132.00	132	40.00	40						
Metal Doors 18 gauge													
3' x 7'	08110 820 1120	2	Ea.	189.00	378	28.00	56						
3' x 7', "B" label	08110 300 0180	12	Ea.	199.00	2388	30.00	360						
For vision lite, add	08110 300 0240	12	Ea.	18.00	216								
Wood Doors - Flush Oak													
3' x 7', "B" 1-1/2 hour	08110 820 4400	9	Ea.	224.00	2016	40.00	360						
3' x 7', "B" 1 hour	08110 820 4400	12	Ea.	237.00	2844	40.00	480						
Hardware													
Locksets	08710 650 1400	8	Ea.	156.00	1248	24.00	192						
Panic Hardware	08710 750 0020	17	Ea.	380.00	6460	48.00	816						
Closers	08710 300 2400	35	Ea.	139.00	4865	40.00	1400						
Subtotals					$ 23,656		$ 4,760						

Figure 9.92

251

CONSOLIDATED ESTIMATE

PROJECT: Office Building
LOCATION:
TAKE OFF BY: ABC
QUANTITIES BY: ABC
PRICES BY:
CLASSIFICATION: Division 8
ARCHITECT:
EXTENSIONS BY:

ESTIMATE NO:
DATE: Jan-02
CHECKED BY: GHI

DESCRIPTION	SOURCE	QUANT	UNIT	MATERIAL DEF COST	TOTAL	LABOR DEF COST	TOTAL	EQUIPMENT DEF COST	TOTAL	SUBCONTRACT COST	TOTAL	TOTAL COST	TOTAL
Division 8: (Cont'd)													
Hardware (Cont'd)													
Push Plate	08710 780 0100	9	Ea.	5.05	45	20.00	180						
Kickplates	08710 550 0010	9	Ea.	14.35	129	16.00	144						
Hinges	08710 520 1400	53	Ea.	35.00	1855								
Special Doors													
Roll Up Grills	08330 720 2100	2	Ea.			SUBCONTRACTED				2275	4550		
Motorized	08330 720 4500	2	Ea.							960	1920		
Entrances and Storefronts													
Front Entrances (2)	08411 140 0400	2	Pair							1625	3250		
Add for black (36%)	08411 140 1500	2	Pair							257	514		
Basement Entrances	08411 140 0400	1	Pair							1625	1625		
Add for black (36%)	08411 140 1500	1	Pair							257	257		
Glazing													
Bathroom Mirrors	08830 100 0200	135	SF							11	1505		
3 @ 3' x 8'													
3 @ 3' x 7'													
Subtotals				$	2,030	$	324			$	13,621		

Figure 9.93

CONSOLIDATED ESTIMATE

PROJECT: Office Building	CLASSIFICATION: Division 8
LOCATION:	ARCHITECT:
TAKE OFF BY: ABC	QUANTITIES BY: ABC
	PRICES BY:

ESTIMATE NO:	
DATE: Jan-02	
CHECKED BY: GHI	

DESCRIPTION	SOURCE	QUANT	UNIT	MATERIAL DEF COST	MATERIAL TOTAL	EXTENSIONS BY: LABOR COST	LABOR DEF TOTAL	EQUIPMENT COST	EQUIPMENT TOTAL	SUBCONTRACT COST	SUBCONTRACT TOTAL	TOTAL COST	TOTAL
Division 8: (Cont'd)													
Window/Curtain Wall	TELEPHONE QUOTE (Including Sales Tax) 08911 290 0050	25060	SF								1091363		
Subtotals											$ 1,091,363		
Sheet 1 Subtotals					$ 23,656		$ 4,760						
Sheet 2 Subtotals					$ 2,030		$ 324				$ 13,621		
Sheet 3 Subtotals											$ 1,091,363		
Division 8 Totals					$ 25,686		$ 5,084				$ 1,104,984		

Figure 9.94

253

		08411	Aluminum Framed Storefront	CREW	DAILY OUTPUT	LABOR-HOURS	UNIT	2002 BARE COSTS				TOTAL INCL O&P	
								MAT.	LABOR	EQUIP.	TOTAL		
100	1500		With 3' high transoms, 6' x 10' opening, clear finish	2 Sswk	5.50	2.909	Opng.	410	99.50		509.50	630	**100**
	1550		Bronze finish		5.50	2.909		440	99.50		539.50	660	
	1600		Black finish	▼	5.50	2.909	▼	520	99.50		619.50	750	
120	0010		**ALUMINUM DOORS** Commercial entrance, no glazing										**120**
	0020		Hardware incls. hinges, push/pull handle, deadlock, cylinders										
	0800		Narrow stile, no glazing, standard hardware, pair of 2'-6" x 7'-0"	2 Carp	1.70	9.412	Pr.	660	282		942	1,175	
	1000		3'-0" x 7'-0", single		3	5.333	Ea.	435	160		595	730	
	1200		Pair of 3'-0" x 7'-0"		1.70	9.412	Pr.	870	282		1,152	1,400	
	1500		3'-6" x 7'-0", single		3	5.333	Ea.	470	160		630	765	
	2000		Medium stile, pair of 2'-6" x 7'-0"		1.70	9.412	Pr.	725	282		1,007	1,250	
	2100		3'-0" x 7'-0", single		3	5.333	Ea.	575	160		735	885	
	2200		Pair of 3'-0" x 7'-0"		1.70	9.412	Pr.	1,100	282		1,382	1,650	
	2300		3'-6" x 7'-0", single	▼	5.33	3.002	Ea.	680	90		770	885	
	5000		Flush panel doors, pair of 2'-6" x 7'-0"	2 Sswk	2	8	Pr.	895	274		1,169	1,475	
	5050		3'-0" x 7'-0", single		2.50	6.400	Ea.	445	219		664	885	
	5100		Pair of 3'-0" x 7'-0"		2	8	Pr.	895	274		1,169	1,475	
	5150		3'-6" x 7'-0", single	▼	2.50	6.400	Ea.	530	219		749	980	
140	0010		**ALUMINUM DOORS & FRAMES** Entrance, narrow stile, including										**140**
	0015		Standard hardware, clear finish, not incl. glass, 2'-6" x 7'-0" opng.	2 Sswk	2	8	Ea.	435	274		709	970	
	0020		3'-0" x 7'-0" opening		2	8		435	274		709	975	
	0030		3'-6" x 7'-0" opening		2	8		450	274		724	990	
	0100		3'-0" x 10'-0" opening, 3' high transom		1.80	8.889		710	305		1,015	1,325	
	0200		3'-6" x 10'-0" opening, 3' high transom		1.80	8.889		700	305		1,005	1,325	
	0280		5'-0" x 7'-0" opening		2	8	▼	745	274		1,019	1,325	
	0300		6'-0" x 7'-0" opening		1.30	12.308	Pr.	725	420		1,145	1,550	
	0400		6'-0" x 10'-0" opening, 3' high transom		1.10	14.545		650	500		1,150	1,625	
	0420		7'-0" x 7'-0" opening		1	16	▼	785	550		1,335	1,850	
	0500		Wide stile, 2'-6" x 7'-0" opening		2	8	Ea.	665	274		939	1,225	
	0520		3'-0" x 7'-0" opening		2	8		655	274		929	1,225	
	0540		3'-6" x 7'-0" opening		2	8		685	274		959	1,250	
	0560		5'-0" x 7'-0" opening		2	8	▼	1,050	274		1,324	1,650	
	0580		6'-0" x 7'-0" opening		1.30	12.308	Pr.	1,000	420		1,420	1,850	
	0600		7'-0" x 7'-0" opening	▼	1	16	"	1,150	550		1,700	2,250	
	1100		For full vision doors, with 1/2" glass, add				Leaf	55%					
	1200		For non-standard size, add					67%					
	1300		Light bronze finish, add					36%					
	1400		Dark bronze finish, add					18%					
	1500		For black finish, add					36%					
	1600		Concealed panic device, add				▼	930			930	1,025	
	1700		Electric striker release, add				Opng.	239			239	263	
	1800		Floor check, add				Leaf	710			710	780	
	1900		Concealed closer, add				"	475			475	520	
	2000		Flush 3' x 7' Insulated, 12"x 12" lite, clear finish	2 Sswk	2	8	Ea.	900	274		1,174	1,475	
600	0010		**STAINLESS STEEL AND GLASS** Entrance unit, narrow stiles										**600**
	0020		3' x 7' opening, including hardware, minimum	2 Sswk	1.60	10	Opng.	4,600	345		4,945	5,675	
	0050		Average		1.40	11.429		4,975	390		5,365	6,175	
	0100		Maximum	▼	1.20	13.333		5,325	455		5,780	6,700	
	1000		For solid bronze entrance units, statuary finish, add					60%					
	1100		Without statuary finish, add				▼	45%					
	2000		Balanced doors, 3' x 7', economy	2 Sswk	.90	17.778	Ea.	6,225	610		6,835	7,950	
	2100		Premium	"	.70	22.857	"	10,700	785		11,485	13,200	
650	0010		**STOREFRONT SYSTEMS** Aluminum frame, clear 3/8" plate glass,										**650**
	0020		incl. 3' x 7' door with hardware (400 sq. ft. max. wall)										

Figure 9.95

TELEPHONE QUOTATION

PROJECT **Office Building**

DATE _____

TIME _____

FIRM QUOTING _____

PHONE (___)

ADDRESS _____

BY _____

ITEM QUOTED _____

RECEIVED BY **EBW**

WORK INCLUDED	AMOUNT OF QUOTATION
Window/Curtain Wall	10 5 25 20 00
25,060 S.F. @ 42.00	
Tax 5% (On Material)	3 8 84 3 00

DELIVERY TIME **18 Weeks**	**TOTAL BID**	10 91 36 3 00

DOES QUOTATION INCLUDE THE FOLLOWING:

If ☐ NO is checked, determine the following:

STATE & LOCAL SALES TAXES	☒ YES	☐ NO	MATERIAL VALUE	
DELIVERY TO THE JOB SITE	☒ YES	☐ NO	WEIGHT	
COMPLETE INSTALLATION	☒ YES	☐ NO	QUANTITY	
COMPLETE SECTION AS PER PLANS & SPECIFICATIONS	☒ YES	☐ NO	DESCRIBE BELOW	

EXCLUSIONS AND QUALIFICATIONS

ADDENDA ACKNOWLEDGEMENT

TOTAL ADJUSTMENTS	
ADJUSTED TOTAL BID	

ALTERNATES

ALTERNATE NO. **1- Custom color**	7 63 10 00
ALTERNATE NO. **2- Grid texture**	1 3 30 55 00
ALTERNATE NO.	
ALTERNATE NO.	
ALTERNATE NO.	
ALTERNATE NO.	
ALTERNATE NO.	

Figure 9.96

or may not be a builder of structures from the ground up—a conventional general contractor. Because of the skills involved and the quality required, subcontractors usually specialize in only one type of finish. Hence, most finish work is subcontracted.

In today's fireproof and fire-resistant types of construction, some finish materials may be the only combustibles used in a building project. Most building codes (and specifications) require strict adherence to maximum fire, flame spread, and smoke generation characteristics. The estimator must be sure that all materials meet the specified requirements. Materials may have to be treated for fire retardancy at an additional cost.

Lathing and Plastering

The different types of plaster work require varied pricing strategies. Large open areas of continuous walls or ceilings will require considerably less labor per unit of area than small areas or intricate work such as archways, curved walls, cornices and at window returns. Gypsum and metal lath are most often used as sub-bases. However, plaster may be applied directly on masonry, concrete, and in some restoration work, wood. In the latter cases, a bonding agent may be specified.

The number of coats of plaster may also vary. Traditionally, a scratch coat is applied to the substrate. A brown coat is then applied two days later, and the finish, smooth coat seven days after the brown coat. Currently, the systems most often used are two-coat and one-coat (imperial plaster on "blueboard"). Textured surfaces, with and without patterns, may be required. All of these variables in plaster work make it difficult to develop "system" prices. Each project, and even areas within each project, must be examined individually.

The quantity takeoff should proceed in the normal construction sequence—furring (or studs), lath, plaster and accessories. Studs, furring and or ceiling suspension systems, whether wood or steel, should be taken off separately. Responsibility for the installation of these items should be made clear. Depending upon local work practices, lathers may or may not install studs or furring. These materials are usually estimated by the piece or linear foot, and sometimes by the square foot. Lath is traditionally estimated by the square yard for both gypsum and metal lath and is done more recently by the square foot. Usually, a 5% allowance for waste is included. Casing bead, corner bead, and other accessories are measured by the linear foot. An extra foot of surface area should be allowed for each linear foot of corner or stop. Although wood plaster grounds are usually installed by carpenters, they should be measured when taking off the plaster requirements. Plastering is also traditionally measured by the square yard. Deductions for openings vary by preference—from zero deduction to 50% of all openings over 2 feet in width. Some estimators deduct a percentage of the total yardage for openings. The estimator should allow one extra square foot of wall area for each linear foot of inside or outside corner located below the ceiling level. Also, double the areas of small radius work. Quantities are determined by measuring surface areas (walls, ceilings). The estimator must consider both the complexity and the intricacy of the work, and in pricing plaster work, should also consider quality. Basically, there are two quality categories:

1. Ordinary—for commercial purposes, and with waves 1/8" to 3/16" in 10 feet, angles and corners fairly true.

2. First Quality – with variations less than 1/16" in 10 feet. Labor costs for first quality work are approximately 20% more than that for ordinary plastering.

Drywall

With the advent of light gauge metal framing, tin snips are as important to the carpenter as the circular saw. Metal studs and framing are usually installed and included by the drywall subcontractor. The estimator should make sure that studs (and other framing – whether metal or wood) are not included twice by different subcontractors. In some drywall systems, such as shaftwall, the framing is integral and installed simultaneously with the drywall panels.

Metal studs are manufactured in various widths (1-5/8", 2-1/2", 3-5/8", 4" and 6") and in various gauges, or metal thicknesses. They may be used for both load-bearing and non-load-bearing partitions, depending on design criteria and code requirements. Metal framing is particularly useful due to the prohibitive use of structural wood (combustible) materials in new building construction. Metal studs, track and accessories are purchased by the linear foot, and usually stocked in 8' to 16' lengths, by 2' increments. For large orders, metal studs can be purchased in any length up to 20'.

For estimating, light gauge metal framing is taken off by the linear foot or by the square foot of wall area of each type. Different wall types—with different stud widths, stud spacing, or drywall requirements—should each be taken off separately, especially if estimating by the square foot.

Metal studs can be installed very quickly. Depending upon the specification, metal studs may have to be fastened to the track with self-tapping screws, tack welds or clips, or may not have to be prefastened. Each condition will affect the labor costs. Fasteners, such as screws, clips and powder-actuated studs are very expensive, though labor-saving. These costs must be included.

Drywall may be purchased in various thicknesses—1/4" to 1"—and in various sizes 2' × 8' to 4' × 20'. Different types include standard, fire-resistant, water-resistant, blueboard, coreboard and pre-finished. There are many variables and possible combinations of sizes and types. While the installation cost of 5/8" standard drywall may be the same as that of 5/8" fire-resistant drywall, the two types (and all other types) should be taken off separately. The takeoff will be used for purchasing, and material costs will vary.

Because drywall is used in such large quantities, current, local prices should always be checked. A variation of a few cents per square foot can become many thousands of dollars over a whole project.

Fire-resistant drywall provides an excellent design advantage in creating relatively lightweight, easy to install firewalls (as opposed to masonry walls). As with any type of drywall partition, the variations are numerous. The estimator must be very careful to take off the appropriate firewalls exactly as specified. (Even more important, the contractor must *build* the firewalls exactly as specified. Liabilities can be great.) For example, a metal stud partition with two layers of 1/2" fire-resistant drywall on each side may constitute a two-hour partition (when all other requirements such as staggered joints, taping, sealing openings, etc. are met). If a one-hour partition is called for, the estimator cannot assume that one layer of 1/2" fire-resistant drywall on each side of a metal stud partition will suffice. Alone, it does not. When left to choose the appropriate assembly (given the rating required), the estimator must be sure that the system has been tested and *approved* for use—by Underwriters Laboratory as well as local building and fire codes and responsible authorities.

In all cases, the drywall (and studs) for firewalls must extend completely from the deck below to the underside of the deck above, covering the area above and around any and all obstructions. All penetrations must be protected.

In the past, structural members—such as beams or columns—to be fireproofed had to be "wrapped" with a specified number of layers of fire-resistant drywall—a very labor-intensive and expensive task. With the advent of spray-on fireproofing, structural members can be much more easily protected. This type of work is usually performed by a specialty subcontractor. Takeoff and pricing are done by square foot of surface area.

When walls are specified for minimal sound transfer, the same continuous, unbroken construction is required, Sound-proofing specifications may include additional accessories and related work. Resilient channels attached to studs, mineral fiber batts and staggered studs may all be used. In order to develop high noise reduction coefficients, double stud walls may be required with sheet lead between double or triple layers of drywall. (Sheet lead may also be required at X-ray installations.) Caulking is required at all joints and seams. All openings must be specially framed with double, "broken" door and window jambs.

Shaftwall, developed for a distinct design advantage, is another drywall assembly which should be estimated separately. Firewalls require protection from both sides of the partition, and hence drywall installation from both sides. Shaftwall, used at vertical openings (elevators, utility chases), can be installed completely from one side. Special track, studs (C-H or double E type) and drywall (usually 1" thick and 2' wide coreboard) are used and should be priced separately from other drywall partition components.

Because of the size and weight of drywall, costs for material handling and loading should be included. Using larger sheets (manufactured up to 4' × 20') may require less taping and finishing, but these sheets may each weigh well in excess of 100 pounds and are awkward to handle. The weight of drywall must also be considered (and distributed) when loading a job on elevated slabs, so that allowable floor loads are not exceeded. All material handling involves costs that must be included.

As with plaster work, open spans of drywall should be priced differently from small, intricate areas that require much cutting and piecing. Similarly, areas with many corners or curves will require higher finishing costs than open walls or ceilings. Corners, both inside and outside, should be estimated by the linear foot, in addition to the square feet of surface area.

Although difficult because of variations, the estimator may be able to develop historical systems or assemblies prices for metal studs, drywall, taping and finishing. When using systems, whether complete or partial, the estimator must be sure that the system, as specified, is exactly the same as the system for which the costs are developed. For example, a cost is developed for 5/8" fire-resistant drywall, taped and finished. The project specifications require a firewall with two layers of the 5/8" drywall on each side of the studs. It could be easy to use the "system" cost for each layer, when only one of the two layers is to be taped and finished. The more detailed the breakdown and delineation, the less chance for error.

Tile and Terazzo

Tile, terrazzo and other hard surface floor and wall finishes are most often, estimated by the square foot. Linear features such as bullnose and cove base are taken off in linear feet. Small areas and individually laid tile patterns should be separately estimated, apart from large, open expanses, and installations of preattached tile sheets which are taken off by the square foot.

In addition to the considerable variation in material prices, the installation method of tile will also have a significant impact on the total cost. Two basic methods are used—"thin" set, with an epoxy-type adhesive, and "mud" set, using a fine-grained mortar. Mud set is approximately 30% more expensive than thin set, but provides a harder and more durable treatment. Hard, cement-like backer board can be used as a base for thin set tile. The more commonly used types of ceramic wall tile are now manufactured in pre-grouted sheets which require less labor expense.

Currently, most terrazzo installed is in the form of manufactured tiles. However, the traditional method of pouring, grinding and rubbing the terrazzo over a concrete base slab is still used. In such cases, an experienced subcontractor should estimate this very labor-intensive work. Accessories, such as embedded decorative strips and grounds, must be included.

In all tile work, surface preparation may be the responsibility of the installer. Especially in renovation, this preparation may be very involved and costly.

Acoustical Treatment

Acoustical treatments may involve sound-absorbing panels on walls, sheet lead within walls or floors, suspended or "dropped" ceilings, or sound attenuation blankets in wall cavities. While sound deadening is primary, it does not have to be the only function.

Most applications of acoustical treatments are for ceilings, whether suspended or attached to the structure above, or concealed spline or lay-in. In most cases, acoustical ceilings are estimated by the square foot. Suspension systems or furring can be estimated by linear feet of pieces or by square feet of surface area. The installation and leveling of suspension grids can be very labor-intensive. The estimator must be sure that adequate points of support are available, spaced often enough to meet specified requirements. If not, attachments for support, such as anchors, toggle bolts, and carrier channels, will be required and the costs included. Invariably, a duct or other obstruction runs just above a main support "T," or runner. Design modification of the grid (or duct) system may be necessary, and the architect should be notified.

In small areas and rooms, leveling a suspension grid may be relatively easy. In large open areas, a laser may be necessary for leveling. Especially for large open areas, the cost of installing a ceiling (whether acoustical or drywall) will depend on how clear and clean the floor space is. Most ceilings are installed from rolling platforms. If the workers are only able to work in certain areas, or if the platform must be hand-lifted over material and debris, installation costs will increase significantly. When developing the preliminary schedule, good planning will help to ensure minimum installation costs for ceilings. Costs for equipment, such as lasers and rolling scaffolding, should be listed separately. For very high installations, scissors or telescoping lifts may be required.

Tile for acoustical ceilings may consist of mineral fiber, wood fiber, fiberglass or metal, and might be cloth-covered, colored, textured and/or patterned. There is little waste when installing grid systems (usually less than 5%), because pieces can be butted and joined. Waste for tile, however, can be as low as 5% for large open areas, to as high as 30%-40%, for small areas and rooms. Waste for tile may depend on grid layout as well as room dimensions. Figure 9.97 demonstrates that for the same size room, the layout of a typical 2' x 4' grid has a significant effect on generated waste of ceiling tile. Since most textures and patterns on ceiling tile are aligned in one direction, pieces cannot be turned 90 degrees (to the specified alignment) to try to reduce waste.

Certain tile types, such as tegular (recessed), require extra labor for cutting and fabrication at edge moldings. Soffits, facias, and "boxouts" should be estimated separately due to extra labor, material waste, and special attachment techniques. Costs should also be added for unusually high numbers of tiles to be specially cut for items such as sprinkler heads, diffusers or telepoles.

While the weight of ceiling tile is not the primary consideration that it is with drywall, some material handling and storage costs will still be incurred and must be included. Acoustical tile is very bulky and cumbersome, as well as fragile, and must be protected from damage before installation.

Flooring

Finish floor coverings include carpet, resilient sheet goods and tile, rubber and vinyl stair treads and risers, and wood finish flooring. When estimating all types of finish flooring, the condition and required preparation of the subfloor must be considered.

Resilient Flooring: Resilient materials are vinyl, rubber and linoleum products in the form of tiles, sheet goods and base. The most commonly used is vinyl composition tile—often mistakenly referred to as vinyl asbestos tile. Resilient flooring is taken off and priced by the square foot. Most resilient flooring is directionally patterned. The construction documents will most likely specify a particular laying pattern. Because of the size of resilient tiles (12" × 12" or 9" × 9"), waste is not necessarily dependent upon room configuration. However, depending upon seam requirements, sheet goods may involve a great deal of waste. Each application should be viewed individually. Cove base and straight base are measured by the linear foot. If the specifications are not clear, the estimator must determine whether the base is to be wrapped around corners, or if premolded corners are required.

Due to the thinness and flexibility of resilient goods, defects in the subfloor easily "telegraph" through the material. Consequently, the subfloor material and the quality of surface preparation are very important. Subcontractors will often make contracts conditional on a smooth, level subfloor. Surface preparation, which can involve chipping and patching, grinding or washing, is often an "extra." Costs should be included to account for some surface preparation which will invariably be required. This is especially true in renovation where, in extreme cases, the floor may have to be leveled with a complete application of special lightweight leveling concrete.

Carpet: There are hundreds of carpeting manufacturers, each with hundreds of products. The specifications for a project will usually include a generic description of the carpeting and a recommended manufacturer. The description will include items such as pile type and weight, backing requirements, and smoke and flame characteristics. Usually this description is so generically specific that even if an approved equal is allowed, no such equal exists. It is best to estimate the carpeting exactly as specified.

Carpeting and padding is manufactured most commonly in 12' widths (sometimes 9' or 15'), and is taken off by the square yard. Costs for installation will vary depending upon the specified method—direct cement (without pad) or stretched (with pad and perimeter tack strips).

The specifications may also require and define the location of seams. Often, butt seams—roll end to roll end—are not allowed. Where seams do occur,

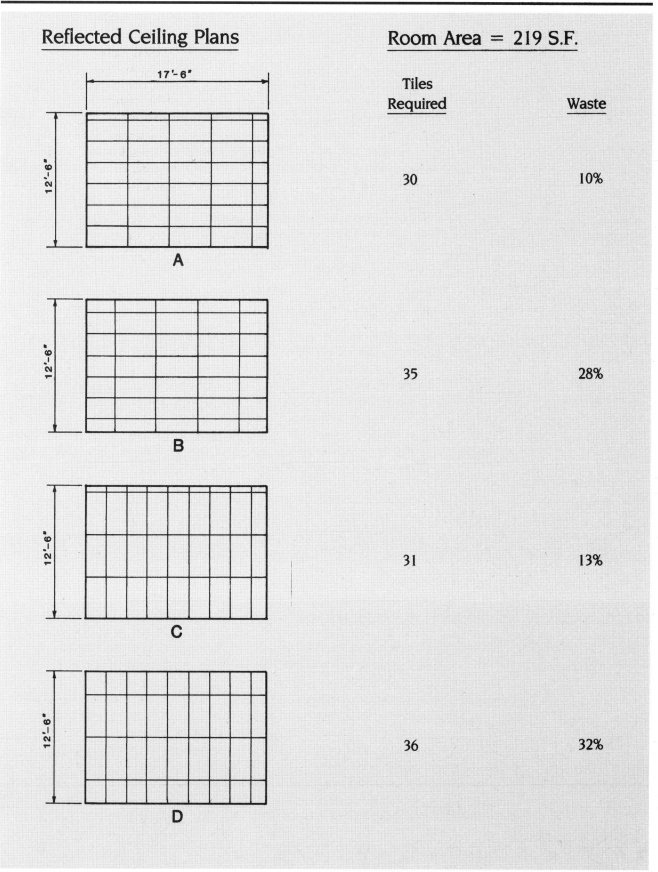

Reflected Ceiling Plans

Room Area = 219 S.F.

	Tiles Required	Waste
A	30	10%
B	35	28%
C	31	13%
D	36	32%

17'-6"

12'-6"

A

12'-6"

B

12'-6"

C

12'-6"

D

Figure 9.97

costs for sewing (by hand) and gluing will differ. When a carpeting subcontractor estimates a job, the whole job is drawn on a floor plan based upon the seaming requirements to determine the quantity of carpet to be purchased. Depending upon the requirements and the configuration of the installation, waste can be high.

Where carpet base is specified, costs must be included for binding, priced by the linear foot. Carpet base may or may not be the same material used on the floor. The specifications should be carefully checked. Carpet tile systems have been developed for ease of repairing damaged areas and for use with under-carpet power and telecommunication systems. Tiles are available in various sizes from 18" to 36" square and are normally taken off and priced by the square foot.

All resilient materials and direct cement carpeting are installed with adhesive. If adhesive materials are estimated separately, costs are per gallon. Coverage depends upon the flooring material and the type of adhesive. Prices developed or quoted for direct glue-down flooring usually include the adhesive.

Wood Flooring: Everyone loves to see a wood floor specified and installed, except for the estimator who has not included all requirements. Wood flooring can be deceptively expensive.

Wood flooring is available in strip, parquet, or block configuration. All types are estimated by the square foot. There are three basic grades of wood flooring: first, second, and third, plus combination grades of "second and better" and "third and better." There are also color grades and special grade labels for different kinds of lumber, such as oak, maple, walnut, pecan, or beech. The estimator should be acquainted with these classifications and the associated price differences. The laying pattern will influence both labor costs and material waste.

Strip wood floors may be used for applications from residences to gymnasiums and large sports complexes. While the wood may be the same, installation methods vary considerably. For residential work, the strips are nailed to wood joist/subfloor systems. For commercial applications, installation is usually over concrete. The strips may be nailed to wood sleepers or attached to steel channels with steel clips. In most cases, resilient materials, pads, and/or sheets are used between wood and concrete in addition to a continuous vapor barrier. Wood expands and contracts considerably with variations in temperature and humidity. Precautions must, therefore, be taken. Since expansion joints are not aesthetically pleasing within wood floors, gaps are used at the perimeter. Costs for hiding this gap, with a wood base or metal angles, must be included. Since large, commercial wood floors are usually installed by a specialty subcontractor, the estimator must be sure that all associated work is taken into account.

Parquet floors are currently made of prefinished, manufactured tiles, installed with adhesive. Durable synthetic finishes make this type of wood flooring suitable for commercial uses and public spaces.

Wood end grain block flooring is still commonly used for industrial applications. Set in, and filled with epoxy or asphalt, this type of flooring is installed by specialty subcontractors using specialized equipment. The estimator must check with all such subcontractors who are familiar with methods and materials to ensure that all requirements are met, and the costs included. For all types of finish flooring, surface treatments may be required after installation. For wood flooring (unless prefinished), sanding and finishing will be necessary.

If the wood is prefinished, as with most resilient goods, waxing and buffing is an additional expense that must be included. Carpeting should be installed after *all* other work is completed. This is rarely the case, however, and invariably, the brand new carpet gets dirty quickly. A minimum of vacuuming and spot cleaning should be anticipated. In extreme cases, steam cleaning of the entire carpet may be necessary. Stretched carpet may require restretching after steam cleaning. The estimator must try to anticipate such items, all of which involve extra cost.

In many projects, and most often with flooring, the architect or designer may not select the final finish material until the project is underway. In such cases, a specified allowance is usually carried in the estimate. The allowance is most often for material *only*, and is specified per unit (square foot for resilient and wood flooring, square yard for carpet), but must be carried in the estimate as a lump sum. The estimator must still determine quantities of materials and installation costs.

Painting: As with most finishes, architects and designers will be particularly insistent about adherence to specifications and specified colors for painting. This is not an area in which to cut corners—in estimating, or performance of the work. The specifications usually will clearly define acceptable materials, manufacturers, preparation and application methods for *each* different type of surface to be painted. Samples may be required for approval by the architect or owner. Quarts of various paint colors may be required before the final color decision is made.

When estimating painting, the materials and methods are included in the specifications. Areas to be painted are usually defined on a Room Finish Schedule and taken off from the plans and elevations in square feet of surface area. Odd-shaped and special items can be converted to an equivalent wall area. The following table includes suggested conversion factors for various types of surfaces:

Balustrades:		1 Side × 4
Blinds:	Plain	Actual area × 2
	Slotted	Actual area × 4
Cabinets:	Including interior	Front area × 5
Downspouts and Gutters:		Actual area × 2
Drop Siding:		Actual area × 1.1
Cornices:	1 Story	Actual area × 2
	2 Story	Actual area × 3
	1 Story Ornamental	Actual area × 4
	2 Story Ornamental	Actual area × 6
Doors:	Flush	Actual area × 1.5
	Two Panel	Actual area × 1.75
	Four Panel	Actual area × 2.0
	Six Panel	Actual area × 2.25 Door
Trim:		LF × 0.5
Fences:	Chain Link	1 side × 3 for both sides
	Picket	1 side × 4 for both sides
Gratings:		1 side × 0.66
Grilles:	Plain	1 side × 2.0
	Lattice	Actual area × 2.0
Moldings:	Under 12" Wide	1 SF per LF
Open Trusses:		Length × Depth × 2.5
Pipes:	Up to 4"	1 SF per LF
	4" to 8"	2 SF per LF

Pipes (cont.):	8" to 12"	3 SF per LF
	12" to 16"	4 SF per LF
	Hangers Extra	
Radiators:		Face area × 7
Sanding and Puttying:	Quality Work	Actual area × 2
	Average Work	Actual area × 0.5
	Industrial	Actual area × 0.25
Shingle Siding:		Actual area × 1.5
Stairs:		No. of risers × 8 widths
Tie Rods:		2 SF per LF
Wainscoting, Paneled:		Actual area × 2
Walls and Ceilings:		Length × Width, no deducts for less than 100 SF
Window Sash:		1 LF of part = 1 SF

While the above factors are used to determine quantities of equivalent wall surface areas, the appropriate, specified application method must be used for pricing.

The choice of application method will have a significant effect on the final cost. Spraying is very fast, but the costs of masking the areas to be protected may offset the savings. Oversized rollers may be used to increase production. Brushwork, on the other hand, is labor-intensive. The specifications often include (or restrict) certain application methods. Typical coverage and labor-hour rates are shown in Figure 9.98 for different application methods.

Finishes — R099 Paints & Coverings

R09910-220   Painting

Item	Coat	One Gallon Covers			In 8 Hours a Laborer Covers			Labor-Hours per 100 S.F.		
		Brush	Roller	Spray	Brush	Roller	Spray	Brush	Roller	Spray
Paint wood siding	prime	250 S.F.	225 S.F.	290 S.F.	1150 S.F.	1300 S.F.	2275 S.F.	.695	.615	.351
	others	270	250	290	1300	1625	2600	.615	.492	.307
Paint exterior trim	prime	400	—	—	650	—	—	1.230	—	—
	1st	475	—	—	800	—	—	1.000	—	—
	2nd	520	—	—	975	—	—	.820	—	—
Paint shingle siding	prime	270	255	300	650	975	1950	1.230	.820	.410
	others	360	340	380	800	1150	2275	1.000	.695	.351
Stain shingle siding	1st	180	170	200	750	1125	2250	1.068	.711	.355
	2nd	270	250	290	900	1325	2600	.888	.603	.307
Paint brick masonry	prime	180	135	160	750	800	1800	1.066	1.000	.444
	1st	270	225	290	815	975	2275	.981	.820	.351
	2nd	340	305	360	815	1150	2925	.981	.695	.273
Paint interior plaster or drywall	prime	400	380	495	1150	2000	3250	.695	.400	.246
	others	450	425	495	1300	2300	4000	.615	.347	.200
Paint interior doors and windows	prime	400	—	—	650	—	—	1.230	—	—
	1st	425	—	—	800	—	—	1.000	—	—
	2nd	450	—	—	975	—	—	.820	—	—

Figure 9.98

Depending upon local work rules, some unions require that painters be paid higher rates for spraying, and even for roller work. In some cases, paint is applied to walls and surfaces with a brush and then rolled for the desired finish. Higher rates also tend to apply for structural steel painting, for high work, and for the application of fire-retardant paints. The estimator should determine which restrictions may be encountered.

The surface preparation of walls, as in the case of floor finishes, may represent a significant cost. Invariably, the painter claims that the walls were not finished properly, and the general contractor contends that a certain amount of wall prep is included in the painting subcontract. This conflict should be resolved at the estimate stage. Also, caulking at door frames and windows—the responsibility similarly argued—must be included.

Wall Covering

Wall coverings are usually estimated by the number of rolls. Single rolls contain approximately 36 S.F.; this figure forms the basis for determining the number of rolls required. Wall coverings are, however, usually sold in double or triple roll bolts.

The area to be covered is measured, length times height of wall above baseboards, in order to get the square footage of each wall. This figure is divided by 30 to obtain the number of single rolls, allowing 6 S.F. of waste per roll. Deduct one roll for every two door openings. Two pounds of dry paste makes about three gallons of ready-to-use adhesive and hangs about 36 single rolls of light to medium weight paper, or 14 rolls of heavy weight paper. Application labor costs vary with the quality, pattern, and type of joint required.

With vinyls and grass cloths requiring no pattern match, a waste allowance of 10% is normal—approximately 3.5 S.F. per roll. Wall coverings that require a pattern match may have about 25%—30% waste, or 9—11 S.F. per roll. Waste can run as high as 50%—60% on wall coverings with a large, bold, or intricate pattern repeat.

Commercial wall coverings are available in widths from 21" to 54", and in lengths from 5-1/3 yards (single roll) to 100 yard bolts. To determine quantities, independent of width, measure the linear (perimeter) footage of walls to be covered. Divide the linear footage by the width of the goods to determine the number of "strips" or drops. Then determine the number of strips per bolt or package by dividing the length per bolt by the ceiling height.

$$\frac{\text{Linear Footage of Walls}}{\text{Width of Goods}} = \text{No. of Strips}$$

$$\frac{\text{Length of Bolt Roll}}{\text{Ceiling Height}} = \frac{\text{No. of Strips (whole no.)}}{\text{Bolt (roll)}}$$

Finally, divide the quantity of strips required by the number of strips per bolt (roll) in order to determine the required amount of material. Use the same waste allowance as above.

Surface preparation costs for wall covering must also be included. If the wall covering is to be installed over new surfaces, the walls must be treated with a wall sizing, shellac or primer coat for proper adhesion. For existing surfaces, scraping, patching, and sanding may be necessary. Requirements will be included in the specifications.

Sample Estimate: Division 9

Most of the finishes for the sample project involve large quantities in wide open areas. For an interior space broken up into many rooms or suites, the estimate would be much more involved, and a detailed Room Finish Schedule should be provided. The estimate sheets for the sample project are shown in Figures 9. 99 to 9.102. On all the estimate sheets for the project, items are entered on every other line. Many more sheets are used this way, but paper is cheap. Invariably, omitted items must be inserted at the last minute and sheet subtotals and division totals will be recalculated as a result. If the sheets are concise, neat and organized, recalculation can be easy, and will involve less chance of error.

Particularly when estimating drywall, the same dimensions can be used to calculate the quantities of different items—in this case, drywall, metal studs, and accessories. When an opening (to be finished with drywall) is deducted from the drywall surface area, perimeter dimensions are used to calculate corner bead. Similarly, at joints with items such as windows, the same dimensions are used for J-bead. The quantity sheet for drywall and associated items is shown in Figure 9.103. Types, sizes, and different applications of drywall and studs are listed (and priced) separately. The square foot quantities of studs and drywall can be used for comparison to check for possible errors. When such a method is used, the estimator must be aware of those partitions which receive drywall only on one side, as opposed to both sides, and those with more than one layer. A comparison for the sample project is performed as follows:

Framing	Stud Area	No. of Sides	No. of Layers	Drywall Area
6"	300 S.F.	2	1	600 S.F.
3-5/8"	3,560 S.F.	2	1	7,120 S.F.
3-5/8"	10,104 S.F.	1	1	10,104 S.F.
1-5/8"	5,640 S.F.	1	2	11,280 S.F.
1-5/8"	1,704 S.F.	1	1	1,704 S.F.
Furring	4,200 S.F.	1	1	4,200 S.F.
	25,508 S.F.			35,008 S.F.

25,508 S.F. = 35,008 − (300 + 3560 + 5640) = 25,508 S.F.

Note that furring is included in the above cross-check calculations as square feet. The estimator must always be aware of units during *any* calculations. When deductions are made from the drywall totals for two-side application and double layers, the drywall and stud quantities should equate.

For the ceiling grid, because of the large, open areas, no allowance for waste is included. Note, however, that the dimensions used are for the exterior of the building (210' × 90' × 3 floors). This includes a small overage. For purchasing purposes, all ceiling perimeter dimensions would be measured separately for edge moldings. A 5% allowance is added to the ceiling tile. The large quantity of light fixtures must be considered and is deducted from the total. Deductions for light fixtures should be based on experience. Carpeting quantities are also derived from the building perimeter measurements, with appropriate deductions for the building core.

Division 10: Specialties

Division 10 includes prefinished, manufactured items that are usually installed at the end of a project when other finish work is complete. Following is a partial list of items that may be included in Division 10:

- Bathroom accessories
- Bulletin and chalkboards

CONSOLIDATED ESTIMATE

PROJECT: Office Building
LOCATION:
TAKE OFF BY: ABC   QUANTITIES BY: ABC
CLASSIFICATION: Division 9
ARCHITECT: PRICES BY: As Shown
ESTIMATE NO:
DATE: Jan-02
CHECKED BY: GHI
EXTENSIONS BY: DEF

SOURCE			DESCRIPTION	QUANT	UNIT	MATERIAL COST	DEF	MATERIAL TOTAL	LABOR COST	LABOR TOTAL	EQUIPMENT COST	EQUIPMENT TOTAL	SUBCONTRACT COST	SUBCONTRACT TOTAL	TOTAL COST	TOTAL
			Division 9: Finishes													
			Framing : Metal Studs													
09110	100	2500	6" - 25 gauge	300	SF	0.31		93	0.51	153						
09110	100	2300	3-5/8" - 25 gauge	13664	SF	0.19		2596	0.50	6832						
09110	100	2000	3-5/8" - 25 gauge	7344	SF	0.16		1175	0.48	3525						
			Accessories													
09270	100	0900	7/8" Furring	31.5	CLF	17.35		547	92.50	2914						
09270	100	1120	J - Bead	34.1	CLF	12.40		423	81.50	2779						
09270	100	0300	Corner Bead	58.4	CLF	10.30		602	60.00	3504						
			Drywall													
09250	700	4050	5/8" F.R. @Columns	5640	SF	0.52		2933	1.60	9024						
09250	700	2150	5/8" F.R. @Core	7720	SF	0.26		2007	0.50	3860						
09250	700	2050	5/8" Standard	16008	SF	0.26		4162	0.50	8004						
09260	800	0300	Shaftwall: @ Elevator	2040	SF	1.11		2264	2.67	5447						
			Subtotals					$ 16,802		$ 46,042						

Figure 9.99

CONSOLIDATED ESTIMATE

PROJECT: Office Building
LOCATION:
TAKE OFF BY: ABC

CLASSIFICATION: Division 9
ARCHITECT: ABC
QUANTITIES BY: ABC

PRICES BY:

EXTENSIONS BY:

ESTIMATE NO:
DATE: Jan-02
CHECKED BY: GHI

DESCRIPTION	SOURCE			QUANT.	UNIT	MATERIAL DEF COST	MATERIAL TOTAL	LABOR COST	LABOR TOTAL	EQUIPMENT DEF COST	EQUIPMENT TOTAL	SUBCONTRACT COST	SUBCONTRACT TOTAL	TOTAL COST	TOTAL TOTAL
Division 9: (Cont'd)															
Fireproofing. @ Beams	07812	600	0400	40500	SF							1.14	46170		
Total Fireproofing													$ 46,170		
Ceramic Tile:															
Walls	09310	100	5400	1584	SF							5.55	8791		
Bull Nose	09310	100	2500	396	LF							7.65	3029		
Cove Base	09310	100	0700	396	LF							8.15	3227		
Floors	09310	100	3300	960	SF							8.00	7680		
Total Ceramic Tile													$ 22,728		
Accoustical Ceiling:															
Grid	09130	100	0050	56700	SF	0.64	36288	0.30	17010						
Tile	09510	700	3740	53735	SF	1.58	84901	0.42	22569						
Total Ceiling							$ 121,189		$ 39,579						

Figure 9.100

CONSOLIDATED ESTIMATE

PROJECT: Office Building
LOCATION:
TAKE OFF BY: ABC
CLASSIFICATION: Division 9
ARCHITECT:
QUANTITIES BY: ABC
PRICES BY: As Shown
EXTENSIONS BY:
ESTIMATE NO:
DATE: Jan-02
CHECKED BY: GHI

DESCRIPTION	SOURCE			QUANT.	UNIT	MATERIAL			LABOR			EQUIPMENT		SUBCONTRACT		TOTAL	
						COST	DEF	TOTAL	COST	DEF	TOTAL	COST	TOTAL	COST	TOTAL	COST	TOTAL
Division 9: (Cont'd)																	
Flooring																	
Carpet	09680	800	3200	5850	SY									33.00	193050		
Cove Base	09658	100	1150	2530	LF									1.61	4073		
Parquet @ Lobby & Elevators	09648	100	6500	1800	SF									7.70	13860		
Total Flooring														$	210,983		
Painting																	
Walls (Drywall)	09910	920	0840	27784	SF									0.51	14170		
Wood Doors	09910	320	1800	21	Ea.									40.50	851		
Metal Doors (Primed)	09910	320	1000	14	Ea.									35.00	490		
Block Walls	09910	920	2880	9846	SF									0.39	3840		
Total Painting														$	19,350		

Figure 9.101

CONSOLIDATED ESTIMATE

PROJECT: Office Building
LOCATION:
TAKE OFF BY: ABC
CLASSIFICATION: Division 9
ARCHITECT:
QUANTITIES BY: ABC
PRICES BY:
EXTENSIONS BY:

ESTIMATE NO:
DATE: Jan-02
CHECKED BY: GHI

SOURCE	DESCRIPTION	QUANT	UNIT	MATERIAL COST	DEF TOTAL	LABOR COST	DEF TOTAL	EQUIPMENT COST	TOTAL	SUBCONTRACT COST	TOTAL	TOTAL COST	TOTAL
	Division 9: (Cont'd)												
	Sheet 1: Drywall & Framing				$ 16,802		$ 46,042						
	Sheet 2: Fireproofing										$ 46,170		
	Ceramic Tile										$ 22,728		
	Accoustical Ceiling				$ 121,789		$ 39,579						
	Sheet 3: Flooring										$ 210,983		
	Painting										$ 19,350		
	Division 9 Totals				$ 138,591		$ 85,621				$ 299,231		

Figure 9.102

QUANTITY SHEET

PROJECT: Office Building
LOCATION:
TAKE OFF BY:

Division 9
ARCHITECT:
EXTENSIONS BY:

DESCRIPTION	NO.	DIMENSIONS L	W	H	Studs	UNIT	5/8" Std. Drywall	UNIT	5/8" F.R. Drywall	UNIT	Acc.	UNIT
Partitions - 25 ga.												
Bath Chase 6"	3	10'		10'	300	SF			600	SF		
Interior 3 5/8"												
Lobby	1	86'		10'	860	SF			1720	SF		
Core	3	90'		10'	2700	SF			5400	SF		
Corner Bead	27	10'									270	LF
Furring 16" O.C.	3	140'		10'			4200	SF			3151	LF
Exterior 3 5/8"	3	560'		10'	10104	SF	10104	SF				
Deduct Windows		496'		4'-5"	10104	SF						
Corner Bead	6	496'									3408	LF
	96	4.5'										
J - Bead	6	496'									3408	LF
	96	4.5'										
Window Returns		3408'	0.5'		1704	SF	1704	SF				
1 5/8" Studs												
Columns 1 5/8"	3	188'		10'	5640	SF						
Unfinished									5640	SF		
Taped									5640	SF		
Corner Bead	3	72'		10'							2160	LF
Quantity Summary												
Studs: 6"					300	SF						
3 5/8"					13664	SF						
1 5/8"					7344	SF						
Drywall: 5/8" F.R.									13360	SF		
Unfinished: 5/8" F.R.									5640	SF		
5/8" Std.							16008	SF				
Corner Bead											5838	LF
J - Bead											3408	LF
Furring											3151	LF

Figure 9.103

- Flagpoles
- Lockers
- Mailboxes
- Partitions
  - Toilet
  - Office
  - Accordion
  - Woven Wire

Items as listed in *Means Building Construction Cost Data* provide a good checklist to help ensure that all appropriate work is included. The prices shown in *Means Building Construction Cost Data* are national averages and do not reflect any particular manufacturer's product or prices.

A thorough review of the drawings and specifications is necessary to be sure that all items are accounted for. The estimator should list each type of item and the recommended manufacturers. Often, no substitutes are allowed. Each type of item is then counted. Takeoff units will vary with different items.

Quotations and bids should be solicited from local suppliers and specialty subcontractors. The estimator must include all appropriate shipping and handling costs. If no historical costs for installation are available, then costs for labor may be taken from cost data books, such as *Means Building Construction Cost Data*, or estimated from scratch. When a specialty item is particularly large, job-site equipment may be needed for placement or installation.

The estimator should pay particular attention to the construction requirements which are necessary to Division 10 work but are included in other divisions. Almost all items in Division 10 require some form of base or backing for proper installation. These requirements may or may not be included in the construction documents, but are usually listed in the manufacturers' recommendations for installation. It is often stated in the General Conditions of the specifications that "the contractor shall install all products according to manufacturers' recommendations"—another catch-all phrase that places responsibility on the contractor (and estimator).

Examples of such items are concrete bases for lockers and supports for ceiling-hung toilet partitions. Concrete bases for lockers cannot be installed until partitions are erected, or at least accurately laid out. The specific locker must be approved by the architect before the exact size of the base is determined. Installation of the concrete often requires a small truckload (hence an extra charge for a minimum order) and hand placement with wheelbarrow and shovel. Similarly, supports for toilet partitions cannot be installed until precise locations are determined. Such supports can be small steel beams, or large angles that must be welded in place. For a room-dividing accordion, folding or telescoping partitions, supports must be strong enough so that no deflection is allowed.

Preparation costs prior to the installation of specialty items may, in some cases, exceed the costs of the items themselves. The estimator must visualize the installation in order to anticipate all of the work and costs.

Sample Estimate: Division 10

The estimate sheet for Division 10 is shown in Figure 9.104. The takeoff and pricing involve a simple counting process. If not included on a schedule, these items should be counted a few times to ensure that all items are included, and in the proper quantities. When estimating for Division 10, it is important to be sure that all backing, supports and blocking are included, usually elsewhere in the estimate. Accessories attached only to drywall with toggles or plastic anchors will work loose very quickly. Installing backing after a wall is finished is very expensive.

There are a few specific manufacturers that are usually specified for bathroom accessories and toilet partitions. A good estimator will find that there are many less expensive but "equal" products (almost exact copies) on the market, and these may be approved by the architects. Smart shopping for such costly items can help to lower the bid.

Division 11: Equipment

Equipment includes permanent fixtures that cause the space to function as designed—i.e., book stacks for libraries, vaults for banks, etc. Often leaving this division until the end of the estimate, the estimator can use the average costs shown in *Means Building Construction Cost Data* for budget pricing purposes. *Means Building Construction Cost Data* can also serve as a checklist to ensure that items have not been omitted.

The construction documents may specify that the owner will purchase equipment directly, and that the contractor will install it. In such cases, the daily output shown in the cost book can be used to complete the labor portion of the estimate. If equipment is furnished by the owner, it is common practice to add about 10% of the materials cost into the estimate. This procedure protects the contractor from the risks, and covers handling costs associated with the materials. Often the contractor is responsible for receipt, storage, and protection of these owner-purchased items until they are installed.

As with specialties in Division 10, equipment must also be evaluated to determine what is required from other divisions for its successful installation, Some possibilities are:

- Concrete
- Miscellaneous Metals
- Rough Carpentry
- Mechanical Coordination
- Electrical Requirements

Division 11 includes equipment that can be packaged and delivered complete or partially assembled by the factory. Also, some items can or must be purchased and installed by an authorized factory representative. The estimator must investigate these variables in order to include adequate costs.

Division 12: Furnishings

Division 12, Furnishings, is best defined as furniture designed for specific uses, such as for dormitories, hospitals, hotels, offices, and restaurants. Window treatments are also included in Division 12. These important furnishings may be listed in the budget estimate to help determine the total financial investment, though they are usually not a part of the actual construction contract. If the furnishings are built-in, then additional labor-hours for unpacking, installing, and clean-up are necessary and must be calculated. In most cases, architects will separately specify and arrange for the purchase and installation of furnishings.

Division 13: Special Construction

The items in this division are specialized subsystems that are usually manufactured or constructed, and installed by specialty subcontractors. The costs shown in *Means Building Construction Cost Data* are for budget purposes only. Final cost figures should be furnished by the appropriate subcontractor after the exact requirements of the project have been specified.

CONSOLIDATED ESTIMATE

PROJECT: Office Building	CLASSIFICATION: Division 10
LOCATION:	ARCHITECT:
TAKE OFF BY: ABC	QUANTITIES BY: ABC PRICES BY: ABC

SHEET NO.   1 of 1

ESTIMATE NO:
DATE: Jan-02
CHECKED BY: GHI

EXTENSIONS BY: DEF   PRICES BY: DEF

SOURCE	DESCRIPTION	QUANT.	UNIT	MATERIAL COST	MATERIAL TOTAL	LABOR COST	LABOR TOTAL	EQUIPMENT COST	EQUIPMENT TOTAL	SUBCONTRACT COST	SUBCONTRACT TOTAL	TOTAL COST	TOTAL
	Division 10: Specialties												
	Bathroom Accessories:												
10820 100 0610	Towel disp./waste receptacle	9	Ea.	345.00	3105	24.00	216						
10820 100 1100	Grab Bar	6	Ea.	26.50	159	12.00	72						
10820 100 4200	Napkin Dispenser	6	Ea.	390.00	2340	16.00	96						
10820 100 3000	Mirrors (Handicapped)	3	Ea.	63.50	191	12.00	36						
10820 100 4600	Soap Dispenser	15	Ea.	42.50	638	12.00	180						
10820 100 5700	Stainless Steel Shelves	6	Ea.	54.50	327	15.00	90						
10820 100 6100	Toilet Tissue Dispenser	18	Ea.	11.75	212	8.00	144						
10820 100 7800	Ash Trays	12	Ea.	75.50	906	13.35	160						
	Toilet Partitions:												
10155 100 2500	Floor Mounted, Handrail	15	Ea.	365.00	5475	80.00	1200						
10155 100 4700	Urinal Screens	3	Ea.	140.00	420	60.00	180						
	Division 10 Total				$ 13,772		$ 2,374						

Figure 9.104

It is a good idea to review this portion of the project with the subcontractor to determine both the exact scope of the work and those items that are not covered by the quotation. If the subcontractor requires services such as excavation, unloading, or other temporary work, then these otherwise excluded items must be included elsewhere in the estimate.

The specialty subcontractor will have more detailed information at hand concerning the system. The more detailed the estimator's knowledge of a system, the easier it will be to subdivide that system into cost components. Each component can be further subdivided into material, labor, and equipment costs that will fully identify the direct cost of the specialty item for future purposes.

Division 14: Conveying Systems

The following systems may be included in this division:

- Correspondence Lifts
- Dumbwaiters
- Elevators
- Escalators and Moving Ramps
- Material Handling Systems
- Pneumatic Tube Systems
- Handicapped Lifts

Because of the specialized construction of the above units, it is almost impossible for the general estimator to price most of this equipment, except in a preliminary, budgetary capacity. Sometimes the plans specify package units that carry standard prices. For general budget pricing, refer to *Means Building Construction Cost Data*. When quotations on specific equipment are received, they should be checked against the specifications to verify that all requirements are met. All required inspections, tests and permits, as well as the responsibility for their costs, should be included in the installed price of each system.

Many of the costs associated with conveying systems are not included in the subcontract bid. When drilling for hydraulic pistons, for example, the preliminary project schedule must be adjusted so that the work can be performed prior to the erection of the superstructure. Drill tailings must also be disposed of, and if ledge or boulders are encountered, drilling costs will increase substantially. Contracts for elevator installations often have more exclusions than inclusions. The estimator must be sure that all associated costs are taken into account.

One of the most commonly underestimated or unanticipated costs is that for the installation of elevator door frames and sills. Typically, frames and sills are supplied by the elevator subcontractor and installed by the general contractor when constructing the elevator shaft walls. At this stage, shop drawings for the elevator usually have not yet been approved. The resulting late installation often involves the cutting and patching of other completed work (concrete floors and block walls). Even if shop drawings are approved, the materials often arrive late. Good planning and proper estimating can help to alleviate or lessen these problems.

Even when determining budget costs for elevators, many variables must be considered. Reference Table R14200-400 from *Means Building Construction Cost Data*, shown in Figure 9.105, indicates the complexity of estimating for elevators.

Sample Estimate: Division 14

The architect's design criteria is apt to be based on the standard elevator "packages" of certain manufacturers. In such cases, firm subcontract bids would be received. For the sample estimate, a budget price is developed and shown in Figure 9.106. The costs would be based on project specifications and are derived from Figure 9.105. If a budget price for elevators is carried in an estimate for bidding, a qualification should be included that the bid may change upon receipt of subcontract prices. Budget prices should be carried only if no specifications are available or obtainable.

Division 15: Mechanical

Unit price estimates for the mechanical and electrical portions of a project should always be performed by the installing subcontractors. Each field requires specialized experience and expertise. Nevertheless, the estimator for the general contractor often requires costs to compare to subcontract bids. In most cases, adjusted square foot costs based on previous, similar projects will suffice. A Systems Estimate may be used if more accuracy is needed. It is essential to understand both the work and the estimating process in order to properly interpret and analyze subcontract costs, and to ensure that all requirements are met. Following are brief discussions of the estimating process for mechanical and electrical work.

When a complete set of mechanical plans and specifications is available, the first step in preparing an estimate is to review all information, making notes on any special or unique requirements. Also, while it is not unusual to see an item on the plans that is not reflected in the specifications, or vice versa, the estimator should make careful note of any contradictions. Such inconsistencies will require resolution before a meaningful estimate can be put together. A review of the other drawings is also advised to determine how the building structure, site layout and other work will affect the mechanical installation. The scales of architectural and engineering drawings should be compared. Plumbing, sprinkler and HVAC designs are frequently prepared by consulting engineers who are not associated with the architects. In these cases, inconsistencies in scale, location, and terminology may result.

One basic feature of almost all mechanical designs is the fact that their different elements can be separated into systems, such as hot water, cold water, fire protection, and heating and/or cooling. Some of these systems may be interconnected or have some common parts, but they are nevertheless distinct. Most systems can be broken down into the following categories: source, conductor or connector, and a terminal unit; for example, water meter—copper tubing—sink faucet. Note that while the sink with faucet or a water closet may be terminal units for the cold water system, they are a source for the drain-waste-vent (DWV) system.

Pre-printed forms are especially useful for preparing the mechanical portion of the estimate. While careful measurements and a count of components are important, they will not compensate for omissions such as forgetting to include pipe insulation. A well-designed form acts as a checklist, a guide for standardization, and a permanent record.

Plumbing

The first step in preparing a plumbing estimate is to visualize the scope of the job by scanning all of the drawings and specifications. The next step is to make a list of the types of materials on a takeoff sheet. This process will help the estimator to remember the various components as they are located on the drawings. For major items, pieces of equipment and fixtures, the quantity sheet can serve as a checklist. The estimator must be sure to include the required

R14200-400   Elevator Cost Development

Requirement: One passenger elevator, five story hydraulic, 2,500 lb. capacity, 12' floor to floor, speed 150 F.P.M., emergency power switching and maintenance contract.

Description	Adjustment		Unit Cost	Total Cost
A. Base Elevator, Hyd Pass, Base Unit, 1500 lb, 100 fpm, 2 Stops, Std. Finish				$39,407
B. Capacity Adjustment (2,500 lb.)	1	Ea.	$1,200	1,200
C. Excess Travel Over Base (4 x 12') = 48' — (12' for Base Unit) =	36	V.L.F.	355	12,780
D. Stops Over Base 5 — (2 for Base Unit) =	3	Stops	3,725	11,175
E. Speed Adjustment (150 F.P.M.)	1	Ea.	1,750	1,750
F. Options:				
1. Intercom Service	1	Ea.	615	615
2. Emergency Power Switching, Automatic	1	Ea.	2,175	2,175
3. Stainless Steel Entrance Doors	5	Ea.	935	4,675
4. Maintenance Contract (12 Months)	1	Ea.	2,675	2,675
5. Position Indicators	2	Ea.	335	670
Total Cost				$77,122

Figure 9.105

CONSOLIDATED ESTIMATE

PROJECT: Office Building	CLASSIFICATION: Division 14	ESTIMATE NO:
LOCATION:	ARCHITECT: As Shown	DATE: Jan-02
TAKE OFF BY: ABC	QUANTITIES BY: ABC PRICES BY: ABC	CHECKED BY: GHI
	EXTENSIONS BY: DEF	

DESCRIPTION	SOURCE			QUANT	UNIT	MATERIAL DEF	COST	TOTAL	LABOR DEF	COST	TOTAL	EQUIPMENT DEF	COST	TOTAL	SUBCONTRACT COST	TOTAL	TOTAL COST	TOTAL
Division 14: Conveying System																		
Pass. Hydraulic - Base	14210	200	2050	1	Ea.										39400	39400		
For 2500 Lbs, add	14210	200	2100	1	Ea.										1200	1200		
Travel 37'	14210	200	2350	37	V.L.F.										470	17390		
Stops (plus 2)	14210	200	2375	1	Stop										3800	3800		
Speed - 100 fpm	14210	200	3400	1	Ea.										925	925		
S.S. Doors	14210	200	3425	1	Ea.										370	370		
Carpet																		
P.L. Walls	14210	200	3625	1	Ea.										925	925		
S.S. Entrance - Doors	14210	200	3650	1	Ea.										925	925		
- Frame																		
Each Elevator															$	64,935		
Elevator				2	Ea.										64935	129870		
Automatic Controls				1	Ea.										3650	3650		
Fire Service																		
Hall Lanterns				8	Ea.										440	3520		
Position Indicators				4	Ea.										335	1340		
Maintenance				2	Ea.										2640	5280		
Division 14 Total															$	143,660		

Figure 9.106

278

labor-only items, such as cleaning, adjusting, purifying, testing, and balancing, since these will not show up on the drawings and have little, if any, material costs.

The easiest way to do a material takeoff is by system, as most of the pipe components of a system will tend to be related as a group—of the same material, class, weight, and grade. For example, a waste system could consist of pipe varying from 3" to 8" in diameter, but would probably be all service weight cast iron, DWV copper, or PVC up to a specified size, and then cast iron for the larger sizes.

Fixtures: Fixture takeoff is usually nothing more than counting the various types, sizes, and styles, and then entering them on a fixture form. It is important, however, that each fixture be fully identified. A common error occurs when parts that must be purchased separately are overlooked; examples are trim, carriers, and flush valves. Equipment like pumps, water heaters, water softeners, and all items not previously counted are also listed at this time. The order in which the takeoff proceeds is not as important as the development of a consistent method. With consistency, the estimator can speed up the process while minimizing the chances of overlooking any item or class of items. Taking off fixtures and equipment before the piping is best for several reasons. Done in this order, the fixture and equipment lists can be given to suppliers for pricing while the estimator is performing the more arduous and time-consuming piping takeoff. Taking off the fixtures and equipment first also gives the estimator a good perspective on the building and its systems. The next step is transferring these figures to the estimate sheet. Figure 9.107 shows relative installation times for different types of fixtures.

Piping: Fixture costs in *Means Building Construction Cost Data* (Figure 9.108) are based on the cost per fixture set in place. The rough-in piping cost, which must be added for each fixture, includes a carrier, if required, some supply, waste and vent pipe, connecting fittings and stops. The lengths of rough-in pipe are

Plumbing Fixture Installation Time

Item	Rough-In	Set	Total Hrs.	Item	Rough-In	Set	Total Hrs.
Bathtub	5	5	10	Shower head only	2	1	3
Bathtub and shower, cast iron	6	6	12	Shower drain	3	1	4
Fire hose reel and cabinet	4	2	6	Shower stall, slate		15	15
Floor drain to 4" diameter	3	1	4	Slop sink	5	3	8
Grease trap, single, cast iron	5	3	8	Test 6 fixtures			14
Kitchen gas range		4	4	Urinal, wall	6	2	8
Kitchen sink, single	4	4	8	Urinal, pedestal or floor	6	4	10
Kitchen sink, double	6	6	12	Water closet and tank	4	3	7
Laundry tubs	4	2	6	Water closet and tank, wall hung	5	3	8
Lavatory wall hung	5	3	8	Water heater, 45 gals. gas, automatic	5	2	7
Lavatory pedestal	5	3	8	Water heaters, 65 gals. gas, automatic	5	2	7
Shower and stall	6	4	10	Water heaters, electric, plumbing only	4	2	6

Figure 9.107

nominal runs which would connect to the larger runs and stacks (to within 10'). The supply runs and DWV runs and stacks must be accounted for in separate entries.

Pipe runs for any type of system consist of straight sections and fittings of various shapes and styles, for various purposes. Depending on the required detail, pipe and fittings can be itemized or included as a percentage of the fixtures. If detail is required, the estimator should measure and record the lengths of each size of pipe. When a fitting is crossed, a list of all fitting types and sizes can be used for counting. Colored pencils can be used to mark runs that have been completed, and to note termination points on the main line where measurements are stopped so that a branch may be taken off. Care must be taken to note changes in pipe material. Since different materials can only meet at a joint, it should become an automatic habit to see that the piping material going into a joint is the same as that leaving the joint.

When summarizing quantities of piping for estimate or purchase, it is good practice to round the totals of each size up to the lengths normally available from the supply house or mill. This method is even more appropriate in larger projects where rounding might be done to the nearest hundred or thousand feet. With this approach, a built-in percentage for scrap or waste is allowed. Rigid copper tubing and rigid plastic pipe are normally supplied in 20' lengths, cast iron soil pipe in either 5' or 10' sections. Steel pipe of 2" diameter or less is furnished in 21' lengths, and pipe of a larger diameter is available in single random lengths ranging from 16' to 22'.

Apart from counting and pricing every pipe and fitting, there are other ways to determine budget costs for plumbing. *Means Building Construction Cost Data* provides costs for fixtures, as well as the associated costs for rough-in of the supply, waste and vent piping as shown in Figure 9.108. The rough-in costs include piping within 10' of the fixture. If these prices are used, further costs must be added for the stacks and mains. Another method involves adding percentages to the cost of the fixtures. Recommended percentages are shown in Figure 9.109. Using information such as that provided in Figures 9.108 and 9.109, the estimator can develop "systems" to be used for budget pricing.

In addition to the cost of pipe installation, the estimator must also consider any associated costs. In many cases, mechanical and electrical subcontractors must dig, by hand, their own under slab trenches. Site utility excavation (and backfill) may also be included, if required. Underground piping may require special wrapping. For interior piping, the estimator must visualize the installation in order to determine how—and to what—the pipe hangers are attached. Overhead installations will require rolling scaffolding. For installations higher than an average of 15', labor costs may be increased by the following suggested percentages:

Ceiling Height	Labor Increase
15' to 20'	10%
20' to 25'	20%
25' to 30'	30%
30' to 35'	40%
35' to 40'	50%
Over 40'	60%

Fire Protection

The takeoff of fire protection systems (sprinklers and standpipes) is very much like that of other plumbing—the estimator should measure the pipe loops and

15418	Resi/Comm/Industrial Fixtures	CREW	DAILY OUTPUT	LABOR-HOURS	UNIT	2002 BARE COSTS				TOTAL INCL O&P	
						MAT.	LABOR	EQUIP.	TOTAL		
600	2100	31" x 22" single bowl R15100-420	Q-1	5.60	2.857	Ea.	247	92.50		339.50	410	600
2200	32" x 21" double bowl		4.80	3.333		287	108		395	480
2300	42" x 21" double bowl		4.80	3.333		470	108		578	685
3000	Stainless steel, self rimming, 19" x 18" single bowl		5.60	2.857		282	92.50		374.50	450
3100	25" x 22" single bowl		5.60	2.857		310	92.50		402.50	480
4000	Steel, enameled, with ledge, 24" x 21" single bowl		5.60	2.857		105	92.50		197.50	255
4100	32" x 21" double bowl	▼	4.80	3.333		143	108		251	320
4960	For color sinks except stainless steel, add					10%				
4980	For rough-in, supply, waste and vent, counter top sinks	Q-1	2.14	7.477	▼	85	242		327	460
5000	Kitchen, raised deck, P.E. on C.I.									
5100	32" x 21", dual level, double bowl	Q-1	2.60	6.154	Ea.	360	199		559	695
5700	For color, add					20%				
5790	For rough-in, supply, waste & vent, sinks	Q-1	1.85	8.649		85	280		365	515
6650	Service, floor, corner, P.E. on C.I., 28" x 28"		4.40	3.636		490	118		608	710
6790	For rough-in, supply, waste & vent, floor service sinks		1.64	9.756		194	315		509	690
7000	Service, wall, P.E. on C.I., roll rim, 22" x 18"		4	4		405	129		534	645
7100	24" x 20"		4	4		445	129		574	685
8600	Vitreous china, 22" x 20"	▼	4	4		380	129		509	615
8960	For stainless steel rim guard, front or side, add					31			31	34
8980	For rough-in, supply, waste & vent, wall service sinks	Q-1	1.30	12.308	▼	305	400		705	940
900	0010	WATER CLOSETS R15100-420										900
0030	For automatic flush, see 15410-300-0972									
0150	Tank type, vitreous china, incl. seat, supply pipe w/stop									
0200	Wall hung, one piece	Q-1	5.30	3.019	Ea.	425	97.50		522.50	615
0400	Two piece, close coupled		5.30	3.019		545	97.50		642.50	745
0960	For rough-in, supply, waste, vent and carrier		2.73	5.861		259	190		449	570
1000	Floor mounted, one piece		5.30	3.019		495	97.50		592.50	690
1100	Two piece, close coupled, water saver CN	▼	5.30	3.019	▼	140	97.50		237.50	300
1960	For color, add					30%				
1980	For rough-in, supply, waste and vent	Q-1	3.05	5.246	Ea.	126	170		296	395
3000	Bowl only, with flush valve, seat									
3100	Wall hung	Q-1	5.80	2.759	Ea.	340	89		429	510
3200	For rough-in, supply, waste and vent, single WC		2.56	6.250		271	202		473	605
3300	Floor mounted		5.80	2.759		315	89		404	485
3350	With wall outlet		5.80	2.759		480	89		569	665
3390	Water closet bowl w/auto self flush, self clean, self sanitizing		4.70	3.404		1,500	110		1,610	1,825
3400	For rough-in, supply, waste and vent, single WC	▼	2.84	5.634	▼	138	182		320	425

15440	Plumbing Pumps										
240	0010	PUMPS, PRESSURE BOOSTER SYSTEM										240
0200	Pump system, with diaphragm tank, control, press. switch									
0300	1 HP pump	Q-1	1.30	12.308	Ea.	3,275	400		3,675	4,200
0400	1-1/2 HP pump		1.25	12.800		3,325	415		3,740	4,275
0420	2 HP pump		1.20	13.333		3,400	430		3,830	4,375
0440	3 HP pump	▼	1.10	14.545		3,450	470		3,920	4,500
0460	5 HP pump	Q-2	1.50	16		3,800	535		4,335	5,000
0480	7-1/2 HP pump		1.42	16.901		4,250	565		4,815	5,525
0500	10 HP pump	▼	1.34	17.910	▼	4,450	600		5,050	5,775
1000	Pump/ energy storage system, diaphragm tank, 3 HP pump									
1100	motor, PRV, switch, gauge, control center, flow switch									
1200	125 lb. working pressure	Q-2	.70	34.286	Ea.	9,550	1,150		10,700	12,200
1300	250 lb. working pressure	"	.64	37.500	"	10,400	1,250		11,650	13,400
400	0010	PUMPS, GRINDER SYSTEM Complete, incl. check valve, tank, std.										400
0020	controls incl. alarm/disconnect panel w/wire. Excavation not included									

Figure 9.108

count fittings, valves, sprinkler heads, alarms, and other components. The estimator then makes note of special requirements, as well as any conditions that would affect job performance and cost.

There are many different types of sprinkler systems; examples are wet pipe, dry pipe, pre-action, and chemical. Each of these types involves a different set of requirements and costs. Figure 9.110, from *Means Mechanical Cost Data*, provides descriptions of various sprinkler systems.

Most sprinkler systems must be approved by Factory Mutual for insurance purposes *before* installation. This requirement means that shop drawings must be produced and submitted for approval. The shop drawings take time and the approval process takes more time. This process can cause serious delays and added expense if it is not anticipated and given adequate consideration at the estimating and scheduling phase.

Square foot historical costs for fire protection systems may be developed for budget purposes. These costs may be based on the relative hazard of occupancy—light, ordinary and extra. A comparison of some requirements of the different hazards is shown in Figure 9.111. Consideration must also be given to special or unusual requirements. For example, many architects specify that sprinkler heads must be located in the center of ceiling tiles. Each head may require extra elbows and nipples for precise location. Recessed heads are more expensive. Special dry pendant heads are required in areas subject to freezing. When installing a sprinkler system in an existing structure, a completely new water service may be required in addition to the existing domestic water service. These are just a few examples of requirements which may necessitate an adjustment of square foot costs.

Heating, Ventilation, and Air Conditioning

As with plumbing, equipment for HVAC should be taken off first so that the estimator is familiarized with the various systems and layouts. While actual

---

Plumbing Approximations for Quick Estimating

Water Control:
Water Meter; Backflow Preventer, Shock Absorbers; Vacuum Breakers; Mixer. .... 10 to 15% of Fixtures

Pipe and Fittings: .... 30 to 60% of Fixtures

> Note: Lower percentage for compact buildings or larger buildings with plumbing in one area.
> Larger percentage for large buildings with plumbing spread out.
> In extreme cases pipe may be more than 100% of fixtures.
> Percentages **do not** include special purpose or process piping.

Plumbing Labor:
1 & 2 Story Residential .... Rough-in Labor = 80% of Materials
Apartment Buildings .... Rough-in Labor = 90 to 100% of Materials
Labor for handling and placing fixtures is approximately 25 to 30% of fixtures.

Quality/Complexity Multiplier (For all installations)
Economy installation, add .... .0 to 5%
Good quality, medium complexity, add .... .5 to 15%
Above average quality and complexity, add .... .15 to 25%

Figure 9.109

R10520-110 Sprinkler Systems (Automatic)

Sprinkler systems may be classified by type as follows:

1. **Wet Pipe System.** A system employing automatic sprinklers attached to a piping system containing water and connected to a water supply so that water discharges immediately from sprinklers opened by a fire.

2. **Dry Pipe System.** A system employing automatic sprinklers attached to a piping system containing air under pressure, the release of which as from the opening of sprinklers permits the water pressure to open a valve known as a "dry pipe valve". The water then flows into the piping system and out the opened sprinklers.

3. **Pre-Action System.** A system employing automatic sprinklers attached to a piping system containing air that may or may not be under pressure, with a supplemental heat responsive system of generally more sensitive characteristics than the automatic sprinklers themselves, installed in the same areas as the sprinklers; actuation of the heat responsive system, as from a fire, opens a valve which permits water to flow into the sprinkler piping system and to be discharged from any sprinklers which may be open.

4. **Deluge System.** A system employing open sprinklers attached to a piping system connected to a water supply through a valve which is opened by the operation of a heat responsive system installed in the same areas as the sprinklers. When this valve opens, water flows into the piping system and discharges from all sprinklers attached thereto.

5. **Combined Dry Pipe and Pre-Action Sprinkler System.** A system employing automatic sprinklers attached to a piping system containing air under pressure with a supplemental heat responsive system of generally more sensitive characteristics than the automatic sprinklers themselves, installed in the same areas as the sprinklers; operation of the heat responsive system, as from a fire, actuates tripping devices which open dry pipe valves simultaneously and without loss of air pressure in the system. Operation of the heat responsive system also opens

approved air exhaust valves at the end of the feed main which facilitates the filling of the system with water which usually precedes the opening of sprinklers. The heat responsive system also serves as an automatic fire alarm system.

6. **Limited Water Supply System.** A system employing automatic sprinklers and conforming to these standards but supplied by a pressure tank of limited capacity.

7. **Chemical Systems.** Systems using halon, carbon dioxide, dry chemical or high expansion foam as selected for special requirements. Agent may extinguish flames by chemically inhibiting flame propagation, suffocate flames by excluding oxygen, interrupting chemical action of oxygen uniting with fuel or sealing and cooling the combustion center.

8. **Firecycle System.** Firecycle is a fixed fire protection sprinkler system utilizing water as its extinguishing agent. It is a time delayed, recycling, preaction type which automatically shuts the water off when heat is reduced below the detector operating temperature and turns the water back on when that temperature is exceeded. The system senses a fire condition through a closed circuit electrical detector system which controls water flow to the fire automatically. Batteries supply up to 90 hour emergency power supply for system operation. The piping system is dry (until water is required) and is monitored with pressurized air. Should any leak in the system piping occur, an alarm will sound, but water will not enter the system until heat is sensed by a firecycle detector.

Area coverage sprinkler systems may be laid out and fed from the supply in any one of several patterns as shown below. It is desirable, if possible, to utilize a central feed and achieve a shorter flow path from the riser to the furthest sprinkler. This permits use of the smallest sizes of pipe possible with resulting savings.

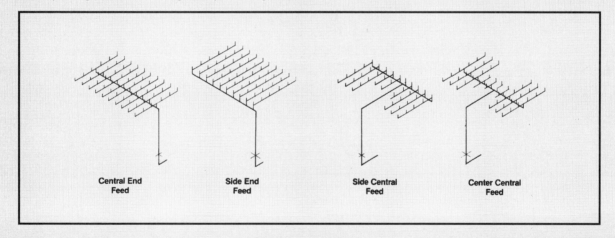

| Central End Feed | Side End Feed | Side Central Feed | Center Central Feed |

Figure 9.110

pieces of equipment must be counted at this time, it is also necessary to note sizes, capacities, controls, special characteristics and features. The weight and size of equipment may be important if the unit is especially large or is going into comparatively close quarters. If hoisting or rigging is not included in the subcontract price, then costs for placing the equipment must be figured and listed. From the equipment totals, other important items, such as motor starters, valves, strainers, gauges, thermometers, traps, and air vents can also be counted.

Sheet metal ductwork for heating, ventilation, and air conditioning is usually estimated by weight. The lengths of the various sizes are measured and recorded on a worksheet. The weight per foot of length is then determined. Figure 9.112 is a conversion chart for determining the weight of ductwork based on size and material. A count must also be made of all duct-associated accessories, such as fire dampers, diffusers, and registers. This count may be done during the duct takeoff. It is usually less confusing, however, to make a separate count. For budget purposes, ductwork, insulation, diffusers and registers may be estimated using the information in Figure 9.113.

The takeoff of heating, ventilation and air conditioning pipe and fittings is accomplished in a manner similar to that of plumbing. In addition to the general, miscellaneous items noted during the review of plans and specifications, the heating, ventilation and air conditioning estimate usually includes the work of subcontractors. All material suppliers and appropriate subcontractors should be notified as soon as possible to verify material availability and pricing and to ensure timely submission of bids. Typical subcontract work includes:

- Balancing
- Controls
- Insulation
- Water Treatment
- Sheet Metal
- Core Drilling

Sprinkler Quantities for Various Size and Types of Pipe

Sprinkler Quantities: The table below lists the usual maximum number of sprinkler heads for each size of copper and steel pipe for both wet and dry systems. These quantities may be adjusted to meet individual structural needs or local code requirements. Maximum area on any one floor for one system is: light hazard and ordinary hazard 52,000 S.F., extra hazardous 25,000 S.F.

Pipe Size	Light Hazard Occupancy		Ordinary Hazard Occupancy		Extra Hazard Occupancy	
Diameter	Steel Pipe	Copper Pipe	Steel Pipe	Copper Pipe	Steel Pipe	Copper Pipe
1"	2 sprinklers	2 sprinklers	2 sprinklers	2 sprinklers	1 sprinklers	1 sprinklers
1-1/4"	3	3	3	3	2	2
1-1/2"	5	5	5	5	5	5
2"	10	12	10	12	8	8
2-1/2"	30	40	20	25	15	20
3"	60	65	40	45	27	30
3-1/2"	100	115	65	75	40	45
4"			100	115	55	65
5"			160	180	90	100
6"			275	300	150	170

Figure 9.111

If similar systems are used repeatedly, it is easy to develop historical square foot costs for budget and comparison purposes. However, with variation in the types and size of HVAC systems, the estimator must use caution when using square foot prices. A way to adjust such relative prices can be to adjust costs based on the quality or complexity of the installation:

Quality/Complexity	Adjustment
Economy/Low	0 to 5%
Good/Medium	5 to 15%
Above Average/High	15 to 25%

Sample Estimate: Division 15

For bidding purposes, firm subcontract prices should always be solicited for all phases of mechanical work. However, the estimator should still be familiar with the project requirements and the basics of mechanical estimating. Only with an understanding of what is involved will the estimator be able to judge the validity of a subcontract quote.

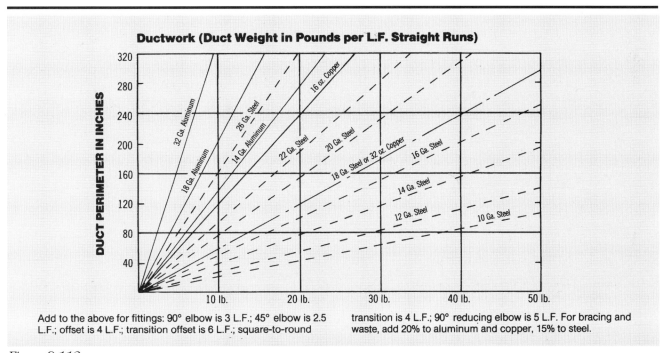

Ductwork (Duct Weight in Pounds per L.F. Straight Runs)

Add to the above for fittings: 90° elbow is 3 L.F.; 45° elbow is 2.5 L.F.; offset is 4 L.F.; transition offset is 6 L.F.; square-to-round transition is 4 L.F.; 90° reducing elbow is 5 L.F. For bracing and waste, add 20% to aluminum and copper, 15% to steel.

Figure 9.112

Ductwork Packages (per ton of cooling)

System	Sheet Metal	Insulation	Diffusers	Return Register
Roof Top Unit Single Zone	120 Lbs.	52 S.F.	1	1
Roof Top Unit Multizone	240 Lbs.	104 S.F.	2	1
Self-contained Air or Water Cooled	108 Lbs.	—	2	—
Split System Air Cooled	102 Lbs.	—	2	—

Figure 9.113

For the purposes of this sample estimate, Figures 9.114 through 9.119 are included to show how a subcontractor might prepare a bid for the HVAC work. Unit prices are from *Means Mechanical Cost Data*. Different types of ductwork installations are separated on the quantity sheet in Figure 9.114. The weight (pounds per linear foot) for each size of duct is derived from the table in Figure 9.120 from *Means Mechanical Cost Data*. The detail of the HVAC estimate reflects the precision required for proper mechanical estimating.

Overhead and profit are added to the bare cost totals using the methods described in Chapters 4 and 7. Ten percent is added to each—material, equipment and subcontracts—for handling and supervision. To labor, an average 50.8% is added. This includes employer-paid taxes and insurance, office overhead, and profit.

While usually only one subcontractor will handle the complete mechanical portion of the project, different subcontractors may separately bid and perform the plumbing, fire protection and HVAC portions of the estimate. When soliciting quotes, the estimator for the general contractor should try to obtain detailed breakdowns of each. The estimator will be able to use these breakdowns for comparing bids if the quote is not as expected, and can thereby discover which areas are "out of line." While subcontractors are often reluctant to provide detailed information, it is to their own benefit if possible errors or omissions which exist in their prices are discovered.

A summary of the subcontract prices is shown in Figure 9.121. Separation and breakdowns as shown can be used to help develop historical costs for future projects, whether they are done by system or square foot.

Division 16: Electrical

A good estimator must be able to visualize the proposed electrical installation from source (service) to end use (fixtures or devices). This process helps to identify each component needed to make the system work. Before starting the takeoff, the estimator should follow some basic steps:

- Read the electrical specifications thoroughly.
- Scan the electrical plans. Check other sections of the plans and specifications for their potential effect on electrical work (special attention should be given to mechanical and site work).
- Check the architectural and structural plans for unique or atypical requirements.
- Check and become familiar with the fixture and power symbols.
- Clarify with the architect or engineer any unclear areas, making sure that the scope of work is understood. Addenda may be necessary to clarify certain items of work so that the responsibility for the performance of all work is defined.
- Immediately contact suppliers, manufacturers, and subsystem specialty contractors in order to get their quotations, sub-bids, and drawings.

Certain information should be taken from complete plans and specifications in order to properly estimate the electrical portion of the work. In addition to floor plans, other sources of information are:

- Power riser diagram
- Panelboard schedule
- Fixture schedule
- Electrical symbol legend
- Reflected ceiling plans
- Branch circuit plans

QUANTITY SHEET

PROJECT: Office Building   Division 15   SHEET NO. 1 of 1

LOCATION: ARCHITECT: ESTIMATE NO:

TAKE OFF BY: EXTENSIONS BY: DATE:

CHECKED BY:

DESCRIPTION	NO.	DIMENSIONS		Insulated Supply	UNIT	Supply not Insulated	UNIT	Return	UNIT	Toilet Exhaust	UNIT
Ductwork											
Supply (Insul.) 3rd											
45 x 14	80	14.8	22	1184	Lbs.						
25 x 14	92	8.4	24	773							
22 x 14	144	7.8	"	1123							
18 x 14	192	6.9	"	1325							
14 x 14	96	6.0	"	576							
12 x 9	96	4.5	"	432							
15 x 14	96	6.2	"	595							
13 x 12	64	5.4	"	346							
12 x 8	96	4.3	"	413							
				(6767)	Lbs.						
Supply (not Insul.) 1st & 2nd											
6767 x 2						(13534)	Lbs.				
Return											
50 x 14	144	16.0	22					2304	Lbs.		
33 x 14	132	11.8	22					1558			
18 x 14	132	6.9	24					911			
								(4773)	Lbs.		
Toilet Exhaust											
18 x 12	18	6.5	24							117	Lbs.
16 x 12	14	6.0	"							84	
12 x 12	14	5.2	"							73	
6 x 6	18	2.6	"							47	
										(321)	Lbs.

Figure 9.114

287

CONSOLIDATED ESTIMATE

PROJECT: Office Building
LOCATION:
TAKE OFF BY: ABC
CLASSIFICATION: Division 15
ARCHITECT:
QUANTITIES BY: ABC
PRICES BY: As Shown
EXTENSIONS BY:

ESTIMATE NO:
DATE: Jan-02
CHECKED BY: GHI

DESCRIPTION	SOURCE			QUANT	UNIT	MATERIAL DEF COST	TOTAL	LABOR DEF COST	TOTAL	EQUIPMENT DEF COST	TOTAL	SUBCONTRACT COST	TOTAL	TOTAL COST	TOTAL
Division 15: Sheet Metal Duct Work															
Supply Duct 3rd Floor (Insul)				6800	Lb.										
Supply Duct (Not Insulated)				14000	Lb.										
Return Duct (Not Insulated)				4800	Lb.										
Toilet Exhaust Duct Work				325	Lb.										
Total Duct Work	15810	600	0580	25925	Lb.	0.31	8037	2.76	71553						
Supply Diffuser 9" x 9"	15850	300	2020	3	Ea.	86.50	260	20.00	60						
Supply Diffuser 15" x 15"	15850	300	2060	132	Ea.	146.00	19272	25.50	3366						
Framed Eggcrate 24" x 24"	15850	500	4040	19	Ea.	58.50	1112	18.70	355						
Splitter Damper, 1 Foot Rod	15820	300	7000	12	Ea.	16.55	199	11.70	140						
Flexible Toilet Exhaust Duct 6"	15810	300	1560	200	LF	1.30	260	1.81	362						
Flexible Toilet Exhaust Duct 4"	15810	300	1520	90	LF	0.98	88	1.40	126						
Exhaust Register 8" X 6"	15850	700	5040	3	Ea.	22.00	66	11.70	35						
Exhaust Register 8" X 4"	15850	700	5020	1	Ea.	22.00	22	10.80	11						
Roof Exhaust fan 4600 CFM	15830	100	7240	1	Ea.	1175	1175	162.00	162						
Double Wall Gavl. Chimney 24"	15550	440	0340	12	LF	147.00	1764	24.50	294						
Roof flashing Collar 24"	15550	440	1170	1	Ea.	247.00	247	65.50	66						
Double Wall Tee 24"	15550	440	1330	1	Ea.	705.00	705	65.50	66						
Tee Cap 24"	15550	440	1590	1	Ea.	48.00	48	37.50	38						
Rain Cap & Screen 24"	15550	440	1880	1	Ea.	680.00	680	39.50	40						
Subtotals						$	33,934	$	76,673							

Figure 9.115

CONSOLIDATED ESTIMATE

PROJECT: Office Building
LOCATION:
TAKE OFF BY: ABC

CLASSIFICATION: Division 15
ARCHITECT:
QUANTITIES BY: ABC
PRICES BY: DEF
EXTENSIONS BY: DEF

ESTIMATE NO:
DATE: Jan-02
CHECKED BY: GHI

Division 15: (Cont'd)

Air Conditioning

DESCRIPTION	SOURCE	QUANT.	UNIT	MATERIAL COST	MATERIAL TOTAL	LABOR COST	LABOR TOTAL	EQUIPMENT COST	EQUIPMENT TOTAL
Chiller 175 Tons	15620 600 1600	1	Ea.	83000	83000	8425	8425		
Fan Coil Unit 30 Tons	15720 200 1600	6	Ea.	7375	44250	1750	10500		
Cooling Tower 175 Tons	15640 400 1900	175	Ton	94.00	16450	6.45	1129		
Thermometer 6", 30 - 140 Deg F	13838 200 4500	4	Ea.	55.50	222	9.05	36		
Dial Pressure Gauge 3-1/2"	13838 200 2300	6	Ea.	21.00	126	9.05	54		
Condenser Water Pump 7-1/2 HP	15180 200 4420	1	Ea.	2000	2000	325.00	325		
Chiller Water Pump 5 HP	15180 200 4410	2	Ea.	2075	4150	325.00	650		
Steel Pipe, Bevel End 4"	15107 620 2110	180	LF	9.00	1620	14.00	2520		
Steel Pipe, Thread & Coupled 2"	15107 620 0610	340	LF	3.18	1081	8.10	2754		
Weld Tee 4"	15107 660 3440	15	Ea.	36.50	548	173.00	2595	26.50	398
Weld Elbow 4"	15107 660 3130	20	Ea.	18.55	371	104.00	2080	16.00	320
Weld Neck Flange 4"	15107 660 6500	14	Ea.	26.00	364	52.00	728	8.00	112
4" Sets, Nuts, Bolts & Gaskets	15107 660 0670	30	Ea.	8.70	261	36.00	1080		
Cast Iron Tee 4"	15550 640 0580	12	Ea.	10.75	129	47.00	564		
Cast Iron Elbow 4"	15107 640 0140	24	Ea.	7.60	182	29.00	696		
Gate Valve OS& Y 4"	15107 200 3680	8	Ea.	355.00	2840	173.00	1384		
Check Valve Wafer 4"	15110 200 6670	3	Ea.	210.00	630	104.00	312		
Gate Valve Bronze 4"	15110 160 3480	6	Ea.	80.50	483	26.00	156		
Balancing Valve Flanged 4"	15110 600 7030	3	Ea.	335.00	1005	173.00	519		
Stop & Balance Valve 2"	15110 600 1080	6	Ea.	120.00	720	36.00	216		
Flanged Y Strainer 4"	15120 840 1060	3	Ea.	155.00	465	174.00	522		
Expansion Tank 31 Gallon ASME	15120 320 1880	1	Ea.	1450	1450	65.00	65		
					$ 162,347		$ 37,310		$ 830

Figure 9.116

289

CONSOLIDATED ESTIMATE

PROJECT: Office Building
LOCATION:
TAKE OFF BY: ABC
CLASSIFICATION: Division 15
ARCHITECT:
QUANTITIES BY: ABC
PRICES BY: As Shown
EXTENSIONS BY:

ESTIMATE NO:
DATE: Jan-02
CHECKED BY: GHI

DESCRIPTION	SOURCE	QUANT	UNIT	MATERIAL COST	MATERIAL TOTAL	LABOR COST	LABOR TOTAL	EQUIPMENT COST	EQUIPMENT TOTAL	SUBCONTRACT COST	SUBCONTRACT TOTAL	TOTAL COST	TOTAL
Division 15: (Cont'd)													
Heating													
Water Heater, Gas CI 3500 MBH	15510 400 3400	1	Ea.	18300	18300	6725	6725						
Electrci Duct Heater	15761 200 0160	1	Ea.	1575	1575	50.00	50						
Expansion Tank 79 Gallon ASME	15120 320 3060	1	Ea.	2175	2175	104.00	104						
Steel Pipe, Bevel End 3"	15107 620 2090	120	LF	6.00	720	12.05	1446	1.86	223				
Stl Pipe, Thread & Coupled 1-1/2	15107 620 0600	320	LF	2.51	803	6.45	2064						
Weld Tee 3"	15107 660 3430	15	Ea.	25.50	383	129.00	1935	20.00	300				
Weld Elbow 3"	15107 660 3120	12	Ea.	10.90	131	74.00	888	11.45	137				
Weld Neck Flange 3"	15107 660 6480	8	Ea.	22.00	176	37.00	296	5.70	46				
3" Sets, Nuts, Bolts & Gaskets	15107 660 0650	16	Ea.	4.80	77	26.00	416						
Cast Iron Tee 1-1/2"	15550 640 0570	12	Ea.	7.70	92	40.00	480						
Cast Iron Elbow 1-1/2"	15107 640 0140	24	Ea.	7.60	182	29.00	696						
Gate Valve OS& Y 3"	15107 200 3660	4	Ea.	250.00	1000	115.00	460						
Check Valve Wafer 3"	15110 200 6670	2	Ea.	165.00	330	64.50	129						
Gate Valve Bronze 1-1/2"	15110 160 3470	6	Ea.	59.00	354	22.00	132						
Balancing Valve Flanged 3"	15110 600 7020	2	Ea.	263.00	526	115.00	230						
Stop & Balance Valve 1-1/2"	15110 600 1070	6	Ea.	102.00	612	26.00	156						
Flanged Y Strainer 3"	15120 840 1040	3	Ea.	84.50	254	116.00	348						
Air Control Fitting 3"	15120 120 0100	3	Ea.	920.00	2760	130.00	390						
Thermometer 6", 30 - 140 Deg F	13838 200 4500	2	Ea.	55.50	111	9.05	18						
Dial Pressure Gauge 3-1/2"	13838 200 2300	4	Ea.	21.00	84	9.05	36						
Hot Water Pump 5 HP	15180 200 4300	2	Ea.	1425	2850	288.00	576						
					$ 33,495		$ 17,575		$ 706				

Figure 9.117

CONSOLIDATED ESTIMATE

PROJECT: Office Building	CLASSIFICATION: Division 15
LOCATION:	ARCHITECT:
TAKE OFF BY: ABC	QUANTITIES BY: ABC PRICES BY: DEF EXTENSIONS BY: DEF

ESTIMATE NO:
DATE: Jan-02
CHECKED BY: GHI

DESCRIPTION	SOURCE	QUANT	UNIT	MATERIAL COST	MATERIAL TOTAL	LABOR COST	LABOR TOTAL	EQUIPMENT COST	EQUIPMENT TOTAL	SUBCONTRACT COST	SUBCONTRACT TOTAL	TOTAL COST	TOTAL
HVAC: Miscellaneous													
Recording Drawings		8	Hour										
Operating Instructions		8	Hour										
Maintenance Manuals		4	Hour										
Cleaning Systems		16	Hour										
		36	Hour			36.20	1303						
Subcontracts													
Insulation											22014		
Balancing											10400		
Controls											47351		
Crane									1706				
						$ 1,303		$ 1,706		$ 79,765			

Figure 9.118

291

CONSOLIDATED ESTIMATE

PROJECT: Office Building　　CLASSIFICATION: Division 15
LOCATION:　　ARCHITECT:
TAKE OFF BY: ABC　　QUANTITIES BY: ABC　　PRICES BY: ABC　　EXTENSIONS BY: DEF

ESTIMATE NO:
DATE: Jan-02
CHECKED BY: GHI

DESCRIPTION	SOURCE	QUANT	UNIT	MATERIAL			LABOR		EQUIPMENT		SUBCONTRACT		TOTAL	
				COST	DEF	TOTAL	COST	TOTAL	COST	TOTAL	COST	TOTAL	COST	TOTAL
Division 15: (Cont'd)														
HVAC: Summary														
Ductwork	Sheet 1					$ 33,934		$ 76,673						
Air Conditioning	Sheet 2					$ 162,347		$ 37,310		$ 830				
Heating	Sheet 3					$ 33,495		$ 17,575		$ 706				
Miscellaneous	Sheet 4							$ 1,303						
Subcontracts	Sheet 4									$ 1,706		$ 79,765		
Material Subtotal						$ 229,776								
Sales Tax					5%	$ 11,489								
Bare Cost Total						$ 241,265		$ 132,861		$ 3,242		$ 79,765		
Overhead & Profit				10%		$ 24,126	50.8%	$ 67,493	10%	$ 324	10%	$ 7,977		
Totals						$ 265,391		$ 200,354		$ 3,566		$ 87,742		$ 557,053
Total Bid to GC														$ 557,053
Division 15 Total														

Figure 9.119

R15810-100 Sheet Metal Calculator (Weight in Lb./Ft. of Length)

Gauge	26	24	22	20	18	16	Gauge	26	24	22	20	18	16
Wt.-Lb./S.F.	.906	1.156	1.406	1.656	2.156	2.656	Wt.-Lb./S.F.	.906	1.156	1.406	1.656	2.156	2.656
SMACNA Max. Dimension – Long Side		30"	54"	84"	85" Up		SMACNA Max. Dimension – Long Side		30"	54"	84"	85" Up	
Sum-2 sides							**Sum-2 Sides**						
2	.3	.40	.50	.60	.80	.90	56	9.3	12.0	14.0	16.2	21.3	25.2
3	.5	.65	.80	.90	1.1	1.4	57	9.5	12.3	14.3	16.5	21.7	25.7
4	.7	.85	1.0	1.2	1.5	1.8	58	9.7	12.5	14.5	16.8	22.0	26.1
5	.8	1.1	1.3	1.5	1.9	2.3	59	9.8	12.7	14.8	17.1	22.4	26.6
6	1.0	1.3	1.5	1.7	2.3	2.7	60	10.0	12.9	15.0	17.4	22.8	27.0
7	1.2	1.5	1.8	2.0	2.7	3.2	61	10.2	13.1	15.3	17.7	23.2	27.5
8	1.3	1.7	2.0	2.3	3.0	3.6	62	10.3	13.3	15.5	18.0	23.6	27.9
9	1.5	1.9	2.3	2.6	3.4	4.1	63	10.5	13.5	15.8	18.3	24.0	28.4
10	1.7	2.2	2.5	2.9	3.8	4.5	64	10.7	13.7	16.0	18.6	24.3	28.8
11	1.8	2.4	2.8	3.2	4.2	5.0	65	10.8	13.9	16.3	18.9	24.7	29.3
12	2.0	2.6	3.0	3.5	4.6	5.4	66	11.0	14.1	16.5	19.1	25.1	29.7
13	2.2	2.8	3.3	3.8	4.9	5.9	67	11.2	14.3	16.8	19.4	25.5	30.2
14	2.3	3.0	3.5	4.1	5.3	6.3	68	11.3	14.6	17.0	19.7	25.8	30.6
15	2.5	3.2	3.8	4.4	5.7	6.8	69	11.5	14.8	17.3	20.0	26.2	31.1
16	2.7	3.4	4.0	4.6	6.1	7.2	70	11.7	15.0	17.5	20.3	26.6	31.5
17	2.8	3.7	4.3	4.9	6.5	7.7	71	11.8	15.2	17.8	20.6	27.0	32.0
18	3.0	3.9	4.5	5.2	6.8	8.1	72	12.0	15.4	18.0	20.9	27.4	32.4
19	3.2	4.1	4.8	5.5	7.2	8.6	73	12.2	15.6	18.3	21.2	27.7	32.9
20	3.3	4.3	5.0	5.8	7.6	9.0	74	12.3	15.8	18.5	21.5	28.1	33.3
21	3.5	4.5	5.3	6.1	8.0	9.5	75	12.5	16.1	18.8	21.8	28.5	33.8
22	3.7	4.7	5.5	6.4	8.4	9.9	76	12.7	16.3	19.0	22.0	28.9	34.2
23	3.8	5.0	5.8	6.7	8.7	10.4	77	12.8	16.5	19.3	22.3	29.3	34.7
24	4.0	5.2	6.0	7.0	9.1	10.8	78	13.0	16.7	19.5	22.6	29.6	35.1
25	4.2	5.4	6.3	7.3	9.5	11.3	79	13.2	16.9	19.8	22.9	30.0	35.6
26	4.3	5.6	6.5	7.5	9.9	11.7	80	13.3	17.1	20.0	23.2	30.4	36.0
27	4.5	5.8	6.8	7.8	10.3	12.2	81	13.5	17.3	20.3	23.5	30.8	36.5
28	4.7	6.0	7.0	8.1	10.6	12.6	82	13.7	17.5	20.5	23.8	31.2	36.9
29	4.8	6.2	7.3	8.4	11.0	13.1	83	13.8	17.8	20.8	24.1	31.5	37.4
30	5.0	6.5	7.5	8.7	11.4	13.5	84	14.0	18.0	21.0	24.4	31.9	37.8
31	5.2	6.7	7.8	9.0	11.8	14.0	85	14.2	18.2	21.3	24.7	32.3	38.3
32	5.3	6.9	8.0	9.3	12.2	14.4	86	14.3	18.4	21.5	24.9	32.7	38.7
33	5.5	7.1	8.3	9.6	12.5	14.9	87	14.5	18.6	21.8	25.2	33.1	39.2
34	5.7	7.3	8.5	9.9	12.9	15.3	88	14.7	18.8	22.0	25.5	33.4	39.6
35	5.8	7.5	8.8	10.2	13.3	15.8	89	14.8	19.0	22.3	25.8	33.8	40.1
36	6.0	7.8	9.0	10.4	13.7	16.2	90	15.0	19.3	22.5	26.1	34.2	40.5
37	6.2	8.0	9.3	10.7	14.1	16.7	91	15.2	19.5	22.8	26.4	34.6	41.0
38	6.3	8.2	9.5	11.0	14.4	17.1	92	15.3	19.7	23.0	26.7	35.0	41.4
39	6.5	8.4	9.8	11.3	14.8	17.6	93	15.5	19.9	23.3	27.0	35.3	41.9
40	6.7	8.6	10.0	11.6	15.2	18.0	94	15.7	20.1	23.5	27.3	35.7	42.3
41	6.8	8.8	10.3	11.9	15.6	18.5	95	15.8	20.3	23.8	27.6	36.1	42.8
42	7.0	9.0	10.5	12.2	16.0	18.9	96	16.0	20.5	24.0	27.8	36.5	43.2
43	7.2	9.2	10.8	12.5	16.3	19.4	97	16.2	20.8	24.3	28.1	36.9	43.7
44	7.3	9.5	11.0	12.8	16.7	19.8	98	16.3	21.0	24.5	28.4	37.2	44.1
45	7.5	9.7	11.3	13.1	17.1	20.3	99	16.5	21.2	24.8	28.7	37.6	44.6
46	7.7	9.9	11.5	13.3	17.5	20.7	100	16.7	21.4	25.0	29.0	38.0	45.0
47	7.8	10.1	11.8	13.6	17.9	21.2	101	16.8	21.6	25.3	29.3	38.4	45.5
48	8.0	10.3	12.0	13.9	18.2	21.6	102	17.0	21.8	25.5	29.6	38.8	45.9
49	8.2	10.5	12.3	14.2	18.6	22.1	103	17.2	22.0	25.8	29.9	39.1	46.4
50	8.3	10.7	12.5	14.5	19.0	22.5	104	17.3	22.3	26.0	30.2	39.5	46.8
51	8.5	11.0	12.8	14.8	19.4	23.0	105	17.5	22.5	26.3	30.5	39.9	47.3
52	8.7	11.2	13.0	15.1	19.8	23.4	106	17.7	22.7	26.5	30.7	40.3	47.7
53	8.8	11.4	13.3	15.4	20.1	23.9	107	17.8	22.9	26.8	31.0	40.7	48.2
54	9.0	11.6	13.5	15.7	20.5	24.3	108	18.0	23.1	27.0	31.3	41.0	48.6
55	9.2	11.8	13.8	16.0	20.9	24.8	109	18.2	23.3	27.3	31.6	41.4	49.1
							110	18.3	23.5	27.5	31.9	41.8	49.5

Example: If duct is 34" x 20" x 15' long, 34" is greater than 30" maximum, for 24 ga. so must be 22 ga. 34" + 20" = 54" going across from 54" find 13.5 lb. per foot. 13.5 x 15' = 202.5 lbs. For S.F. of surface area 202.5 ÷

1.406 = 144 S.F.
Note: Figures include an allowance for scrap.

Figure 9.120

CONSOLIDATED ESTIMATE

PROJECT: Office Building
LOCATION:
TAKE OFF BY: ABC
QUANTITIES BY: ABC
CLASSIFICATION: Division 15
ARCHITECT:
PRICES BY: DEF
EXTENSIONS BY: DEF
CHECKED BY: GHI

ESTIMATE NO:
DATE: Jan-02
SHEET NO. 1 of 1

DESCRIPTION	SOURCE	QUANT	UNIT	MATERIAL COST	MATERIAL TOTAL	LABOR COST	LABOR TOTAL	EQUIPMENT COST	EQUIPMENT TOTAL	SUBCONTRACT COST	SUBCONTRACT TOTAL	TOTAL COST	TOTAL TOTAL
Division 15: Mechanical													
	SUBCONTRACTOR QUOTATION (Including Sales Tax)												
Plumbing													
Base Price											$ 173,423		
Pipe Insulation (HW only)											$ 4,205		
Total Plumbing											$ 177,628		
Fire Protection:													
Standpipes											$ 51,656		
Sprinklers											$ 87,532		
Total Fire Protection											$ 139,188		
HVAC: (Includes the following)													
Base Price													
Insulation													
Balancing													
Controls													
Crane													
Total HVAC											$ 557,053		
Division 15 Total											$ 873,869		

Figure 9.121

294

- Fire alarm and telephone riser diagrams
- Special systems

Performing a fixture and device takeoff can be a good way to become familiar with the proposed electrical installation. Take off one bay, section, or floor at a time. Mark each fixture and device with colored pencils and list them as you proceed. Check the site plans for exterior fixtures.

Fixtures should be taken off using the fixture schedule in conjunction with the reflected ceiling plan. Fixture counts should have outlet boxes, plaster rings, or Greenfield with connectors and fixture wire if needed. Include fixture supports as needed.

Separate quantity sheets should be used for each of the major categories of electrical work. By keeping each system on different sheets, the estimator will find it easy to isolate various costs. This format becomes a reference for purchasing and cost control. It also provides a breakdown of items that can be submitted to the general contractor when billing.

While making takeoffs of each category, identify all other required components of the systems. Determine the materials that will be part of a quotation or sub-bid from a supplier, including additional items to complete the vendor's package.

Switchgear

Material costs for large equipment such as Switchgear motor control centers and associated items should be obtained from suppliers or manufacturers. When determining the installation cost for large equipment (for any division) the estimator must consider a number of factors to be sure that all work is included:

- Access and placement
- Uncrating
- Rigging and setting
- Pads and anchors
- Leveling and shimming
- Assembly of components
- Connections
- Temporary protection requirements

While pads and anchors for large equipment may be well designated on the plans, supports for smaller equipment, such as panels and transformers, may not be well defined. For example, floor to ceiling steel supports may be required. The same consideration must be given to large cable troughs and conduits. Special support requirements may be necessary. If so, they must be included, whether they are the responsibility of the electrical or general contractor. For elevated installation of equipment, labor costs should be adjusted. Suggested adjustments follow:

Ceiling Height	Labor Adjustment
10' – 15'	+ 15%
15' – 25'	+ 30%
Over 25'	+ 35%

Ducts and Cable Trays

Bus ducts and cable trays should be estimated by component:

- Type and size of duct or tray
- Material (aluminum or galvanized)
- Hangers
- Fittings

The installation of under floor ducts systems (and conduit) must be scheduled after placement of any reinforcing at the bottom of slabs, but before installation of the upper steel. Without proper coordination, delays can occur. The estimate and preliminary schedule should reflect associated costs. Because of the high cost of tray and duct systems, the takeoff should be as accurate as possible. Fittings should not be deducted from straight lengths. In this way, an adequate allowance for waste should be provided. For high ceiling installations, labor costs for bus duct and cable trays should be adjusted accordingly. Suggested adjustments are:

Ceiling Height	Labor Adjustment
15' – 20'	+ 10%
20' – 25'	+ 20%
25' – 30'	+ 25%
30' – 35'	+ 30%
35' – 40'	+ 35%
Over 40'	+ 40%

Feeders

Feeder conduit and wire should be carefully taken off using a scale or printed dimensions. This is a more accurate method than the use of a rotometer. Large conduit and wire are expensive and require a considerable amount of labor. Accurate quantities are important. Switchboard locations should be marked on each floor before measuring horizontal runs. The distance between floors should be marked on riser plans and added to the horizontal for a complete feeder run. Conduits should be measured at right angles to the structure unless it has been determined that they can be routed directly. Elbows, terminations, bends, or expansion joints should also be taken off at this time. Under-slab and high ceiling installations should be treated the same as in the case of ducts and trays. The estimator should also consider the weight of material for high ceiling installations. Extra workers may be required for heavier components. Typical weights are shown in Figure 9.122. If standard feeder takeoff sheets are used, the wire column should reflect longer lengths than the conduit. This is because of added amounts of wire used in the panels and switchboard to make connections. If the added length is not shown on the plans, it can be determined by checking a manufacturer's catalog. Conduit, cable supports, and accessories should all be totalled at this time.

Branch Circuits

Branch circuits may be taken off using a rotometer. The estimator should take care to start and stop accurately at boxes. Start with two wire circuits and mark with colored pencil as items are taken off. Add about 5% to conduit quantities for normal waste. On wire, add 10% to 12% overage to make connections. Add conduit fittings, such as locknuts, bushings, couplings, expansion joints, and fasteners. Two conduit terminations per box are average. Figure 9.123 shows prices for conduit as presented in *Means Building Construction Cost Data*. Note that a minimum amount of fittings are included in the costs.

Conduit Weight Comparisons (Lbs. per 100 ft.)

Type	1/2"	3/4"	1"	1-1/4"	1-1/2"	2"	2-1/2"	3"	3-1/2"	4"	5"	6"
Rigid Aluminum	28	37	55	72	89	119	188	246	296	350	479	630
Rigid Steel	79	105	153	201	249	332	527	683	831	972	1314	1745
Intermediate Steel (IMC)	60	82	116	150	182	242	401	493	573	638		
Electrical Metallic Tubing (EMT)	29	45	65	96	111	141	215	260	365	390		
Polyvinyl Chloride, Schedule 40	16	22	32	43	52	69	109	142	170	202	271	350
Polyvinyl Chloride Encased Burial						38		67	88	105	149	202
Fibre Duct Encased Burial						127		164	180	206	400	511
Fibre Duct Direct Burial						150		251	300	354		
Transite Encased Burial						160		240	290	330	450	550
Transite Direct Burial						220		310		400	540	640

Weight Comparisons of Common Size Cast Boxes in Lbs.

Size NEMA 4 or 9	Cast Iron	Cast Aluminum	Size NEMA 7	Cast Iron	Cast Aluminum
6" x 6" x 6"	17	7	6" x 6" x 6"	40	15
8" x 6" x 6"	21	8	8" x 6" x 6"	50	19
10" x 6" x 6"	23	9	10" x 6" x 6"	55	21
12" x 12" x 6"	52	20	12" x 6" x 6"	100	37
16" x 16" x 6"	97	36	16" x 16" x 6"	140	52
20" x 20" x 6"	133	50	20" x 20" x 6"	180	67
24" x 18" x 8"	149	56	24" x 18" x 8"	250	93
24" x 24" x 10"	238	88	24" x 24" x 10"	358	133
30" x 24" x 12"	324	120	30" x 24" x 10"	475	176
36" x 36" x 12"	500	185	30" x 24" x 12"	510	189

Size Required and Weight (Lbs./1000 L.F.) of Aluminum and Copper THW Wire by Ampere Load

Amperes	Copper Size	Aluminum Size	Copper Weight	Aluminum Weight
15	14	12	24	11
20	12	10	33	17
30	10	8	48	39
45	8	6	77	52
65	6	4	112	72
85	4	2	167	101
100	3	1	205	136
115	2	1/0	252	162
130	1	2/0	324	194
150	1/0	3/0	397	233
175	2/0	4/0	491	282
200	3/0	250	608	347
230	4/0	300	753	403
255	250	400	899	512
285	300	500	1068	620
310	350	500	1233	620
335	400	600	1396	772
380	500	750	1732	951

Weight (Lbs./L.F.) of 4 Pole Aluminum and Copper Bus Duct by Ampere Load

Amperes	Aluminum Feeder	Copper Feeder	Aluminum Plug-In	Copper Plug-In
225			7	7
400			8	13
600	10	10	11	14
800	10	19	13	18
1000	11	19	16	22
1350	14	24	20	30
1600	17	26	25	39
2000	19	30	29	46
2500	27	43	36	56
3000	30	48	42	73
4000	39	67		
5000		78		

Figure 9.122

Wiring devices should be entered with plates, boxes, and plaster rings. Calculate stub ups or drops for wiring devices. Switches should be counted and multiplied by the distance from switch to ceiling.

Receptacles are handled similarly, depending on whether they are wired from the ceiling or floor. In many cases, there are two conduits going to a receptacle box as you feed in and out to the next outlet. Some estimators let rotometers overrun outlets purposely as an adjustment for vertical runs; this practice can, however, lead to inaccuracies.

In metal stud partitions, it is often acceptable to go horizontally from receptacle to receptacle. In wood partitions, horizontal runs are also the usual practice. In suspended ceilings, the specifications should be checked to see if straight runs are allowed. If the space above the ceiling is used as a return air plenum, conduit is usually required for all wiring.

Motors, safety switches, starters, and controls, should each include power lines, junction boxes, supports, wiring troughs, and wire terminations. Short runs of Greenfield or Sealtite with appropriate connectors and wire should be added for motor wiring.

For large installations, the economy of scale may have a definite impact on the electrical costs. If large quantities of a particular item are installed in the same general area, certain deductions can be made for labor. Suggested deductions include:

	Quantity	Labor Deduction
Under floor ducts, bus ducts, conduit, cable systems:	150 to 250 LF	−10%
	250 to 350 LF	−15%
	350 to 500 LF	−20%
	Over 500 LF	−25%
Outlet boxes:	25 to 50	−15%
	50 to 75	−20%
	75 to 100	−25%
	Over 100	−30%
Wiring devices:	10 to 25	−20%
	25 to 50	−25%
	50 to 100	−30%
	Over 100	−35%
Lighting fixtures:	25 to 50	−15%
	50 to 75	−20%
	75 to 100	−25%
	Over 100	−30%

The estimator for the general contractor may require only a budget cost for the purpose of verifying subcontractor bids. As described in the section on mechanical estimating, square foot or systems costs developed from similar, past projects can also be useful for electrical estimating.

Sample Estimate: Division 16

While electrical work should be accurately estimated by a subcontractor for bidding purposes, the estimator for the general contractor should understand the work involved. Complete plans and specifications will include a number of schedules and diagrams that will help the estimator to determine quantities. Figures 9.124 and 9.125 are the lighting fixture and panelboard schedules,

16131 | Cable Trays

		Description	CREW	DAILY OUTPUT	LABOR-HOURS	UNIT	MAT.	LABOR	EQUIP.	TOTAL	TOTAL INCL O&P	
105	0010	**CABLE TRAY LADDER TYPE** w/ ftngs & supports, 4" dp., to 15' elev.										105
	0160	Galvanized steel tray										
	0170	4" rung spacing, 6" wide	2 Elec	98	.163	L.F.	8.15	5.80		13.95	17.60	
	0200	12" wide		86	.186		9.80	6.60		16.40	20.50	
	0400	18" wide		82	.195		11.40	6.90		18.30	23	
	0600	24" wide		78	.205		13.10	7.25		20.35	25.50	
	3200	Aluminum tray, 4" deep, 6" rung spacing, 6" wide		134	.119		10.25	4.23		14.48	17.55	
	3220	12" wide		124	.129		11.45	4.57		16.02	19.45	
	3230	18" wide		114	.140		12.75	4.98		17.73	21.50	
	3240	24" wide		106	.151		14.90	5.35		20.25	24.50	
	9980	Allow. for tray ftngs., 5% min.-20% max.										

16132 | Conduit & Tubing

		Description	CREW	DAILY OUTPUT	LABOR-HOURS	UNIT	MAT.	LABOR	EQUIP.	TOTAL	TOTAL INCL O&P	
205	0010	**CONDUIT** To 15' high, includes 2 terminations, 2 elbows and										205
	0020	11 beam clamps per 100 L.F. [R16132 -210]										
	0300	Aluminum, 1/2" diameter	1 Elec	100	.080	L.F.	1.22	2.84		4.06	5.60	
	0500	3/4" diameter		90	.089		1.65	3.15		4.80	6.50	
	0700	1" diameter		80	.100		2.21	3.55		5.76	7.75	
	1000	1-1/4" diameter		70	.114		2.86	4.05		6.91	9.20	
	1030	1-1/2" diameter		65	.123		3.53	4.36		7.89	10.40	
	1050	2" diameter		60	.133		4.68	4.73		9.41	12.20	
	1070	2-1/2" diameter		50	.160		7.30	5.65		12.95	16.45	
	1100	3" diameter	2 Elec	90	.178		9.75	6.30		16.05	20	
	1130	3-1/2" diameter		80	.200		12.05	7.10		19.15	24	
	1140	4" diameter		70	.229		14.60	8.10		22.70	28	
	1750	Rigid galvanized steel, 1/2" diameter	1 Elec	90	.089		1.65	3.15		4.80	6.50	
	1770	3/4" diameter		80	.100		1.94	3.55		5.49	7.45	
	1800	1" diameter		65	.123		2.73	4.36		7.09	9.50	
	1830	1-1/4" diameter		60	.133		3.66	4.73		8.39	11.05	
	1850	1-1/2" diameter		55	.145		4.37	5.15		9.52	12.50	
	1870	2" diameter		45	.178		5.80	6.30		12.10	15.80	
	1900	2-1/2" diameter		35	.229		9.65	8.10		17.75	23	
	1930	3" diameter	2 Elec	50	.320		12.20	11.35		23.55	30.50	
	1950	3-1/2" diameter		44	.364		14.90	12.90		27.80	35.50	
	1970	4" diameter		40	.400		17.80	14.20		32	40.50	
	2500	Steel, intermediate conduit (IMC), 1/2" diameter	1 Elec	100	.080		1.28	2.84		4.12	5.65	
	2530	3/4" diameter		90	.089		1.54	3.15		4.69	6.40	
	2550	1" diameter		70	.114		2.12	4.05		6.17	8.40	
	2570	1-1/4" diameter		65	.123		2.80	4.36		7.16	9.60	
	2600	1-1/2" diameter		60	.133		3.22	4.73		7.95	10.60	
	2630	2" diameter		50	.160		4.15	5.65		9.80	13	
	2650	2-1/2" diameter		40	.200		8.25	7.10		15.35	19.65	
	2670	3" diameter	2 Elec	60	.267		10.60	9.45		20.05	26	
	2700	3-1/2" diameter		54	.296		12.95	10.50		23.45	30	
	2730	4" diameter		50	.320		15	11.35		26.35	33.50	
	5000	Electric metallic tubing (EMT), 1/2" diameter	1 Elec	170	.047		.37	1.67		2.04	2.90	
	5020	3/4" diameter **CN**		130	.062		.57	2.18		2.75	3.88	
	5040	1" diameter		115	.070		.98	2.47		3.45	4.76	
	5060	1-1/4" diameter		100	.080		1.47	2.84		4.31	5.85	
	5080	1-1/2" diameter		90	.089		1.86	3.15		5.01	6.75	
	5100	2" diameter		80	.100		2.37	3.55		5.92	7.90	
	5120	2-1/2" diameter		60	.133		5.60	4.73		10.33	13.20	
	5140	3" diameter	2 Elec	100	.160		6.05	5.65		11.70	15.15	
	5160	3-1/2" diameter		90	.178		7.70	6.30		14	17.90	
	5180	4" diameter		80	.200		8.65	7.10		15.75	20	

Figure 9.123

respectively, for the building project. While such schedules may not always provide quantities, they can serve as a checklist to ensure that all types are counted.

Figure 9.126 is the quantity sheet for lighting. Note that associated boxes, devices and wiring are taken off at the same time as the fixtures and that the takeoff is done by floor. By performing the estimate in this way, quantities for

Type	Manufacturer & Catalog #	Fixture	Type	Lamps		Watts	Mounting	Remarks
				Qty	Volts			
A	Meansco #7054	2' x 4' Troffer	F-40 CW	4	277	40	Recessed	Acrylic Lens
B	Meansco #7055	1' x 4' Troffer	F-40 CW	2	277	40	Recessed	Acrylic Lens
C	Meansco #7709	6" x 4'	F-40 CW	1	277	40	Surface	Acrylic Wrap
D	Meansco #7710	6" x 8' Strip	F96T12 CW	1	277	40	Surface	
E	Meansco #7900A	6" x 4'	F-40	1	277	40	Surface	Mirror Light
F	Kingston #100A	6" x 4'	F-40	1	277	40	Surface	Acrylic Wrap
G	Kingston #110C	' Strip	F-40 CW	1	277	40	Surface	
H	Kingston #3752	Wallpack	HPS	1	277	150	Bracket	W/Photo Cell
J	Kingston #201-202	Floodlight	HPS	1	277	400	Surface	2' Below Fascia
K	Kingston #203		HPS	1	277	100	Wall Bracket	
L	Meansco #8100	Exit Light	1-13W 20W	1	120	13	Surface	
			T6-1/2	2	6½	20		
M	Meansco #9000	Battery Unit	Sealed Beam	2	12	18	Wall Mount	12 Volt Unit

Figure 9.124

Panel	Main Breaker	Main	Lugs	Amps	Circ.	Breakers		Location	Volts
						#	Type		
PP-1	No	3-P	Wire	100	20	(20)	IP-20 A	Basement	120/208
PP-2	No	3-P	4 Wire	225	42	(42)	IP-20 A	1st Floor	120/208
PP-3	No	3-P	4 Wire	225	42	(42)	IP-20 A	2nd Floor	120/208
PP-4	No	3-P	4 Wire	225	42	(42)	IP-20 A	3rd Floor	120/208
LP-1	No	3-P	4 Wire	225	42	(42)	IP-20 A	Basement	277/480
LP-2	No	3-P	4 Wire	225	42	(42)	IP-20 A	1st Floor	277/480
LP-3	No	3-P	4 Wire	225	42	(42)	IP-20 A	2nd Floor	277/480
LP-4	No	3-P	4 Wire	225	42	(42)	IP-20 A	3rd Floor	277/480

Figure 9.125

purchase are easily determined and changes can be calculated quickly. For example, in a typical project, lighting fixtures are often changed or substituted, either because of aesthetic decisions or unavailability. If quantities and costs are isolated, adjustments are easily made.

The estimate sheets for lighting—as might be prepared by an electrical subcontractor—are shown in Figures 9.127 and 9.128. Prices are from *Means Electrical Cost Data*. By including the associated costs for boxes, devices and wiring with the lighting fixtures and the other major components, the electrical estimator is able to develop system costs for comparison to other projects, and for cross-checking. Since this is only part of the electrical work, overhead and profit are not likely to be added until the end of the electrical estimate. However, if a breakdown of costs is provided as part of the quotation, each portion of that breakdown would include overhead and profit. Such is the case in the estimate sheet for Division 16 as prepared by the general contractor, in Figure 9.129. For each system, the associated costs for devices and wiring are included, in addition to overhead and profit.

Estimate Summary

At this point in the estimating process, the estimate is complete for Divisions 2 through 16. All vendor quotations and subcontractors' bids should be in hand (ideally, but not necessarily realistic); and all costs should be determined for the work to be done "in-house." All costs known at this time should be entered on the Estimate Summary sheet as shown in Figure 9.130. Based on the subtotal of these costs, the items in Division 1 which are dependent on job costs can then be calculated.

The Project Overhead Summary is shown in Figures 9.131 and 9.132. In addition to those items related to job costs, there are also certain project overhead items that depend on job duration. A preliminary schedule is required in order to calculate these time-related costs. Using the procedure described in Chapter 5, a preliminary precedence schedule is prepared for the project as shown in Figure 9.133. Note that only those items that affect the project duration are calculated and included in the total time (247 working days). A more extensive and detailed project schedule is completed when the construction contract is awarded. The corresponding preliminary bar schedule is shown in Figure 9.134.

Throughout the estimate for all divisions, the estimator should note all items which should be included as project overhead as well as those which may affect the project schedule. The estimator must be especially aware of items which are implied but may not be directly stated in the specifications. Some of the "unspecified" costs which may be included are for the preparation of shop drawings, as-built drawings and the final project schedule. Costs for general job safety—such as railings, safety nets, fire extinguishers—must also be included if these items are required. If costs for job safety are not listed in the estimate, the items may not be installed on the job—until an OSHA fine, or worse, an accident.

Costs for shop drawings are mentioned above because the time involved in their preparation, submission and approval must also be considered when preparing the schedule. Construction delays are often blamed on the fact that shop drawings or "cuts" weren't submitted early enough, or approved in a timely fashion. The estimator must base such scheduling decisions on experience.

PROJECT Office Building **ARCHITECT** ___
LOCATION ___ **OWNER** ___

SECTION Lighting **QUANTITY SHEET** **DATE** ___
TAKE OFF BY PHD **EXTENSIONS BY** ___ **CHECKED BY** ___

NO	DESCRIPTION	DIMENSIONS	Garage	1st Flr	2nd Flr	3rd Flr	Penthouse	Sub Total	Adjustments	Total
Fixtures										
A	2'x4' Troffer, 4L		1	257	244	244		725		725 Ea.
B	1'x4' Troffer, 2L		6	10	6	6		28		28
C	6'x4' Acrylic		4	9	—	8		29		29
D	6'x8' Strip		53					53		53
E	6'x8' Mirror Lite		6	3	3	3		9		9
G	4" Shop Surf		6	1	1	1		27		27
H	150w Wall Pack		11	2	—	—	18	2		2
J								-10		-10
K	70w Wall Pack		1	1			1	3		3
L	Exit Light		6	8	4	4	1	23		23
N	Remote Head		1	1	1	1				4
M	Emergency Lite		3	3	3	3	1	13		13

Type F is deleted – Use Type C

NO	DESCRIPTION	DIMENSIONS	Garage	1st Flr	2nd Flr	3rd Flr	Penthouse	Sub Total	Adjustments	Total
Boxes & Devices										
O	4" Round Junction Box	Steel	8	128	129	129		394		394 Ea.
O	4" Round Cover	Steel	8	128	129	129		394		394
S₁	Switch Box	1 Gang Steel	1	5	5	5	4	20		20
S₁	S.P. Switch 20A	Single Pole Sw.	6	8	4	4	1	23		23
	4" Sq. Box		5	5	5	5	4	20		20
	Switch Cover		4	8	4	4	1	23		23
	Plaster Rings 4"									

NO	DESCRIPTION	DIMENSIONS	Garage	1st Flr	2nd Flr	3rd Flr	Penthouse	Sub Total	Adjustments	Total
Raceways										
	1/2" EMT		1460	2070	1880	1880	940	8250 LF	850 10%	9100 LF
	1/2" Conn.	Set screw type	190	264	242	242	114	1052 Ea.		1052 Ea.
	1/2" Conn.	Compression Type	118	216	24	24	12	104 Ea.		104 Ea.
	5/8" Greenfield		70	1295	1330	1330		4025 LF	405 10%	4430 LF
	3/8" Greenfield Conn.		32	474	488	488	1	1482 Ea.		1482 Ea.

NO	DESCRIPTION	DIMENSIONS	Garage	1st Flr	2nd Flr	3rd Flr	Penthouse	Sub Total	Adjustments	Total
Conductors										
	#12 THHN / Copper		4650	10095	9630	9650	2820	36825 LF	3683 LF ⇒	405 CLF

Figure 9.126

302

CONSOLIDATED ESTIMATE

PROJECT: Office Building
LOCATION:
TAKE OFF BY: ABC
CLASSIFICATION: Division 16
ARCHITECT:
QUANTITIES BY: ABC
PRICES BY: As Shown
ESTIMATE NO:
DATE: Jan-02
CHECKED BY: GHI

DESCRIPTION	SOURCE	QUANT.	UNIT	MATERIAL COST	MATERIAL DEF TOTAL	LABOR COST	LABOR TOTAL	EQUIPMENT COST	EQUIPMENT TOTAL	SUBCONTRACT COST	SUBCONTRACT TOTAL	TOTAL COST	TOTAL TOTAL
Division 16: Electrical													
Lighting													
Fixtures													
Type													
A 2' x 4' w/ 4 Lamps	16510 440 0600	725	Ea.	56.50	40963	60.50	43863						
B 1' x 4' w/ 2 Lamps	16510 440 0200	28	Ea.	46.50	1302	50.00	1400						
C 6" x 4' w/ 1 Lamp	16510 440 2020	29	Ea.	66.00	1914	35.50	1030						
D 6" x 8' w/ 1 Lamp	16510 440 2600	53	Ea.	39.50	2094	42.50	2253						
E 6" x 4' Mirror Light	16510 440 6900	9	Ea.	84.50	761	35.50	320						
G 4' Surface Mount	16510 440 2200	27	Ea.	26.50	716	33.50	905						
H 150 W HPS Wall Pack	16520 300 1170	2	Ea.	228.00	456	71.00	142						
K 70 W HPS Wall Pack	16520 300 1160	3	Ea.	185.00	555	71.00	213						
L Exit Light, Single Face	16530 320 0080	23	Ea.	37.50	863	35.50	817						
M Remote Head	16530 320 0780	4	Ea.	21.00	84	10.60	42						
N Emergency Battery Unit	16530 320 0500	13	Ea.	110.00	1430	71.00	923						
Raceways:													
1/2" EMT	16132 205 5000	9080	Ea.	0.37	3360	1.67	15164						
Set Screw Connectors	16132 205 6500	1052	Ea.	0.53	558	2.36	2483						
Compression Box Connectors	16132 205 8800	104	Ea.	2.00	208	2.36	245						
1/2" Flexible Metal Steel Conduit	16132 320 0050	4430	Ea.	0.23	1019	1.42	6291						
Connectors	16132 320 0420	1482	Ea.	0.95	1408	2.84	4209						
Subtotals				$	57,687	$	80,297						

Figure 9.127

CONSOLIDATED ESTIMATE

PROJECT:	Office Building		CLASSIFICATION:	Division 16	
LOCATION:			ARCHITECT:		ESTIMATE NO:
TAKE OFF BY:	ABC	QUANTITIES BY: ABC	PRICES BY: As Shown	EXTENSIONS BY:	DATE: Jan-02
					CHECKED BY: GHI

DESCRIPTION	SOURCE	QUANT.	UNIT	MATERIAL COST	MATERIAL DEF TOTAL	LABOR COST	LABOR TOTAL	EQUIPMENT COST	EQUIPMENT DEF TOTAL	SUBCONTRACT COST	SUBCONTRACT TOTAL	TOTAL COST	TOTAL TOTAL
Division 16: (Cont'd)													
Lighting: (Cont'd)													
Conductors:													
THHN copper #12	16120 900 1200	405	CLF	6.20	2511	26.00	10530						
Boxes & Devices:													
Box - 4" Octagon	16136 600 0020	394	Ea.	1.48	583	14.20	5595						
4" Blank Cover	16136 600 0250	394	Ea.	0.70	276	4.43	1745						
Box - 4" Square	16136 600 0150	23	Ea.	2.05	47	14.20	327						
Plaster Rings	16136 600 0300	23	Ea.	1.14	26	4.43	102						
Switch Box	16136 600 0650	20	Ea.	2.31	46	10.50	210						
Single Pole Switch 20 A	16140 910 0500	20	Ea.	6.50	130	10.50	210						
Switch Plate	16140 910 2600	20	Ea.	1.80	36	3.55	71						
Subtotals					$ 3,655		$ 18,790						
Sheet 1 subtotals					$ 57,687		$ 80,297						
Sheet 2 subtotals					$ 3,655		$ 18,790						
Material Subtotal					$ 61,342								
Sales Tax				5%	$ 3,067								
Bare Cost Total					$ 64,409		$ 99,087						
Overhead & Profit				10%	$ 6,441	49.2%	48,751	10%		10%			
Totals					$ 70,850		$ 147,838					$	218,688
Total Lighting Bid to GC												$	218,688

Figure 9.128

PROJECT: Office Building   **CLASSIFICATION:** Division 16

LOCATION: **ARCHITECT:**

TAKE OFF BY: ABC   **QUANTITIES BY:** ABC   **PRICES BY:**   **EXTENSIONS BY:**

SHEET NO. 1 of 1

ESTIMATE NO:

DATE: Jan-02

CHECKED BY: GHI

DESCRIPTION	SOURCE	QUANT.	UNIT	MATERIAL			LABOR			EQUIPMENT			SUBCONTRACT			TOTAL	
				DEF COST		TOTAL	DEF COST		TOTAL	DEF COST		TOTAL	COST		TOTAL	COST	TOTAL
Division 16: Electrical																	
	SUBCONTRACTOR QUOTATION (Including Sales Tax)																
Lighting													$		218,688		
Service & Distribution													$		146,467		
Branch Circuits																	
Incl. Under Carpet System													$		168,849		
Motors													$		38,782		
Fire Alarm													$		52,488		
Miscellaneous Fees													$		10,675		
Division 16 Total													$		635,949		

Figure 9.129

CONDENSED ESTIMATE SUMMARY

PROJECT	Office Building				SHEET NO.		
LOCATION		TOTAL AREA / VOLUME			ESTIMATE NO:		
ARCHITECT As Shown		COST PER S.F. / C.F.			DATE: Jan/2002		
PRICES BY: DEF		EXTENSIONS BY: DEF			NO. OF STORIES		
					CHECKED BY: GHI		

DIV.	DESCRIPTION	Mat'l	Labor	Equip.	Sub w/ Tax	Sub w/o Tax	Total
1	General Requirements						
2	Site Construction	$136,574	$44,206	$35,427	$38,802		
3	Concrete	$312,367	$240,339	$8,832			
4	Masonry	$23,458	$28,045	$4,882			
5	Metals					$370,465	
6	Wood & Plastics	$6,279	$2,799				
						$87,519	
7	Thermal & Moisture Protection	$9,064	$6,954				
8	Doors & Windows	$25,686	$5,084		$1,091,363	$13,621	
9	Finishes	$138,591	$85,621			$299,231	
10	Specialties	$13,772	$2,374				
11	Equipment						
12	Furnishings						
13	Special Construction						
14	Conveying Systems					$143,660	
15	Mechanical				$873,869		
16	Electrical				$635,949		
	Subtotals						
	Sales Tax - Mat. & Equip. 5%						
	Subcontract 2.5%						
	Overhead & Profit 10%M, 56.3%L 10%E, 10%S						
	Subtotal						
	Bond ($12/M)						
	Contingency 2%						
	Adjustments						
	TOTAL BID						

Figure 9.130

PROJECT
OVERHEAD SUMMARY

PROJECT: Office Building								ESTIMATE NO:
LOCATION:			ARCHITECT:					DATE: Jan-02
QUANTITIES BY: ABC PRICES BY: DEF			EXTENSIONS BY: DEF			CHECKED BY: GHI		

DESCRIPTION	QUANTITY	UNIT	MATERIAL/EQUIP.		LABOR		TOTAL COST	
			UNIT	TOTAL	UNIT	TOTAL	UNIT	TOTAL
Job Organization: Superintendent	49	Week			1355	66395		
Project Manager								
Timekeeper & Material Clerk	40	Week			790	31600		
Clerical								
Safety, Watchman & First Aid								
Travel Expense: Superintendent								
Project Manager								
Engineering: Layout (3 person crew)	10	Day			780	7800		
Inspection / Quantities								
Drawings								
CPM Schedule								
Testing: Soil	1	LS		9750				
Materials								
Structural								
Equipment: Cranes								
Concrete Pump, Conveyor, Etc.								
Elevators, Hoists								
Freight & Hauling								
Loading, Unloading, Erecting, Etc.								
Maintenance								
Pumping								
Scaffolding								
Small Power Equipment / Tools	0.5	%						
Field Offices: Job Office, Trailer	11	Mo	300	3300				
Architect / Owner's Office								
Temporary Telephones	11	Mo	65	715				
Utilities								
Temporary Toilets	11	Mo	152.00	1672				
Storage Areas & Sheds								
Temporary Utilities: Heat								
Light & Power	567	CSF	9.88	5602				
PAGE TOTALS			$	21,039	$	105,795		

Figure 9.131

DESCRIPTION	QUANTITY	UNIT	MATERIAL/EQUIP.		LABOR		TOTAL COST	
			UNIT	TOTAL	UNIT	TOTAL	UNIT	TOTAL
Totals Brought Forward				$21,039		$105,795		
Winter Protection: Temp. Heat/Protection	56700	SF	0.45	25515	0.45	25515		
Snow Plowing								
Thawing Materials								
Temporary Roads	750	SY	3.13	2348	1.67	1253		
Signs & Barricades: Site Sign								
Temporary Fences	1	LS	130.00	130				
Temporary Stairs, Ladders & Floors	1350	LF	2.26	3051	3.75	5063		
Photographs								
Clean Up								
Dumpster	40	Week	425.00	17000				
Final Clean Up	56.7	MSF	6.03	342	37.00	2098		
Continuous - One Laborer	45	Week			938	42210		
Punch List	0.2	%		9380				
Permits: Building	1	%		46900				
Misc.								
Insurance: Builders Risk - Additional Rider	1	%		46900				
Owner's Protective Liability								
Umbrella								
Unemployment Ins. & Social Security								
Bonds								
Performance								
Material & Equipment								
Main Office Expense								
Special Items								
Totals:				$172,604		$181,933		

Figure 9.132

308

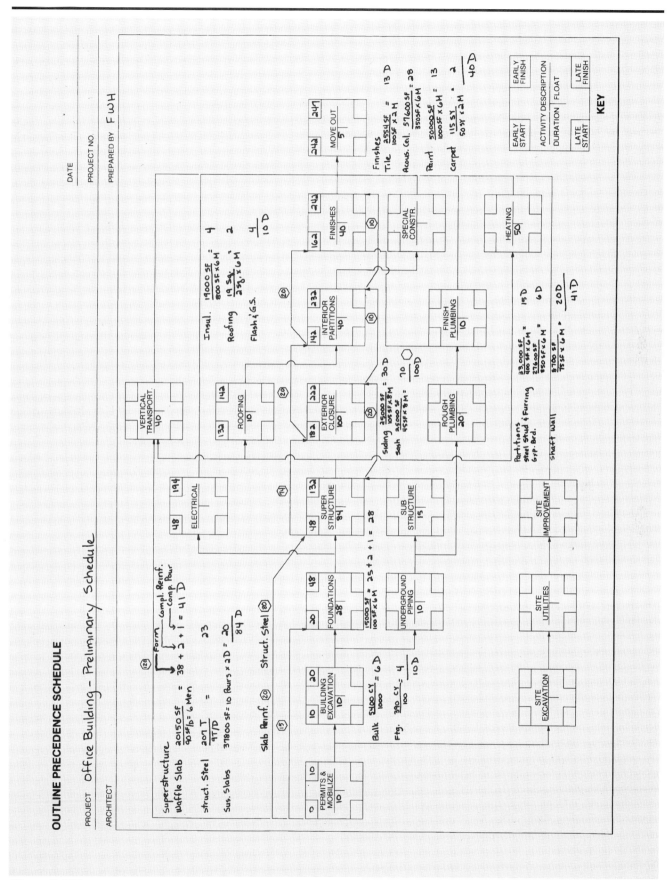

Figure 9.133

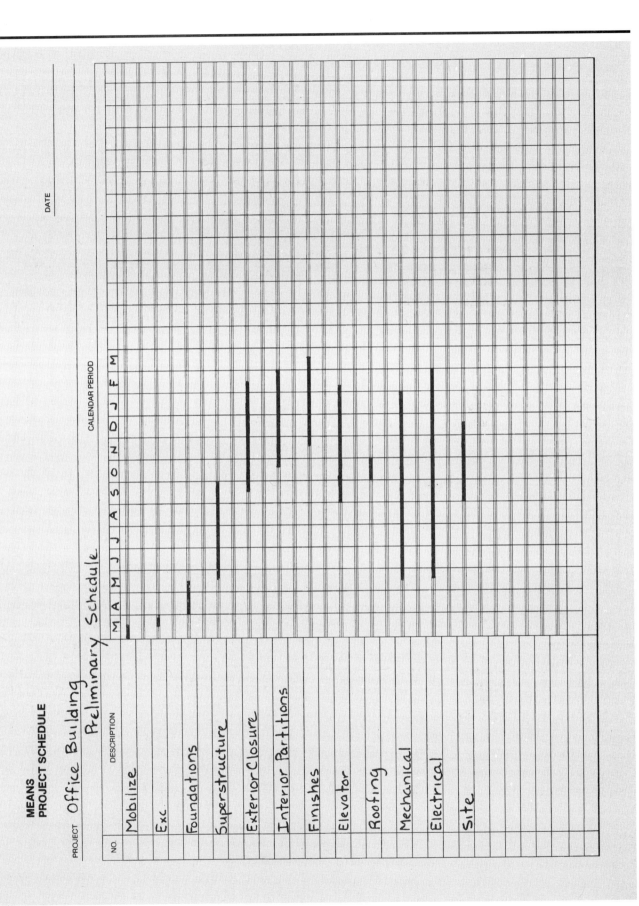

Figure 9.134

310

At this point, all divisions for the building project have been estimated. The preliminary schedule and project overhead summary are complete. Finally, the Estimate Summary sheet can be completed and the final number can be derived. The completed Estimate Summary sheet for the sample estimate is shown in Figure 9.135. Note that the difference between the subtotal of bare costs for all divisions and the "Total Bid" figure is roughly $1,036,370—almost 20%. This amount is the sum of all the indirect costs. (See Chapters 4 and 7.)

Sales tax (in this example, 5%) is added to the total bare costs of materials. An additional 10%—for handling and supervision—is added to the material (including tax), equipment and subcontract costs. This is standard industry practice. Bare labor costs, however, are treated differently. As discussed in Chapter 7, office overhead expenses may be more directly attributable to labor. Consequently, the percentage markup for labor (56.3%) includes employer-paid taxes and insurance, office overhead and profit. This relationship is shown in Figure 9.136. (Profit may or may not be listed as a separate item.) Since all trades are included in the bare labor subtotal, the percentage used is the average for all skilled workers. The requirements for bonds are determined from the specifications and their costs are obtained from local bonding agencies. Finally, a contingency is added if the job warrants. This decision is made when the estimator's role comes to an end and the job of the bidder begins. The decision to add a contingency is often based on the relative detail of the plans and specifications, experience, knowledge of the market and competition, and most likely—a gut feeling. The estimate total, with all indirect costs included, may still not be the same number as that which is *finally* submitted as the bid.

Bid day in a contractor's office is usually a frantic, hectic time. Many subcontractors delay telephoning/faxing quotations until the last possible moment in order to prevent the possibility of bid shopping. Subs to subcontractors also withhold quotes in the same way and for the same reason. This situation, in many instances, forces the prime contractor to use a budget or even a "ballpark" figure so that the bid sheets can be tabulated to obtain a final number. The importance of the estimator's reviewing all disciplines of the estimate cannot be overstressed. The chaotic atmosphere of bid day can only be compensated by thoroughness, precision and attention to detail during the whole preceding estimating process. If an estimate is poorly prepared and put together, flaws will be amplified at these late stages.

Many general or prime contractors or construction managers use an adjustment column to arrive at a "quote" in the last hour before bid is due. Such adjustments might appear as follows:

Structural Steel	+ $10,000
Mechanical	− 35,000
Partitions	− 10,000
Electrical (Use 2nd Bidder)	+ 6,000
Total Deduct	$29,000

These deductions or additions may be obtained from last minute phone calls, faxed quotes, error corrections or discovered omissions. The final number, or bid price, is, in many cases, arrived at by a principal or senior member of the firm. Bidding strategies are discussed in Chapter 6, but the final quotation is usually determined by personal judgment, good or bad.

Factors that may affect a final decision are: the risk involved, competition from other bidders, thoroughness of the plans and specs, and above all, the

CONDENSED ESTIMATE SUMMARY

PROJECT Office Building					ESTIMATE NO:		
LOCATION		TOTAL AREA / VOLUME			DATE: Jan/2002		
ARCHITECT As Shown		COST PER S.F. / C.F.			NO. OF STORIES		
PRICES BY: DEF		EXTENSIONS BY: DEF			CHECKED BY: GHI		

DIV.	DESCRIPTION	Mat'l	Labor	Equip.	Sub w/ Tax	Sub w/o Tax	Total
1	General Requirements	$172,604	$181,933				
2	Site Construction	$136,574	$44,206	$35,427	$38,802		
3	Concrete	$312,367	$240,339	$8,832			
4	Masonry	$23,458	$28,045	$4,882			
5	Metals					$370,465	
6	Wood & Plastics	$6,279	$2,799				
7	Thermal & Moisture Protection	$9,064	$6,954			$87,519	
8	Doors & Windows	$25,686	$5,084		$1,091,363	$13,621	
9	Finishes	$138,591	$85,621			$299,231	
10	Specialties	$13,772	$2,374				
11	Equipment						
12	Furnishings						
13	Special Construction						
14	Conveying Systems					$143,660	
15	Mechanical				$873,869		
16	Electrical				$635,949		
	Subtotals	$838,395	$597,355	$49,141	$2,639,983	$914,496	$5,039,370
	Sales Tax - Mat. & Equip. 5% Subcontract 2.5%	$41,920				$22,862	$64,782
	Overhead & Profit 10%M, 56.3%L 10%E, 10%S	$83,840	$336,311	$4,914	$263,998	$91,450	$780,512
	Subtotal	$964,154	$933,666	$54,055	$2,903,981	$1,028,808	$5,884,665
	Bond ($12/M)						$70,616
	Contingency 2%						$117,693
	Adjustments						$6,072,974
	TOTAL BID						$6,073,000

Figure 9.135

Installing Contractor's Overhead & Profit

Below are the **average** installing contractor's percentage mark-ups applied to base labor rates to arrive at typical billing rates.

Column A: Labor rates are based on union wages averaged for 30 major U.S. cities. Base rates including fringe benefits are listed hourly and daily. These figures are the sum of the wage rate and employer-paid fringe benefits such as vacation pay, employer-paid health and welfare costs, pension costs, plus appropriate training and industry advancement funds costs.

Column B: Workers' Compensation rates are the national average of state rates established for each trade.

Column C: Column C lists average fixed overhead figures for all trades. Included are Federal and State Unemployment costs set at 7.0%; Social Security Taxes (FICA) set at 7.65%; Builder's Risk Insurance costs set at 0.34%; and Public Liability costs set at 1.55%. All the percentages except those for Social Security Taxes vary from state to state as well as from company to company.

Columns D and E: Percentages in Columns D and E are based on the presumption that the installing contractor has annual billing of $1,500,000 and up. Overhead percentages may increase with smaller annual billing. The overhead percentages for any given contractor may vary greatly and depend on a number of factors, such as the contractor's annual volume, engineering and logistical support costs, and staff requirements. The figures for overhead and profit will also vary depending on the type of job, the job location, and the prevailing economic conditions. All factors should be examined very carefully for each job.

Column F: Column F lists the total of Columns B, C, D, and E.

Column G: Column G is Column A (hourly base labor rate) multiplied by the percentage in Column F (O&P percentage).

Column H: Column H is the total of Column A (hourly base labor rate) plus Column G (Total O&P).

Column I: Column I is Column H multiplied by eight hours.

		A		B	C	D	E	F		G	H	I
		Base Rate Incl. Fringes		Workers' Comp. Ins.	Average Fixed Overhead	Overhead	Profit	Total Overhead & Profit			Rate with O & P	
Abbr.	Trade	Hourly	Daily					%	Amount		Hourly	Daily
Skwk	Skilled Workers Average (35 trades)	$30.95	$247.60	16.8%	16.5%	13.0%	10%	56.3%	$17.40		$48.35	$386.80
	Helpers Average (5 trades)	22.75	182.00	18.5		11.0		56.0	12.75		35.50	284.00
	Foreman Average, Inside ($.50 over trade)	31.45	251.60	16.8		13.0		56.3	17.70		49.15	393.20
	Foreman Average, Outside ($2.00 over trade)	32.95	263.60	16.8		13.0		56.3	18.55		51.50	412.00
Clab	Common Building Laborers	23.45	187.60	18.1		11.0		55.6	13.05		36.50	292.00
Asbe	Asbestos/Insulation Workers/Pipe Coverers	33.45	267.60	16.2		16.0		58.7	19.65		53.10	424.80
Boil	Boilermakers	36.25	290.00	14.7		16.0		57.2	20.75		57.00	456.00
Bric	Bricklayers	30.50	244.00	16.0		11.0		53.5	16.30		46.80	374.40
Brhe	Bricklayer Helpers	23.50	188.00	16.0		11.0		53.5	12.55		36.05	288.40
Carp	Carpenters	30.00	240.00	18.1		11.0		55.6	16.70		46.70	373.60
Cefi	Cement Finishers	28.70	229.60	10.6		11.0		48.1	13.80		42.50	340.00
Elec	Electricians	35.45	283.60	6.7		16.0		49.2	17.45		52.90	423.20
Elev	Elevator Constructors	37.10	296.80	7.7		16.0		50.2	18.60		55.70	445.60
Eqhv	Equipment Operators, Crane or Shovel	32.35	258.80	10.6		14.0		51.1	16.55		48.90	391.20
Eqmd	Equipment Operators, Medium Equipment	31.20	249.60	10.6		14.0		51.1	15.95		47.15	377.20
Eqlt	Equipment Operators, Light Equipment	29.80	238.40	10.6		14.0		51.1	15.25		45.05	360.40
Eqol	Equipment Operators, Oilers	26.65	213.20	10.6		14.0		51.1	13.60		40.25	322.00
Eqmm	Equipment Operators, Master Mechanics	32.80	262.40	10.6		14.0		51.1	16.75		49.55	396.40
Glaz	Glaziers	30.00	240.00	13.8		11.0		51.3	15.40		45.40	363.20
Lath	Lathers	28.75	230.00	11.1		11.0		48.6	13.95		42.70	341.60
Marb	Marble Setters	30.10	240.80	16.0		11.0		53.5	16.10		46.20	369.60
Mill	Millwrights	31.75	254.00	10.6		11.0		48.1	15.25		47.00	376.00
Mstz	Mosaic & Terrazzo Workers	29.25	234.00	9.8		11.0		47.3	13.85		43.10	344.80
Pord	Painters, Ordinary	27.15	217.20	13.8		11.0		51.3	13.95		41.10	328.80
Psst	Painters, Structural Steel	27.90	223.20	48.4		11.0		85.9	23.95		51.85	414.80
Pape	Paper Hangers	27.10	216.80	13.8		11.0		51.3	13.90		41.00	328.00
Pile	Pile Drivers	29.80	238.40	24.9		16.0		67.4	20.10		49.90	399.20
Plas	Plasterers	28.10	224.80	15.8		11.0		53.3	15.00		43.10	344.80
Plah	Plasterer Helpers	23.70	189.60	15.8		11.0		53.3	12.65		36.35	290.80
Plum	Plumbers	35.95	287.60	8.3		16.0		50.8	18.25		54.20	433.60
Rodm	Rodmen (Reinforcing)	34.25	274.00	28.3		14.0		68.8	23.55		57.80	462.40
Rofc	Roofers, Composition	26.60	212.80	32.6		11.0		70.1	18.65		45.25	362.00
Rots	Roofers, Tile & Slate	26.75	214.00	32.6		11.0		70.1	18.75		45.50	364.00
Rohe	Roofers, Helpers (Composition)	19.80	158.40	32.6		11.0		70.1	13.90		33.70	269.60
Shee	Sheet Metal Workers	35.10	280.80	11.7		16.0		54.2	19.00		54.10	432.80
Spri	Sprinkler Installers	36.20	289.60	8.7		16.0		51.2	18.55		54.75	438.00
Stpi	Steamfitters or Pipefitters	36.20	289.60	8.3		16.0		50.8	18.40		54.60	436.80
Ston	Stone Masons	30.65	245.20	16.0		11.0		53.5	16.40		47.05	376.40
Sswk	Structural Steel Workers	34.25	274.00	39.8		14.0		80.3	27.50		61.75	494.00
Tilf	Tile Layers	29.15	233.20	9.8		11.0		47.3	13.80		42.95	343.60
Tilh	Tile Layers Helpers	23.35	186.80	9.8		11.0		47.3	11.05		34.40	275.20
Trlt	Truck Drivers, Light	24.30	194.40	14.9		11.0		52.4	12.75		37.05	296.40
Trhv	Truck Drivers, Heavy	25.00	200.00	14.9		11.0		52.4	13.10		38.10	304.80
Sswl	Welders, Structural Steel	34.25	274.00	39.8		14.0		80.3	27.50		61.75	494.00
Wrck	*Wrecking	23.45	187.60	41.2		11.0		78.7	18.45		41.90	335.20

*Not included in averages

Figure 9.136

313

years of experience—the qualification to make such a judgment. Some firms are so scientific and calculating that every quoted price includes a lucky number.

A great deal of success in bidding can be attributed to the proper choice of jobs to bid. A contracting firm can go broke estimating every available job. The company must be able to recognize which jobs are too risky and when the competition is too keen, while not overlooking those which can be profitable. Again, knowledge of the marketplace and *experience* are the keys to successful bidding.

The primary purpose of this text has not been to tell the reader how much an item will cost, but instead, how to develop a consistent and thorough approach to the estimating process. If such a pattern is developed, employing consistency, attention to detail, experience and above all, common sense, accurate costs will follow.

If an estimate is thorough, organized, neat, and concise, the benefits go beyond winning contracts. The information and data that is developed will be useful throughout a project—for purchasing, change orders, cost accounting and control, and development of historical costs.

Appendix A

Appendix A

The following is a multiple-page form, called Means' SPEC-AID, which was developed to aid the estimator in preparing all parts of a building project estimate. Each of the sixteen MasterFormat divisions is represented by a complete and thorough checklist. These lists are used to record project requirements and to help ensure inclusion of all items in an estimate.

R.S. Means has published two complete books of reproducible forms, checklists and aids to the estimator. The first, entitled *Means Forms for Building Construction Professionals*, includes the SPEC-AID. The second, *Means Forms for Contractors*, is a collection of forms for estimating, bidding and scheduling as well as project administrative tasks, personnel management. Both forms books show sample uses and include a full explanation of each type of form.

SPEC-AID

OUTLINE SPECIFICATION
AND
QUESTIONAIRE
FOR

PROJECT _____

ADDRESS _____

CITY/STATE/ZIP _____

PREPARED BY _____

_____

DATE_____

TABLE OF CONTENTS

INSTRUCTIONS FOR USE OF SPEC-AID©

A. A primary use of SPEC-AID© is to provide construction planners with a detailed checklist that will facilitate concurrent precise development of specifications with drawings. (It will ensure that all planning groups are talking about the same type and quality of building.)

B. DIVISION 1 is a General Project Description which outlines code requirements and basic building parameters. It is usually completed by the project manager, architect or engineer, having responsible knowledge of the appropriate necessary information.

C. DIVISION 2 through 16 are arranged according to the CSI MasterFormat. Section items should be completed by designers or other technical personnel as appropriate. This should be done as early in the design stage as possible.

D. When an item or product is specifically selected or the drawings are detailed around it, identify it by name, type, number, manufacturer, and reference. Use additional sheets as required.

E. The PROJECT SCHEDULE on page 320 should be one of the first items considered. It can be refined, added to, and revised as the project requirements develop. It is used to show lead time, to coordinate systems, and to visually indicate major project components by events (design, bid, begin construction, etc.), date and duration.

F. The PROJECT SKETCH on page 321 is a set of grid lines to facilitate schematic representation of project phases, column lines, etc. Use of this feature helps to define the scope of the project.

G. The SYSTEM ESTIMATE can be summarized on pages 351 and 352. The systems headings can be coordinated with the PROJECT SCHEDULE on page 320. By filling in the columns (Quantity, Unit, Unit Price, Total Cost, Cost/S.F., and % of Bldg.), each estimated aspect of the project is shown at a glance. By keeping a complete copy of SPEC-AID© for each project on file, valuable preliminary estimating quantities and costs as well as specifications will be at your fingertips for future reference.

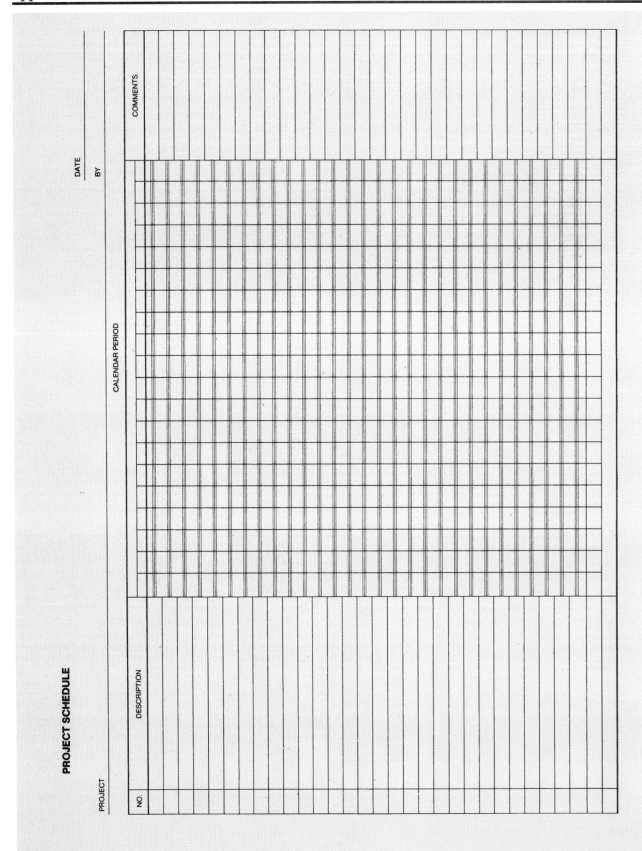

PROJECT SCHEDULE

PROJECT

DATE

BY

CALENDAR PERIOD

COMMENTS

NO.

DESCRIPTION

SPEC-AID

PROJECT _____ DATE _____

ADDRESS _____

CITY/STATE/ZIP _____ PREPARED BY: _____

PROJECT SKETCH   1″ = _____ Ft.

SPEC-AID

DATE _____

DIVISION 1: GENERAL

PROJECT _____ LOCATION _____

Owner _____ **Architect** _____ **Project Mgr.** _____

Engineer: Structural _____ Plumbing _____

H.V.A.C. _____ Electrical _____

Contractor: General _____ Structural _____

Mechanical _____ Electrical _____

Building Type _____

Building Capacities: _____

Quality ☐ Economy ☐ Average ☐ Good ☐ Luxury Describe _____

Size

Ground Floor Area _____ S.F.

Supported Levels (No.) _____ x Area/Level _____ S.F.

Supported Levels (No.) _____ x Area/Level _____ S.F.

Below Grade Area _____ S.F.

Other Area _____ S.F.

TOTAL GROSS AREA _____ S.F.

Floor to Floor Height: Maximum _____ Minimum _____ Average _____

Floor to Ceiling Height: Maximum _____ Minimum _____ Average _____

Floor System Depth: Maximum _____ Minimum _____ Average _____

Building Codes ☐ City _____ ☐ County _____

☐ State _____ ☐ National _____

☐ Other _____ Seismic Zone _____

Zoning ☐ Residential ☐ Commercial ☐ Industrial ☐ None ☐ Other _____

Design Criteria Live Loads: Roof _____ psf. End Walls _____ psf. Window Openings _____%

Supported Floor _____ psf. Side Walls _____ psf. Window Openings _____%

Ground Floor _____ psf.   WIND PRESSURE

Corridors _____ psf. _____ psf. from _____ to _____ ft.

Balconies _____ psf. _____ psf. from _____ to _____ ft.

Allow for Partitions _____ psf. _____ psf. from _____ to _____ ft.

Miscellaneous _____ psf. _____ psf. from _____ to _____ ft.

Comments _____

Typical Bay Spacing _____

Structural Frame ☐ Concrete ☐ Steel ☐ Wood ☐ Wall Bearing ☐ Other _____

Describe _____

Fireproofing ☐ None ☐ Columns _____ Hours ☐ Girders _____ Hours ☐ Beams _____ Hours ☐ Floor _____ Hours

Estimating Budget Estimate Due _____ 19 _____ Schematic Estimate Due _____ 19 ____

Preliminary Estimate Due _____ 19 ____ Final Estimate Due _____ 19 ____ at ____% Working Drawings

Labor Market ☐ Highly Competitive ☐ Normal ☐ Non-Competitive ☐ Unreliable ☐ Union ☐ Non-union

Describe _____

Taxes Tax exempt ☐ No ☐ Yes State ____% County ____% City ____% Other ____%

Bond ☐ Not Required ☐ Required _____

Bidding Date _____ Start Date _____ Construction Duration _____ Months

☐ Open Competitive ☐ Selected Competitive ☐ Negotiated ☐ Filed Bids _____

Contract ☐ Single ☐ Multiple Describe _____

Multiple Type assigned to General Contractor ☐ No ☐ Yes _____

SPEC-AID

DATE _____

DIVISION 2: SITEWORK

PROJECT _____   LOCATION _____

Demolition Site: ☐ No ☐ Yes Allowance _____ ☐ Separate Contract
  Interior: ☐ No ☐ Yes ☐ Allowance _____ ☐ Separate Contract
  Removal From Site: ☐ No ☐ Yes Dump Location _____ Distance _____
Topography ☐ Level ☐ Moderate Grades ☐ Steep Grades Describe _____
Subsurface Exploration ☐ Borings ☐ Test Pits ☐ USDA Maps ☐ Other _____
  Performed by: ☐ Owner ☐ Engineer ☐ Contractor _____
Site Area: Total _____ Acres to Clear _____ Acres To Thin _____ Acres Open _____ Acres
Clearing and Grubbing: ☐ No ☐ Light ☐ Medium ☐ Heavy _____
Topsoil: ☐ No ☐ Strip ☐ Stockpile ☐ Dispose on Site ☐ Dispose off Site _____ Miles ☐ Furnish
  Existing _____ Inches Deep Final Depth _____ Inches Describe _____

Soil Type: ☐ Gravel ☐ Sand ☐ Clay ☐ Silt ☐ Rock ☐ Peat ☐ Other _____
  Rock Expected: ☐ No ☐ Ledge ☐ Boulders ☐ Hardpan ☐ Describe _____
  How Paid _____
  Ground Water Expected: ☐ No ☐ Yes Depth or Elevation _____
  Disposal by ☐ Pumping ☐ Wells ☐ Wellpoints ☐ Other _____
Excavation: ☐ Grade and Fill on Site ☐ Dispose off Site _____ Miles ☐ Borrow Expected _____ Miles
  Quantity Involved _____
  Describe _____
  Sheeting Required: ☐ No ☐ Yes Describe _____
  Protect Existing Structures: ☐ No ☐ Yes Describe _____
Backfill: ☐ No ☐ Yes Area _____ Material _____ Inches Deep _____ % Compaction
  Landscape Area ☐ No ☐ Yes Area _____ Material _____ Inches Deep _____ % Compaction
  Building Area ☐ No ☐ Yes Area _____ Material _____ Inches Deep _____ % Compaction
  Source of Materials _____
Water Control: ☐ Ditching ☐ Sheet Piling ☐ Pumping ☐ Wells ☐ Wellpoints ☐ Pressure Grouting
  ☐ Chemical Grouting ☐ Other _____
  Describe _____
Termite Control: ☐ No ☐ Yes Describe _____
Special Considerations: _____

Piles: ☐ No ☐ Yes ☐ Friction ☐ End Bearing ☐ Concrete ☐ Pipe, Empty ☐ Pipe, Concrete Filled ☐ Steel
  ☐ Step Tapered ☐ Tapered Thin Shell ☐ Wood ☐ Capacity _____ Tons
  Size _____ Length _____ Number Required _____
Caissons: ☐ No ☐ Yes ☐ Cased ☐ Uncased Capacity _____
  Size _____ Length _____ Number Required _____
Pressure Injected Footings: ☐ No ☐ Yes ☐ Cased ☐ Uncased Capacity _____
  Size _____ Length _____ Number Required _____
Special Considerations: _____

Storm Drains: ☐ No ☐ Yes ☐ Asbestos Cement ☐ Bituminous Fiber ☐ Concrete ☐ Corrugated Metal ☐ _____
  Size and Length _____
  Headwall: ☐ No ☐ Yes Type _____ Number _____
  Catch Basins: ☐ No ☐ Yes ☐ Block ☐ Brick ☐ Concrete ☐ Precast Size _____ Number _____
  Manholes: ☐ No ☐ Yes ☐ Block ☐ Brick ☐ Concrete ☐ Precast Size _____ Number _____
Building Sub Drains: ☐ No ☐ Yes ☐ Type _____ Length _____
French Drains: ☐ No ☐ Yes Size _____ Length _____
Trenches: Swales, etc.: ☐ No ☐ Yes Describe _____
Rip Rap: ☐ No ☐ Yes Describe _____
Special Considerations: _____

SPEC-AID

DIVISION 2: SITEWORK

Water Supply Existing Main: ☐ No ☐ Yes Location _____ Size _____
   Service Piping: By Utility ☐ By Others ☐ This Contract _____ Size _____
   Wells: ☐ No ☐ Yes ☐ By Others ☐ This Contract _____ Capacity _____
   Water Pumping Station: ☐ No ☐ Yes Type _____ Capacity _____
Sewers: ☐ No ☐ Yes ☐ By Others ☐ Asbestos Cement ☐ Concrete ☐ Plastic ☐ Vitrified Clay ☐ _____
   Manholes: ☐ No ☐ Yes ☐ Block ☐ Brick ☐ Concrete ☐ Precast Size _____ Number _____
   Sewage Pumping Station: ☐ No ☐ Yes Type _____ Capacity _____
Sewage Treatment: ☐ No ☐ Yes ☐ By Others ☐ This Contract ☐ Septic Tank ☐ Package Treatment Plant
   Describe _____
Special Considerations: _____

Driveways: ☐ No ☐ Yes ☐ By Others ☐ Bituminous ☐ Concrete ☐ Gravel ☐ _____ Thickness _____
Parking Area: ☐ No ☐ Yes ☐ By Others ☐ Bituminous ☐ Concrete ☐ Gravel ☐ _____ Thickness _____
   Base Course: ☐ No ☐ Yes ☐ By Others ☐ Gravel ☐ Stone ☐ _____ Thickness _____
Curbs: ☐ No ☐ Yes ☐ By Others ☐ Bituminous ☐ Concrete ☐ Granite ☐ _____ Size _____
Parking Bumpers: ☐ No ☐ Yes ☐ By Others ☐ Concrete ☐ Timber _____
Painting Lines: ☐ No ☐ Yes ☐ By Others ☐ Paint ☐ Thermo Plastic ☐ Traffic Lines ☐ Stalls ☐ _____
Guard Rail: ☐ No ☐ Yes ☐ By Others ☐ Cable ☐ Steel ☐ Timber _____
Sidewalks: ☐ No ☐ Yes ☐ By Others ☐ Bituminous ☐ Brick ☐ Concrete ☐ Stone _____
   Width _____ Thickness _____
Steps: ☐ No ☐ Yes ☐ Brick ☐ Concrete ☐ Stone ☐ Timber _____
Signs: ☐ No ☐ Yes ☐ Stock ☐ Custom _____
Traffic Signals: ☐ No ☐ Yes ☐ By Others _____
Special Considerations: _____

Fencing: ☐ No ☐ Yes ☐ By Others ☐ Chain link ☐ Aluminum ☐ Steel ☐ Other _____
   Height _____ Length _____ Gates _____
Fountains: ☐ No ☐ Yes ☐ By Others _____
Planters: ☐ No ☐ Yes ☐ By Others ☐ Asbestos Cement ☐ Concrete ☐ Fiberglass ☐ _____
Playground Equipment: ☐ No ☐ Yes ☐ By Others ☐ Benches _____ ☐ Bleachers _____
   ☐ Bike Rack _____ ☐ Goal Posts _____ ☐ Posts _____
   ☐ Running Track _____
   ☐ See Saw _____ ☐ Shelters _____ ☐ Slides _____
   ☐ Swings _____ ☐ Whirlers _____ ☐ _____
Playing Fields: ☐ No ☐ Yes ☐ By Others _____
Railroad Work: ☐ No ☐ Yes ☐ By Others Weight _____ lb. per L.Y. ☐ New ☐ Relay Length _____
   Turnout: ☐ No ☐ Yes _____ ☐ Bumpers _____ ☐ Derails _____
   Wheel Stops _____ ☐ Others _____
Retaining Walls: ☐ No ☐ Yes ☐ By Others ☐ Gravity Concrete ☐ Cantilever Concrete ☐ Steel Bin ☐ Cribbing
   ☐ Timber ☐ Other _____ Height _____ Length _____
Irrigation System: ☐ No ☐ Yes ☐ By Others _____
Tennis Courts: ☐ No ☐ Yes ☐ By Others Type _____ Number _____
Trash Closures: ☐ No ☐ Yes Size _____
Lawns & Planting: ☐ No ☐ Yes ☐ By Others _____ Allowance _____
   Topsoil: ☐ No ☐ Yes ☐ By Others Depth _____ Inches Source _____
   Shrubs: ☐ No ☐ Yes ☐ By Others Describe _____ Allowance _____
   Trees: ☐ No ☐ Yes ☐ By Others Describe _____ Allowance _____
   Seeding: ☐ No ☐ Yes ☐ By Others Describe _____
   Sodding: ☐ No ☐ Yes ☐ By Others Describe _____ Thickness _____
   ☐ Ground Cover _____ ☐ Edging _____ ☐ Mulching _____
Special Considerations: _____

SPEC-AID

DATE _____

DIVISION 3: CONCRETE

PROJECT _____ LOCATION _____

Foundations Bearing on: ☐ Rock   ☐ Earth   ☐ Piles   ☐ Caissons   ☐ Other _____

Bearing Capacity _____

Footings Pile Caps: ☐ No   ☐ Yes _____ psi   Size _____

Forms _____ Reinforcing _____ Waterproofing _____

Spread Footings: ☐ No   ☐ Yes ____ psi   Size _____ Soil Bearing Capacity _____

Forms _____ Reinforcing _____ Waterproofing _____

Continuous Footings: ☐ No   ☐ Yes _____ psi   Size _____

Forms _____ Reinforcing _____ Waterproofing _____

Grade Beams: ☐ No   ☐ Yes _____ psi   Size _____

Forms _____ Reinforcing _____ Waterproofing _____

Piers: ☐ No   ☐ Yes _____ psi   Size _____

Forms _____ Reinforcing _____ Finish _____

Anchor Bolts: ☐ No   ☐ Yes   Size _____

Grout Column Base Plates: ☐ No   ☐ Yes _____

Underslab Fill: ☐ No   ☐ Yes   Material _____ Depth _____

Vapor Barrier: ☐ No   ☐ Yes   Material _____ Thickness _____

Perimeter Insulation: ☐ No   ☐ Yes   Material _____ Dimensions _____

Slab on Grade: ☐ No   ☐ Yes _____ psi   Thickness _____

Forms: ☐ Cold Keyed   ☐ Expansion   ☐ Other _____ Spacing _____

Reinforcing: ☐ No   ☐ Mesh   ☐ Bars _____ Type _____

Finish: ☐ Screed   ☐ Darby   ☐ Float   ☐ Broom   ☐ Trowel   ☐ Granolithic _____

Special Finish: ☐ No   ☐ Hardner   ☐ Colors   ☐ Abrasives   ☐ _____

Columns: ☐ No   ☐ Round   ☐ Square   ☐ Rectangular   ☐ Precast   ☐ Steel   ☐ Encased Steel   ☐ Lightweight

_____ psi   Size _____

Forms: ☐ Optional   ☐ Framed Plywood   ☐ Plywood   ☐ Fiber Tube   ☐ Steel   ☐ Round Fiberglass   ☐ _____

Reinforcing: ☐ No ☐ Square Tied ☐ Spirals Grade _____ Bar Sizes _____ Type Splice _____

Finish: ☐ Break Fins   ☐ Rubbed   ☐ Other _____

Elevated Slab System: ☐ No   ☐ Flat Plate   ☐ Flat Slab   ☐ Domes   ☐ Pans   ☐ Beam & Slab   ☐ Lift Slab   ☐ Composite

☐ Floor Fill   ☐ Roof Fill   ☐ Standard Weight   ☐ Lightweight   Concrete Strength _____

Forms: ☐ Optional   ☐ Plywood   ☐ Other _____ Ceiling Height _____

Reinforcing: ☐ Mesh   ☐ Bars   Grade _____ Size: _____

Post-tension: ☐ No   ☐ Simple Spans   ☐ Continuous   Spans _____ Depth _____

☐ Grouted   ☐ Ungrouted   Perimeter Conditions _____

Slab Finish: ☐ Screed   ☐ Darby   ☐ Float   ☐ Broom   ☐ Trowel   ☐ Granolithic   ☐ _____

Special Finish: ☐ No   ☐ Hardener   ☐ Colors   ☐ Abrasives   ☐ _____

Ceiling Finish: ☐ No   ☐ Break Fins   ☐ Rubbed   ☐ Other _____

Beams: ☐ No   ☐ Steel   ☐ Encased Steel   ☐ Precast   ☐ Regular Weight   ☐ Lightweight   ☐ Steel Composite _____

Description: _____

Forms: ☐ Optional   ☐ Framed Plywood   ☐ Plywood   ☐ Steel   ☐ Other _____ Ceiling Height _____

Reinforcing: ☐ Conventional   ☐ Post-tension   ☐ Simple Span   ☐ Continuous Spans _____ Depth _____

☐ Grouted   ☐ Ungrouted   Perimeter Conditions _____

Walls: ☐ No ☐ Precast ☐ Tilt up ☐ Regular Weight ☐ Lightweight _____ psi Thickness _____

Forms: ☐ Optional   ☐ Framed Plywood   ☐ Plywood   ☐ Steel   ☐ Slipform   ☐ _____

Reinforcing: ☐ No   ☐ Bars   Grade _____ Clear Height _____

Finish: ☐ No   ☐ Break Fins   ☐ Rubbed   ☐ _____

Stairs: ☐ No   ☐ Precast   ☐ Ground Cast   ☐ Form Cast   ☐ Pan Fill Treads   ☐ _____

Forms: ☐ Plywood   ☐ Steel   ☐ Prefab Steel, Left in Place   ☐ _____

Reinforcing: ☐ Conventional   Grade _____

Finish: ☐ All Surfaces   ☐ Treads   ☐ Risers   ☐ Abrasives   ☐ Nosings _____

Page 5 of 30

SPEC-AID

DIVISION 3: CONCRETE

Reinforcing Splices: ☐ No ☐ Yes ☐ Lap Type ☐ Compression Only ☐ 125% Yield ☐ Full Tension
☐ Horizontal ☐ Vertical ☐ Special _____

Gunite: ☐ No ☐ Yes _____

Cast in Place Special Considerations: _____
_____
_____
_____

Copings: ☐ No ☐ Yes Size _____ Finish _____

Curbs: ☐ No ☐ Yes Size _____ Finish _____

Joists: ☐ No ☐ Yes Live load _____ psf. Span _____ Size _____
Describe _____

Lift Slab: ☐ No ☐ Yes No. Slabs _____ Thickness _____ Inches Columns Spacing _____
Story Height _____ ☐ Conventional Reinforcing ☐ Post-tension _____

Lintels: ☐ No ☐ Yes ☐ Doors ☐ Windows ☐ Other _____
☐ Conventional Reinforcing ☐ Prestressed _____

Prestressed Precast Floors: ☐ No ☐ Yes ☐ Plank ☐ Multiple Tee Depth _____ Span _____ Width _____
Roofs: ☐ No ☐ Yes ☐ Plank ☐ Double Tee ☐ Single Tee Depth _____ Span _____ Width _____
Supporting Beams: ☐ No ☐ Yes ☐ Cast in Place ☐ Precast Describe _____

Columns: ☐ No ☐ Yes ☐ Steel ☐ Cast in Place ☐ Precast Size _____
Walls: ☐ No ☐ Yes ☐ Multiple Tee ☐ Other Thickness _____ Height _____ Width _____

Stairs: ☐ No ☐ Yes ☐ Treads Only ☐ Tread & Riser Units ☐ Complete Stairs ☐ Other _____
Describe _____

Tilt Up Walls: ☐ No ☐ Yes Size _____ Finish _____

Wall Panels: ☐ No ☐ Yes ☐ Insulated ☐ Regular Weight ☐ Lightweight Panel Size _____
Finish: ☐ Gray ☐ White ☐ Exposed Aggregate ☐ Other _____
Reinforcing: ☐ Conventional ☐ Prestressed ☐ Plain ☐ Galvanized _____
Erection: ☐ No. Stories _____ Maximum Lift _____ Overhangs, etc. _____

Window Section: ☐ No ☐ Yes ☐ Size _____ Finish _____

Window Sills: ☐ No ☐ Yes Size _____ Finish _____

Precast Special Considerations: _____
_____
_____
_____
_____

Concrete Decks: ☐ No ☐ Yes ☐ Cast in Place ☐ Plank ☐ Topping ☐ Cement Fiber ☐ Channel Slab ☐ _____
Depth _____ Span _____ Sub Purlins _____ Roof Pitch _____

Concrete Fill: ☐ No ☐ Yes ☐ Regular Weight ☐ Lightweight Type _____ Depth _____

Formboard: ☐ No ☐ Yes ☐ Type _____ Depth _____ Spans _____
Sub Purlins: ☐ No ☐ Yes Span _____ Describe _____

Gypsum Decks Floor Plank: ☐ No ☐ Yes Depth _____ Span _____ Underlayment _____
Roofs: ☐ No ☐ Yes ☐ Cast in Place ☐ Plank Depth _____ Span _____ Pitch _____

Other Cementitious Decks: ☐ No ☐ Yes Describe _____

Cementitious Decks: Special Considerations _____
_____
_____

General Notes: Concrete Section _____
_____
_____

SPEC-AID

DATE _____

DIVISION 4: MASONRY

PROJECT _____ LOCATION _____

Exterior Walls: ☐ No ☐ Yes ☐ Load Bearing ☐ Non Load Bearing Story Height _____
Describe _____

Interior Walls: ☐ No ☐ Yes ☐ Load Bearing ☐ Non Load Bearing Ceiling Height _____
Describe _____

Mortar: ☐ Optional ☐ Type K ☐ Type 0 ☐ Type N ☐ Type S ☐ Type M ☐ Thinset ☐ _____
☐ Colors _____ ☐ Other _____

Cement Brick: ☐ No ☐ Yes ☐ Solid ☐ Cavity ☐ Veneer ☐ _____
Describe _____
_____ Compressive Strength _____ psi. ASTM No. _____
Size _____ Bond _____ Joints _____ Reinforcing _____ Ties _____

Common Brick: ☐ No ☐ Yes ☐ Solid ☐ Cavity ☐ Veneer ☐ _____
Describe _____
_____ Compressive Strength _____ psi. ASTM No. _____
Size _____ Bond _____ Joints _____ Reinforcing _____ Ties _____

Face Brick: ☐ No ☐ Yes ☐ Solid ☐ Cavity ☐ Veneer ☐ _____
Describe _____
_____ Compressive Strength _____ psi. ASTM No. _____
Size ☐ Standard ☐ Jumbo ☐ Norman ☐ Roman ☐ Engineer ☐ Double ☐ _____
Allowance: ☐ No ☐ Yes $ _____ per M Delivered ☐ Unglazed ☐ Single Glazed ☐ Double Glazed
Bond: ☐ Running ☐ Common ☐ English ☐ Flemish ☐ Stack ☐ _____ Headers Every _____ Course.
Joints: ☐ Concave ☐ Struck ☐ Flush ☐ Raked ☐ Weathered ☐ Stripped ☐ _____
Reinforcing: ☐ No ☐ Yes Describe _____
Wall Ties: ☐ No ☐ Yes Describe _____

Anchor Bolts: ☐ No ☐ Yes Size _____
Chimneys: ☐ No ☐ Yes ☐ Regular Brick ☐ Radial Brick Size _____
Columns: ☐ No ☐ Yes Size _____
Control Joints: ☐ No ☐ Yes Spacing _____ Material _____
Copings ☐ No ☐ Yes ☐ Concrete ☐ Stone ☐ _____ Describe _____
Fire Brick: ☐ No ☐ Yes ☐ Low Duty ☐ High Duty Describe _____
Fireplaces: ☐ No ☐ Yes Describe _____
Accessories _____
Flooring: ☐ No ☐ Yes ☐ Laid Flat ☐ Laid on Edge ☐ Pattern _____
☐ Regular ☐ Acid Resisting Describe _____
Insulating Brick: ☐ No ☐ Yes Describe _____
Insulation: ☐ No ☐ Yes ☐ Board ☐ Poured ☐ Sprayed Material _____
Thickness _____ Describe _____
Lintels: ☐ No ☐ Yes ☐ Block ☐ Precast ☐ Steel Describe _____
Masonry Restoration: ☐ No ☐ Yes ☐ Cut ☐ Recaulk ☐ Repoint ☐ Stucco Finish ☐ _____
Sand Blast: ☐ No ☐ Yes Describe _____
Steam Clean: ☐ No ☐ Yes Describe _____
Piers: ☐ No ☐ Yes Size _____
Pilasters: ☐ No ☐ Yes Size _____
Refractory Work: ☐ No ☐ Yes Describe _____
Simulated Brick: ☐ No ☐ Yes Material _____ Describe _____
Steps: ☐ No ☐ Yes Describe _____
Vent Box: ☐ No ☐ Yes ☐ Aluminum ☐ Bronze Size _____
Weep Holes: ☐ No ☐ Yes Spacing _____ Describe _____
Window Sills and Stools: ☐ No ☐ Yes ☐ Brick ☐ Concrete ☐ Stone ☐ _____
Describe _____

SPEC-AID

DIVISION 4: MASONRY

Concrete Block: ☐ No ☐ Yes ☐ Exterior ☐ Interior ☐ Regular Weight ☐ Lightweight ☐ Solid ☐ Hollow
☐ Load Bearing ☐ Non Load Bearing Describe _____
_____ Compressive Strength _____ psi. ASTM No. _____
Size: _____
Finish: ☐ Regular ☐ Ground ☐ Ribbed ☐ Glazed ☐ _____
Bond: ☐ Common ☐ Stack ☐ Other _____ Headers Every _____ Course.
Joints: ☐ Concave ☐ Struck ☐ Flush ☐ Raked ☐ Weathered ☐ Stripped ☐ _____
Reinforcing: ☐ No ☐ Yes _____ Strips Every _____ Course.
Wall Ties: ☐ No ☐ Yes Describe _____
Bond Beams: ☐ No ☐ Yes Size _____ Describe _____
Reinforcing _____ Grout _____
Lintels: ☐ No ☐ Yes ☐ Precast ☐ Steel ☐ Block Size _____ Describe _____
Reinforcing _____ Grout _____
Columns: ☐ No ☐ Yes Size _____
Pilasters: ☐ No ☐ Yes Size _____
Glass Block: ☐ No ☐ Yes ☐ Size _____ Type _____
Special Block _____
Describe _____
Glazed Concrete Block: ☐ No ☐ Yes ☐ Solid ☐ Hollow Type _____
☐ Non Reinforced ☐ Reinforced _____ Strips Every _____ Course.
Describe _____
Gypsum Block: ☐ No ☐ Yes ☐ Solid ☐ Hollow Thickness _____ Describe _____

Grouting: ☐ No ☐ Yes ☐ Block Cores ☐ Bond Beams ☐ Cavity Walls ☐ Door Frames ☐ Lintels ☐ _____
Describe _____
Insulation: ☐ No ☐ Yes ☐ Board ☐ Poured ☐ Sprayed Material _____
Thickness _____ Describe _____
Solar Screen: ☐ No ☐ Yes Describe _____
Special Block: ☐ No ☐ Yes Describe _____
☐ Parge Block ☐ Clean Cavity ☐ Spandrel Flashing _____
Special Considerations: _____
Ceramic Veneer: ☐ No ☐ Yes Describe _____
Structural Facing Tile: ☐ No ☐ Yes ☐ 6T Series ☐ 8W Series ☐ Other _____
Describe _____
Terra Cotta: ☐ No ☐ Yes ☐ Floors ☐ Partitions ☐ Fireproofing ☐ Load Bearing ☐ Non Load Bearing
Describe _____
Ashlar Stone: ☐ No ☐ Yes Type _____ Thickness _____
Describe _____
Rubble Stone: ☐ No ☐ Yes ☐ Coarsed ☐ Uncoarsed Type _____ Thickness _____
Describe _____
Cut Stone: ☐ No ☐ Yes ☐ Granite ☐ Limestone ☐ Marble ☐ Sand Stone ☐ Slate ☐ _____
☐ Base _____ ☐ Columns _____ ☐ Coping _____
☐ Curbs _____ ☐ Facing Panels _____
☐ Flooring _____ ☐ Showers _____
☐ Soffits _____ ☐ Stair Treads _____
☐ Stairs _____ ☐ Thresholds _____
☐ Window Sills _____ ☐ Window Stools _____
☐ _____
Simulated Stone: ☐ No ☐ Yes Material _____ Describe _____
Special Stone _____
General Notes: Masonry _____
_____
_____

Page 8 of 30

SPEC-AID

DATE _____

DIVISION 5: METALS

PROJECT _____ LOCATION _____

Design Criteria: In Division 1 _____

Typical Bay Spacings: _____

Floor to Ceiling Heights: _____

Beam Depths: _____

Roof Slope: ☐ Flat ☐ Other _____

Eave Height: _____

Anchor Bolts: ☐ No ☐ By Others ☐ Yes Describe _____ Number _____

Base Plates: ☐ No ☐ By Others ☐ Yes Describe _____

_____ Number _____

Metal Decking Floors: ☐ No ☐ By Others ☐ Cellular ☐ Non Cellular ☐ Painted ☐ Galv. Depth _____

☐ Acoustical ☐ Ventilating Gauge _____ Describe _____

Roof Deck: ☐ No ☐ By Others ☐ Cellular ☐ Non Cellular ☐ Painted ☐ Galv. Depth _____

☐ Acoustical ☐ Ventilating Gauge _____ Describe _____

Structural System: ☐ Wall Bearing ☐ Free Standing ☐ Simple Spans ☐ Continuous Spans _____

☐ Conventional Design ☐ Plastic Design ☐ Field Welded ☐ Field Bolted ☐ Composite Design

☐ Other _____

Type Steel _____ Grade _____

Estimated Weights: Beams _____ Roof Frames _____ Adjustable Spandrel Angles _____

Girders _____ Girts _____ Hanger Pods _____ Bracing _____

Columns _____ Connections _____ Other _____

Paint: Shop ☐ No ☐ Yes _____ Coats Material _____

Field Paint ☐ No ☐ Yes ☐ Brush ☐ Roller ☐ Spray _____ Coats Material _____

Galvanizing: ☐ No ☐ Yes Thickness _____

Other: _____

Fireproofing: ☐ No ☐ Yes ☐ Beams ☐ Columns ☐ Decks ☐ Other _____ Rating _____ Hr.

☐ Concrete Encasement ☐ Spray ☐ Plaster ☐ Drywall ☐ _____

Describe _____

Open Web Joists: ☐ No ☐ Yes ☐ H Series ☐ LH Series ☐ _____

Estimated Weights: _____

Bridging: ☐ No ☐ Yes ☐ Bolted ☐ Welded ☐ Pod Bridging _____

Paint: Shop ☐ Standard ☐ Special _____ Coats Field Paint ☐ No ☐ Yes Describe _____

Light Gauge Joists: ☐ No ☐ Yes Describe _____

Light Gauge Framing: ☐ No ☐ Yes Describe _____

Special Considerations: _____

_____

_____

_____

Fasteners: Expansion Bolts _____ High Strength Bolts _____

Machine Screws _____ Machinery Anchors _____

Nails _____ Roof Bolts _____

Sheet Metal Screws _____ Studs _____

Timber Connectors _____ Toggle Bolts _____

Welded Studs _____ Other _____

_____

_____

SPEC-AID

DIVISION 5: METALS

Area Walls: ☐ No ☐ Yes _____ Gratings _____ Caps _____

Bumper Rails: ☐ No ☐ Yes _____

Canopy Framing: ☐ No ☐ Yes _____

Checkered Plate: ☐ No ☐ Trench Covers ☐ Pit Covers ☐ Platforms ☐ _____

Columns: ☐ No ☐ Aluminum ☐ Steel ☐ Square ☐ Rectangular ☐ Round ☐ _____

Construction Castings: ☐ No ☐ Chimney Specialties _____ ☐ Column Bases _____

☐ Manhole Covers _____ ☐ Wheel Guards _____

☐ _____

Corner Guards: ☐ No ☐ Yes _____

Crane Rail: ☐ No ☐ Yes _____

Curb Angles: ☐ No ☐ Straight ☐ Curved _____

Decorative Covering: ☐ No ☐ Stock Sections ☐ Custom Sections _____

Doors _____ Walls _____

Door Frames: ☐ No ☐ Yes _____ ☐ Protection: ☐ No ☐ Yes _____

Expansion Joints Ceilings: ☐ No ☐ Yes _____ Cover Plates: ☐ No ☐ Yes _____

Floors: ☐ No ☐ Yes _____ Cover Plates: ☐ No ☐ Yes _____

Walls: ☐ No ☐ Yes _____ Cover Plates: ☐ No ☐ Yes _____

Fire Escape: ☐ No ☐ Yes _____ Size _____

Stairs _____ Ladders _____ Cantilever _____

Floor Grating: ☐ No ☐ Aluminum ☐ Steel ☐ Fiberglass ☐ Platforms ☐ Stairs ☐ _____

Type _____ Weight _____

Special Finish _____

Ladders: ☐ No ☐ Aluminum ☐ Steel ☐ _____

☐ With Cage ☐ No Cage _____ ☐ Inclined Type _____

Lamp Posts: ☐ No ☐ Yes _____

Lintels: ☐ No ☐ Yes ☐ Plain ☐ Built-up ☐ Painted ☐ Galvanized _____

_____

Louvers: ☐ No ☐ Yes _____

Manhole Covers: ☐ No ☐ Yes _____

Mat Frames: ☐ No ☐ Yes _____

Overhead Supports: ☐ No ☐ Toilet ☐ Partitions ☐ _____

Pipe Bumpers: ☐ No ☐ Yes _____

Pipe Supports: ☐ No ☐ Yes _____

Railings: ☐ No ☐ Yes ☐ Aluminum ☐ Steel ☐ Pipe ☐ _____

Balconies: ☐ No ☐ Yes _____

Stairs: ☐ No ☐ Yes _____

Wall: ☐ No ☐ Yes _____

Solar Screens: ☐ No ☐ Yes _____

Stairs: ☐ No ☐ Yes ☐ Aluminum ☐ Steel ☐ Stock ☐ Custom ☐ _____

Size _____ Landings _____

Spiral: ☐ No ☐ Aluminum ☐ Steel ☐ Stock ☐ Custom ☐ _____

Pre-erected: ☐ No ☐ Yes _____

Stair Treads: ☐ No ☐ Yes _____

Trench Covers: ☐ No ☐ Yes _____

Weather Vanes: ☐ No ☐ Yes _____

Window Guards: ☐ No ☐ Bars ☐ Woven Wire ☐ _____

Wire: ☐ No ☐ Yes _____

Wire Rope: ☐ No ☐ Yes _____

Special Considerations: _____

_____

_____

SPEC-AID
DIVISION 6: CARPENTRY

DATE _____

PROJECT _____ LOCATION _____

Framing: Type Wood _____ Fiber Stress _____ psi.

Beams: ☐ Single ☐ Built Up _____ Grade _____
Bracing: ☐ No ☐ Let In ☐ _____
Bridging: ☐ Steel ☐ Wood _____
Canopy Framing: _____
Columns _____ Fiber Stress _____ psi.
Door Bucks: ☐ No ☐ Treated ☐ Untreated _____
Floor Planks _____ Grade _____
Furring: ☐ Metal ☐ Wood _____
Grounds: ☐ No ☐ Casework ☐ Plaster ☐ On Wood ☐ On Masonry _____
Joists: ☐ No ☐ Floor ☐ Ceiling _____ Grade _____
Ledgers: ☐ No ☐ Bolted ☐ Nailed _____
Lumber Treatment: ☐ No ☐ Creosote ☐ Salt Treated ☐ Fire Retardant _____
☐ Kiln Dry _____
Nailers: ☐ No ☐ Treated ☐ Untreated _____
Plates: _____ **Platform Framing:** _____
Plywood Treatment: ☐ No ☐ Salt Treated ☐ Fire Retardant _____
Posts & Girts: _____
Rafters: ☐ No ☐ Ordinary ☐ Hip _____
Roof Cants: ☐ No ☐ Yes _____ **Roof Curbs:** ☐ No ☐ Yes _____
Roof Decks: ☐ No ☐ Yes _____ Inches Thick
Roof Purlins: ☐ No ☐ Yes _____
Roof Trusses: ☐ No ☐ Timber Connectors ☐ Nailed ☐ Glued Spaced _____ O.C. Span _____ feet
Sheathing, Roof: ☐ No ☐ Plywood ☐ Boards ☐ Wood Fiber ☐ Gypsum _____
  Wall: ☐ No ☐ Plywood ☐ Boards ☐ Wood Fiber ☐ Gypsum _____
Siding Hardboard: ☐ Plain ☐ Primed ☐ Stained _____
  Particle Board: _____ Wood Fiber _____
  Plywood: ☐ Cedar ☐ Fir ☐ Redwood ☐ Marine ☐ Natural ☐ Stained ☐ Plastic Faced ☐ _____
  Wood: ☐ Cedar ☐ Redwood ☐ White Pine ☐ Bevel ☐ Board & Batten ☐ Channel ☐ T & C ☐ Shiplap _____
  ☐ Natural ☐ Stained _____
Sills: _____ **Sleepers:** _____
Soffits: ☐ No ☐ Open ☐ Vented ☐ Plywood _____
Stressed Skin Plywood Box Beams: _____ Depth _____
  Floor Panels: ☐ No ☐ Yes _____ Depth _____
  Roof Panels: ☐ No ☐ Straight ☐ Curved _____ Depth _____
  Folded Plate: ☐ No ☐ Yes _____ Depth _____
Studs: ☐ No ☐ Yes _____ Grade _____
Subfloor: ☐ No ☐ Plywood ☐ Boards ☐ Wood Fiber ☐ _____
Suspended Ceiling Framing: ☐ No ☐ Yes _____
Underlayment: ☐ No ☐ Particle Board ☐ Plywood ☐ Wood Fiber ☐ Hardboard _____
Special Considerations: _____

Laminated Framing: ☐ Beams ☐ Straight ☐ Curved _____ Span _____
  ☐ Bowstring Trusses ☐ Radial Arch ☐ Tudor Arch ☐ Columns _____
  Span _____ Height _____
  ☐ Industrial Grade ☐ Premium Grade ☐ Exterior Glue ☐ Stain ☐ Varnish ☐ Treated ☐ _____
Laminated Roof Deck: ☐ No ☐ Yes _____ Thickness _____
Special Considerations: _____

SPEC-AID

DIVISION 6: CARPENTRY

Base: ☐ No   ☐ One Piece   ☐ Built up   ☐ Pine   ☐ Hardwood _____

Cabinets: ☐ No   ☐ Corner   ☐ Kitchen   ☐ Toilet Room   ☐ Other _____
   ☐ Stock   ☐ Custom _____
   ☐ Unfinished   ☐ Prefinished _____
   Base Cabinets: ☐ Softwood   ☐ Hardwood   ☐ Drawer Units _____
   Wall Cabinets: ☐ Softwood   ☐ Hardwood _____
   Tall Cabinets: ☐ Softwood   ☐ Hardwood _____
   Special: _____

Casings: ☐ No   ☐ Doors   ☐ Windows   ☐ Beams   ☐ Others _____
   ☐ Softwood   ☐ Hardwood _____

Ceiling Beams: ☐ No   ☐ Cedar   ☐ Pine   ☐ Fir   ☐ Plastic _____

Chair Rail: ☐ No   ☐ Pine   ☐ Other _____

Closets: ☐ No   ☐ Pole   ☐ Shelf   ☐ Prefabricated _____

Columns: ☐ No   ☐ Square   ☐ Round   ☐ Solid   ☐ Built up   ☐ Hollow   ☐ Tapered _____
   Diameter _____ Height _____

Convector Covers: ☐ No   ☐ Yes _____

Cornice: ☐ No   ☐ 1 Piece   ☐ 2 Piece   ☐ 3 Piece   ☐ Pine   ☐ Cedar   ☐ Other _____

Counter Tops: ☐ No   ☐ Plastic   ☐ Ceramic Tile   ☐ Marble   ☐ Suede Finish   ☐ Other _____
   ☐ Stock   ☐ Custom _____
   ☐ No Splash   ☐ Square Splash   ☐ Cover Splash _____
   ☐ Self Edge   ☐ Stainless Edge   ☐ Aluminum Edge _____
   Special _____

Cupolas: ☐ No   ☐ Stock   ☐ Custom   ☐ Wood   ☐ Fiberglass   ☐ Square   ☐ Octagonal   Size _____
   ☐ Aluminum Roof   ☐ Copper Roof   ☐ Other _____

Doors and Frames: See Division 8 _____

Door Moldings: ☐ No   ☐ Yes _____

Door Trim: ☐ No   ☐ Yes _____

Fireplace Mantels: ☐ No   ☐ Beams   ☐ Moldings _____
   Size _____

Moldings: ☐ No   ☐ Softwood   ☐ Hardwood   ☐ Metal   ☐ Other _____

Paneling Hardboard: ☐ No   ☐ Tempered   ☐ Untempered   ☐ Pegboard   ☐ Plastic Faced _____
   Plywood, Unfinished: ☐ No   ☐ Veneer Core   ☐ Lumber Core   Grade _____ Thick _____
   Plywood, Prefinished: ☐ No   ☐ Stock   ☐ Architectural Finish _____
   Size _____
   Wood Boards: ☐ No   ☐ Softwood   ☐ Hardwood _____

Railings: ☐ No   ☐ Stock   ☐ Custom   ☐ Softwood   ☐ Hardwood _____
   ☐ Stairs   ☐ Balcony   ☐ Porch   ☐ Wall   ☐ Other _____

Shelving: ☐ No   ☐ Prefinished   ☐ Unfinished   ☐ Stock   ☐ Custom   ☐ Plywood   ☐ Particle Board   ☐ Boards _____
   ☐ Book Shelves _____   ☐ Linen Shelves _____
   ☐ Storage Shelves _____   ☐ Other _____

Stairs: ☐ No   ☐ Prefabricated   ☐ Built in Place   ☐ Softwood   ☐ Hardwood _____
   ☐ Box   ☐ Open   ☐ Circular _____

Thresholds: ☐ No   ☐ Interior   ☐ Exterior _____

Wainscot: ☐ No   ☐ Boards   ☐ Plywood   ☐ Moldings _____

Windows and Frames: See Division 8 _____

Window Trim: ☐ No   ☐ Yes _____

Special Considerations: _____

SPEC-AID DATE _____

DIVISION 7: MOISTURE PROTECTION

PROJECT _____ LOCATION _____

Bentonite: □ No □ Panels □ Granular _____
Bituminous Coating: □ No □ Brushed □ Sprayed □ Troweled □ 1 Coat □ 2 Coat □ Protective Board _____
Building Paper: □ No □ Asphalt □ Polyethylene □ Rosin □ Kraft □ Foil Backed □ _____
□ Roof Deck Vapor Barrier _____
Caulking: □ No □ Gun Grade □ Knife Grade □ Plain □ Colors _____
□ Doors □ Windows □ _____
Cementitious: □ No □ 1 Coat □ 2 Coat Thickness _____ Inches Mix _____
Control Joints: _____ **Expansion Joints:** _____
Elastomeric Waterproofing: □ No □ EPDM □ Neoprene □ PVC □ Urethane □ _____
Liquid Waterproofing: □ No □ Silicone □ Stearate □ _____
Membrane Waterproofing: □ 1 Ply □ 2 Ply □ 3 Ply □ Felt □ Fabric □ Elastomeric □ _____
Metallic Coating: □ No □ Walls _____ in. Thick □ Floors _____ in. Thick _____
Preformed Vapor Barrier: □ No □ Yes _____
Sealants: □ No □ Butyl □ Polysulfide □ PVC □ Urethane □ _____
□ Doors □ Windows □ _____
Special Waterproofing _____

Building Insulation: Rigid: □ No □ Fiberglass □ Polystyrene □ Urethane □ _____
Non Rigid: □ No □ Fiberglass □ Mineral Fiber □ Vermiculite □ Perlite □ _____
Form Board: □ No □ Acoustical □ Asbestos Cement □ Fiberglass □ Gypsum □ Mineral Fiber □ Wood Fiber
□ Other _____ □ Sub Purlins _____ Span _____
Masonry Insulation: □ No □ Cavity Wall □ Block Cores □ Poured □ Foamed Type _____
Perimeter Insulation: □ No □ Yes Type _____ Thickness _____
Roof Deck Insulation: □ No □ Fiberboard □ Fiberglass □ Foamglass □ Polystyrene □ Urethane □ _____
Thickness _____ □ **Cants** _____ Size _____
Sprayed: □ No □ Fibrous □ Cementitious □ Urethane □ _____
Special Insulation _____

Shingles: Aluminum: □ No □ Yes _____ Asbestos: □ No □ Yes _____
Asphalt: □ No □ Class C □ Class A □ _____ Weight _____ lb. per Sq. _____
Clay Tile: □ No □ Plain □ Glazed □ Spanish □ _____ Weight _____ lb. per Sq. _____
Concrete Tile: □ No □ Yes _____ Porcelain Enamel: □ No □ Yes _____
Slate: □ No □ Yes Type _____ Color _____ Exposure _____
Wood: □ No □ Roofing □ Siding □ Fire Retardant Type _____ Grade _____ Exposure _____
Shingle Underlayment: □ No □ Asbestos □ Asphalt □ _____ Weight _____
Special Shingles: _____

Aluminum: □ No □ Roofing □ Siding □ Painted □ Insulated □ Sandwich □ _____
Thickness _____ Type _____
Asbestos Cement: □ No □ Roofing □ Siding □ Flat □ Corrugated □ Natural □ Painted □ Sandwich
□ Fire Rated Thickness _____ Type _____
Epoxy Panels: □ No □ Solid □ Plywood Back □ Hardboard Back □ Exposed Aggregate □ _____
Fiberglass Panels: □ No □ Roofing □ Siding □ Flat □ Corrugated □ _____ Thickness _____
Metal Facing Panels: □ No □ Field Assembled □ Factory Made Insulation _____
Outside Face _____ Inside Face _____
Protected Metal: □ No □ Roofing □ Siding Type _____ Gauge _____
Steel: □ No □ Roofing □ Siding □ Painted □ Galvanized □ Insulated □ Sandwich □ _____
Type _____ Gauge _____
Vinyl Siding: □ No □ Plain □ Insulated Type _____
Special Roofing & Siding: _____

SPEC-AID

DIVISION 7: MOISTURE PROTECTION

Built Up Roofing: □ No □ Tar & Gravel □ Asphalt & Gravel □ Felt □ Mineral Surface □ Aggregate
□ 1 Ply □ 2 Ply □ 3 Ply □ 4 Ply □ 5 Ply □ Bonded _____ years Roof Pitch _____ Type Deck _____
Underlayment: □ No □ Rosin Paper □ Vapor Barrier □ _____

Elastic Sheet Roofing: □ No □ Butyl □ Neoprene □ _____ Thickness _____
Describe _____

Fluid Applied Roofing: □ No □ Hypalon Neoprene □ Silicone □ Vinyl □ _____ Thickness _____
Describe _____

Roll Roofing: □ No □ Smooth □ Granular _____ Weight _____ lbs. per Sq.

Special Membrane Roofing: _____
_____

Downspouts: □ No □ Aluminum □ Copper □ Lead Coated Copper □ Galvanized Steel □ Stainless Steel
□ Steel Pipe □ Vinyl □ Zinc Alloy □ Stock □ Custom □ _____ Size _____
Describe _____

Expansion Joints: □ No □ Roof □ Walls □ No Curbs □ Curbs □ Rubber □ Metallic □ _____
Describe _____

Fascia: □ No □ Yes Describe _____ Thickness _____

Flashing: □ No □ Aluminum □ Asphalt □ Copper □ Fabric □ Lead □ Lead Coated Copper □ PVC □ Rubber
□ Stainless Steel □ Terne □ Zinc Alloy □ Paper Backed □ Mastic Backed □ Fabric Backed □ _____
Describe _____ Thickness _____
_____

Gravel Stop: □ No □ Aluminum □ Copper □ PVC □ Stainless Steel □ _____
□ With Fascia □ No Fascia □ Natural □ Painted Thickness _____ Face Height _____

Gutters: □ No □ Aluminum □ Copper □ Lead Coated Copper □ Galvanized Steel □ Stainless Steel
□ Vinyl □ Wood □ Zinc Alloy □ _____ Thickness _____
□ Box Type □ K Type □ Half Round □ Stock □ Custom □ _____ Size _____

Louvers: □ No □ Yes _____

Mansard: □ No □ Yes _____ Thickness _____

Metal Roofing: □ No □ Copper □ Copper Bearing Steel □ Lead □ Lead Coated Copper □ Stainless Steel
□ Terne □ Zinc Alloy □ _____ Size _____ Thickness _____
□ Standing Seam □ Flat Seam □ Batten Seam □ _____ Weight _____ lbs. per Sq.

Underlayment: □ No □ 15 lb. Felt □ 30 lb. Felt □ Rosin Paper □ _____

Reglet: □ No □ Aluminum □ Copper □ Galvanized Steel □ Stainless Steel □ Zinc Alloy □ _____
Thickness _____ Counter Flashing: □ No □ Yes _____ Thickness _____

Soffit: □ No □ Yes _____ Thickness _____

Special Sheet Metal Work: _____
_____

Ceiling Hatches: □ No □ Steel □ Galvanized □ Painted □ Aluminum □ _____
Size _____

Roof Drains: □ No □ In Plumbing □ Yes _____

Roof Hatches: □ No □ Steel □ Galvanized □ Painted □ Aluminum □ _____
□ Insulated □ Not Insulated □ With Curbs □ No Curbs □ _____ Size _____

Smoke Hatches: □ No □ Yes _____

Snow Guards: □ No □ Yes _____

Skylights: □ No □ Domes □ Vaulted □ Ridge Units □ Field Fabricated □ Glass □ Plastic □ Single
□ Double □ Sandwich Panels □ With Curbs □ No Curbs □ _____ Size _____

Smoke Vents: □ No □ Yes _____

Skyroofs: □ No □ Yes _____

Ventilators: □ No □ In Ventilating □ Stationary □ Spinners □ Motorized _____

Special Roof Accessories _____
_____

Page 14 of 30

SPEC-AID   DATE _____

DIVISION 8: DOORS, WINDOWS & GLASS

PROJECT _____   LOCATION _____

Hollow Metal Frames: ☐ No   ☐ Baked Enamel   ☐ Galvanized   ☐ Porcelain Enamel _____

Hollow Metal Doors: ☐ No _____ ☐ Core _____ ☐ Labeled _____

Aluminum Frames: ☐ No   ☐ Clear   ☐ Bronze   ☐ Black _____

Aluminum Doors and Frames: ☐ No   ☐ Yes _____ Frames _____

Wood Frames: ☐ No   ☐ Exterior   ☐ Interior   ☐ Custom   ☐ With Sill   ☐ Vinyl Covered   ☐ Pine   ☐ Oak

Wood Doors: ☐ No _____ Core   ☐ Labeled _____ Frames _____

Interior Door Frames: ☐ No   ☐ Aluminum   ☐ Hollow Metal   ☐ Steel   ☐ Wood   ☐ Prehung   ☐ Stock

☐ Custom   ☐ _____

Custom Doors: ☐ No   ☐ Swing   ☐ Bi-Passing   ☐ Bi-Folding _____ Frames _____

Accordion Folding Doors: ☐ No   ☐ Yes _____ Frames _____

Acoustical Doors: ☐ No   ☐ Yes _____ Decibels _____ Frames _____

Cold Storage: ☐ No   ☐ Manual   ☐ Power   ☐ Sliding   ☐ Hinged _____

Counter Doors: ☐ No   ☐ Aluminum   ☐ Steel   ☐ Wood _____ Frames _____

Dark Room Doors: ☐ N0   ☐ Revolving   ☐ 2 Way   ☐ 3 Way _____

Floor Opening Doors: ☐ No   ☐ Aluminum   ☐ Steel   ☐ Single   ☐ Double   ☐ Commercial   ☐ Industrial _____

Glass Doors: ☐ No   ☐ Sliding   ☐ Swing _____ Frames _____

Hangar Doors: ☐ No   ☐ Bi-Fold   ☐ Other   ☐ Electric _____

Jalousie Doors: ☐ No   ☐ Plain Glass   ☐ Tempered Glass _____

Kalamein: ☐ No   ☐ Yes   ☐ Labeled _____ Frames _____

Kennel Doors: ☐ No   ☐ 2 Way Swing _____

Overhead Doors: ☐ No   ☐ Regular Duty   ☐ Heavy Duty   ☐ Stock   ☐ Custom   ☐ One Piece   ☐ Sectional

☐ Manual   ☐ Electric   ☐ Aluminum   ☐ Fiberglass   ☐ Steel   ☐ Wood   ☐ Hardboard   ☐ Commercial

☐ Residential   Size _____

Rolling Doors Exterior: ☐ No   ☐ Manual   ☐ Electric   ☐ Labeled _____

Rolling Doors Interior: ☐ No   ☐ Manual   ☐ Electric   ☐ Labeled _____ Frames _____

Rolling Grilles: ☐ No   ☐ Manual   ☐ Electric   ☐ Aluminum   ☐ Steel _____

Service Door Frames: ☐ No   ☐ Aluminum   ☐ Hollow Metal   ☐ Steel   ☐ Wood   ☐ Stock   ☐ Custom _____

Service Doors: ☐ No   ☐ Stock   ☐ Custom _____   ☐ Transoms _____   ☐ Sidelights _____

☐ Aluminum _____   ☐ Hollow Metal _____   ☐ Core _____   ☐ Kalamein _____

☐ Steel _____   ☐ Wood _____   ☐ Core _____   ☐ Labeled _____   ☐ Special Finish _____

Shock Absorbing Doors: ☐ No   ☐ Flexible   ☐ Rigid _____ Frames _____

Sliding Doors: ☐ No   ☐ Glazed   ☐ Unglazed Aluminum   ☐ Steel   ☐ Wood _____

Swing Doors: ☐ No   ☐ Single   ☐ Double _____

Telescoping Door: ☐ No   ☐ Manual   ☐ Electric _____

Tinclad Doors: ☐ No   ☐ Manual   ☐ Electric _____

Vault Front Doors: ☐ No   ☐ Stainless Steel   ☐ Time Lock   ☐ 1 Hr. Test   ☐ 2 Hr. Test   ☐ 4 Hr. Test _____

Special Exterior Doors: ☐ No   ☐ Yes _____

Special Interior Doors: ☐ No   ☐ Yes _____

Balanced Doors: ☐ No   ☐ Economy   ☐ Premium   ☐ Aluminum   ☐ Stainless Steel _____

Revolving Doors: ☐ No   ☐ Stock   ☐ Custom   ☐ Manual   ☐ Electric   ☐ Diameter _____

Entrance Units: ☐ No   ☐ Aluminum   ☐ Bronze   ☐ Glass   ☐ Hollow Metal   ☐ Stainless Steel   ☐ Wood   ☐ Steel

☐ Stock   ☐ Custom   ☐ Balanced   ☐ Sidelights   ☐ Transoms   Special Finish _____

Entrance Frames: ☐ No   ☐ Aluminum   ☐ Hollow Metal   ☐ Steel   ☐ Wood   ☐ Stainless Steel   ☐ Stock

☐ Custom _____

Store Fronts: ☐ No   ☐ Sliding   ☐ Fixed   ☐ Institutional Grade   ☐ Monumental Grade   ☐ Commercial Grade _____

Windows: _____ % of Exterior Walls _____

Projected: ☐ No   ☐ Glazed   ☐ Unglazed   ☐ Aluminum   ☐ Steel   ☐ Wood _____

Single Hung: ☐ No   ☐ Glazed   ☐ Unglazed   ☐ Aluminum   ☐ Steel   ☐ Wood _____

Sliding: ☐ No   ☐ Glazed   ☐ Unglazed   ☐ Aluminum   ☐ Steel   ☐ Wood _____

Security Windows: ☐ No   ☐ Yes _____

SPEC-AID

DIVISION 8: DOORS, WINDOWS & GLASS

Casement: ☐ No ☐ Fixed _____% Vented ☐ Aluminum ☐ Steel ☐ Wood _____

Picture Window: ☐ No ☐ Glazed ☐ Unglazed ☐ Aluminum ☐ Steel ☐ Wood _____

Double Hung: ☐ No ☐ Glazed ☐ Unglazed ☐ Aluminum ☐ Steel ☐ Wood _____

Special Windows: ☐ No ☐ Yes _____

Screens: ☐ No ☐ Aluminum ☐ Steel ☐ Wood _____

Finish Hardware Allowance: ☐ No ☐ Yes _____

Exterior Doors _____

Interior Doors _____

Automatic Openers: ☐ No ☐ 1 Way ☐ 2 Way ☐ Double Door ☐ Activating Carpet

Automatic Operators: ☐ No ☐ Sliding ☐ Swing ☐ Controls _____

Bumper Plates: ☐ No ☐ U Channel ☐ Teardrop _____

Door Closers: ☐ No ☐ Regular ☐ Fusible Link ☐ Concealed ☐ Heavy Use _____

Door Stops: ☐ No ☐ Yes _____

Floor Checks: ☐ No ☐ Single Acting ☐ Double Acting _____

Hinges: ☐ No ☐ Butt ☐ Pivot ☐ Spring ☐ Frequency _____

Kick Plates: ☐ No ☐ Yes _____

Lock Set: ☐ No ☐ Cylindrical ☐ Mortise ☐ Heavy Duty ☐ Commercial ☐ Residential _____

Panic Device: ☐ No ☐ Yes ☐ Exit Only ☐ Exit & Entrance _____

Push-Pull Device: ☐ No ☐ Yes ☐ Bronze ☐ Aluminum ☐ Other _____

Cabinet Hardware: ☐ No ☐ Yes _____

Window Hardware: ☐ No ☐ Yes _____

Special Hardware: ☐ No ☐ Yes _____

Threshold: ☐ No ☐ Yes _____

Weather Stripping Doors: ☐ No ☐ Zinc ☐ Bronze ☐ Stainless Steel ☐ Spring Type ☐ Extruded Sections

Windows: ☐ No ☐ Zinc ☐ Bronze _____

Acoustical Glass: ☐ No ☐ Yes _____ Thickness _____

Faceted Glass: ☐ No ☐ Yes _____ Thickness _____

Glazing: ☐ No ☐ Putty ☐ Flush ☐ Bead ☐ Gasket ☐ Butt ☐ Riglet ☐ _____

Insulated Glass: ☐ No ☐ Standard ☐ Non-Standard _____ Thickness _____

Laminated Glass: ☐ No ☐ Yes _____ Thickness _____

Mirrors: ☐ No ☐ Plate ☐ Sheet ☐ Transparent ☐ Incl. Frames ☐ No Frames ☐ _____

Door Type _____ Wall Type _____

Obscure Glass: ☐ No ☐ Yes _____ Thickness _____

Plate Glass: ☐ No ☐ Clear ☐ Tinted ☐ Tempered _____ Thickness _____

Plexiglass: ☐ No ☐ Masked ☐ Unmasked _____ Thickness _____

Polycarbonate: ☐ No ☐ Masked ☐ Unmasked _____ Thickness _____

Reflective: ☐ No ☐ Clear ☐ Tinted _____ Thickness _____

Sand Blasted: ☐ No ☐ Yes _____ Thickness _____

Sheet or Float Glass: ☐ No ☐ Clear ☐ Gray _____ Thickness _____

Spandrel Glass: ☐ No ☐ Plain ☐ Insulated ☐ Sandwich _____ Thickness _____

Stained Glass: ☐ No ☐ Yes _____

Vinyl Glazing: ☐ No ☐ Yes _____ Thickness _____

Window Glass: ☐ No ☐ DSA ☐ DBS ☐ Tempered _____ Thickness _____

Wire Glass: ☐ No ☐ Yes _____ Thickness _____

Special Glazing: ☐ No ☐ Yes _____

Curtain Walls: ☐ No ☐ Yes _____

Window Walls: ☐ No ☐ Yes _____

SPEC-AID

DATE _____

DIVISION 9: FINISHES

PROJECT _____ LOCATION _____

Furring: Ceiling: ☐ No ☐ Wired Direct ☐ Suspended _____

Partitions: ☐ No ☐ Load Bearing ☐ Non Load Bearing _____ Thickness _____

Walls: ☐ No ☐ Yes _____

Gypsum Lath: ☐ No ☐ Walls ☐ Ceilings ☐ Regular ☐ Foil Faced ☐ Fire Resistant ☐ Moisture Resistant _____

_____ Thickness _____

Metal Lath: ☐ No ☐ Diamond ☐ Rib ☐ _____ Weight _____

☐ Painted ☐ Galvanized ☐ Paper Backed _____

☐ Walls ☐ Ceilings ☐ Suspended ☐ Partitions ☐ Load Bearing ☐ Non Load Bearing _____

Drywall Finishes: ☐ Taped & Finished ☐ Thin Coat Plaster ☐ Prime Coat ☐ Electric Heat Compound ☐ _____

Mountings: ☐ Nailed ☐ Screwed ☐ Laminated ☐ Clips ☐ _____

Beams: ☐ No _____ Layers _____ Thickness _____

Ceilings: ☐ No ☐ Standard ☐ Fire Resistant ☐ Water Resistance _____ Thickness _____

Columns: ☐ No _____ Layers _____ Thickness _____

Partitions: ☐ No ☐ Wood Studs ☐ Steel Studs _____ Layers _____ Thickness _____

Prefinished: ☐ No ☐ Standard ☐ Fire Resistance _____ Thickness _____

Soffits: ☐ No _____ Layers _____ Thickness _____

Sound Deading Board: ☐ No Type _____ Thickness _____

Walls: ☐ No _____ Layers _____ Thickness _____

Plaster Finishes: ☐ 1 Coat ☐ 2 Coat ☐ 3 Coat ☐ Gypsum ☐ Perlite ☐ Vermiculite ☐ Wood ☐ _____

Beams: ☐ No _____ Ceilings: ☐ _____

Columns: ☐ No _____ Soffits: ☐ No _____

Partitions: ☐ No ☐ Wood Studs ☐ Steel Studs ☐ Solid ☐ Hollow _____

Walls: ☐ No _____

Special Plaster: _____

Sprayed Acoustical: ☐ No ☐ Yes _____ Thickness _____

Fireproofing: ☐ No ☐ Yes _____ Thickness _____

Stucco: ☐ No ☐ On Mesh ☐ Masonry _____

Cast Stone: ☐ No ☐ Glazed ☐ Unglazed ☐ Waxed _____ Thickness _____

Ceramic Tile Base: ☐ No ☐ Cove ☐ Sanitary ☐ _____ Set _____ Height _____

Floors: ☐ No ☐ _____ Set ☐ Natural Clay ☐ Porcelain ☐ Conductive _____ Color Group _____

Walls: ☐ No ☐ _____ Set ☐ Interior ☐ Exterior ☐ Glazed ☐ Crystalline Glazed ☐ _____

☐ Unmounted ☐ Backmounted _____

Panels: ☐ No ☐ Yes _____

Glass Mosaics: ☐ No ☐ Yes _____ Color Group _____

Metal Tile: ☐ No ☐ Aluminum ☐ Copper ☐ Stainless Steel _____

Plastic Tile: ☐ No ☐ Yes _____ Thickness _____

Quarry Tile Base: ☐ No ☐ Cove ☐ Sanitary _____ Height _____

Floor ☐ No _____ Set Size _____ Color _____

Stairs: ☐ No ☐ Treads ☐ Risers _____

Wainscot: ☐ No _____ Set Size _____

Cast In Place Terrazzo Base: ☐ No ☐ Yes _____ Curb: ☐ No ☐ Yes _____

Floor: ☐ No ☐ Bonded ☐ Unbonded ☐ Gray Cement ☐ White Cement ☐ Conventional ☐ Venetian _____

☐ Conductive ☐ Monolithic ☐ Epoxy ☐ _____

Divider Strips: ☐ No ☐ Brass ☐ Zinc _____ Spacing _____

Stairs: ☐ No ☐ Yes _____ Wainscot: ☐ No ☐ Yes _____

Precast Terrazzo Base: ☐ No ☐ Yes Curb: ☐ No ☐ Yes _____

Floor Tiles: ☐ No Size _____ Thickness _____

Stairs: ☐ No ☐ Treads ☐ Risers ☐ Stringers ☐ Landings _____

Wainscot: ☐ No ☐ Yes _____ Thickness _____

Page 17 of 30

337

SPEC-AID

DIVISION 9: FINISHES

Barriers: ☐ No ☐ Aluminum ☐ Foil ☐ Mesh ☐ Lead ☐ Leaded Vinyl _____

Acoustical Barriers: ☐ No ☐ Yes _____

Ceilings: ☐ No ☐ Boards ☐ Tile ☐ Cemented ☐ Stapled ☐ On Suspension ☐ _____

☐ Fiberglass ☐ Mineral Fiber ☐ Wood Fiber ☐ Metal Pan _____

☐ Fire Rated ☐ Ventilating ☐ _____ Ceiling Height _____

☐ Luminous Panels _____ ☐ Access Panels _____

Suspension System: ☐ No ☐ T Bar ☐ Z Bar ☐ Carrier Channels ☐ _____

Strip Lighting: ☐ No ☐ Yes _____ Foot Candles _____

Special Acoustical _____

Brick Flooring: ☐ No ☐ Yes _____

Carpet ☐ No ☐ Yes ☐ With Padding ☐ With Backing ☐ _____ Allowance _____

Type: ☐ Acrylic ☐ Nylon ☐ Polypropylene ☐ Wool ☐ Tile ☐ _____ Face Weight _____

Padding: ☐ No ☐ Yes _____ Backing: ☐ No ☐ Yes _____

Composition Flooring: ☐ No ☐ Acrylic ☐ Epoxy ☐ Mastic ☐ Neoprene ☐ Polyester ☐ _____

☐ Regular Duty ☐ Heavy Duty _____ Thickness _____

Concrete Floor Topping: ☐ No ☐ In Concrete ☐ Yes _____

Resilient Floors: Base: ☐ No ☐ Rubber ☐ Vinyl _____ Height _____

Asphalt Tile: ☐ No ☐ Yes _____ Color Group _____

Conductive Tile: ☐ No ☐ Yes _____ Thickness _____

Cork Tile: ☐ No ☐ Yes _____ Thickness _____

Linoleum: ☐ No ☐ Yes _____ Thickness _____

Polyethylene: ☐ No ☐ Yes _____

Polyurethane: ☐ No ☐ Yes _____ Thickness _____

Rubber Tile: ☐ No ☐ Yes _____ Thickness _____

Vinyl: ☐ Sheet ☐ Tile _____ Thickness _____

Vinyl Asbestos Tile: ☐ No ☐ Yes _____ Color Group _____

Stair Covering: ☐ No ☐ Risers ☐ Treads ☐ Landings ☐ Nosings ☐ Rubber ☐ Vinyl _____

Steel Plates: ☐ No ☐ Cement Bed ☐ Epoxy Bed _____

Wood Floor: ☐ No ☐ Block ☐ Strip ☐ Parquetry ☐ Unfinished ☐ Prefinished ☐ Stock ☐ Custom _____

Fir: ☐ No ☐ Flat Grain ☐ Vertical Grain _____ Size _____

Gym: ☐ No ☐ Yes Type _____

Maple: ☐ No ☐ Yes Grade _____ Size _____

Oak: ☐ No ☐ Red ☐ White Grade _____ Size _____

Other: ☐ _____ Grade _____ Size _____

Finish Required: ☐ No ☐ Yes _____

Wood Block Floor: ☐ No ☐ Creosoted ☐ ☐ Natural _____ Thickness _____

Special Coatings: ☐ No ☐ Floor ☐ Wall _____

Painting: ☐ No ☐ Regular ☐ Fireproof ☐ Fire Retardant ☐ Brush ☐ Roller ☐ Spray _____

Casework: ☐ No _____ Coats _____ Ceilings: ☐ No _____ Coats _____

Doors: ☐ No _____ Coats _____ Trim: ☐ No _____ Coats _____

Walls, Exterior: ☐ No _____ Coats _____ Interior Walls: ☐ No _____ Coats _____

Windows: ☐ No _____ Coats _____ Piping: ☐ No _____ Coats _____

Other: _____

Structural Steel: ☐ No ☐ Yes _____ Miscellaneous Metals: ☐ No ☐ Yes _____

Wall Covering: ☐ No ☐ Cork Tile _____ ☐ Metal Foil _____

☐ Flexible Wood Veneers _____ ☐ Vinyl _____ Weight _____

Wall Paper _____ Vinyl _____ Murals _____

Other _____

Guards: Corner: ☐ No ☐ Rubber ☐ Steel ☐ Vinyl _____

Wall: ☐ No ☐ Rubber ☐ Steel ☐ Vinyl _____

SPEC-AID

DATE _____

DIVISION 10: SPECIALTIES

PROJECT _____ LOCATION _____

Bathroom Accessories: □ No □ Curtain Rod _____ □ Dispensers _____ □ Grab Bar _____
□ Hand Dryer _____ □ Medicine Cabinet _____ □ Mirror _____ □ Robe Hook _____
□ Soap Dispenser _____ □ Shelf _____ □ Tissue Dispenser _____ □ Towel Bar _____
□ Tumbler Holder _____ □ Wall Urn _____ □ Waste Receptical _____ □ _____
Bulletin Board: □ No □ Cork □ Vinyl Cork □ Unbacked □ Backed □ Stock □ Custom
□ Tan □ Framed □ No Frames □ Changeable Letter □ _____ Thickness _____
_____

Canopies: □ No □ Free Standing □ Wall Hung □ Stock □ Custom _____
Chalkboard: □ No □ Cement Asbestos □ Hardboard □ Metal _____ Ga. □ Slate _____ in. Thick □ Tempered Glass _____
□ Treated Plastic □ _____ □ Unbacked □ Backed with _____
□ No Frames □ Frames □ Chalk Tray □ Map Rail □ _____
□ Built in Place □ Prefabricated _____
□ Portable □ Reversible □ Swing Wing □ Sliding Panel _____
_____

Chutes Linen: □ No □ Aluminum □ Aluminized Steel □ Stainless Steel □ _____ Ga. Diameter _____
□ Bottom Collector □ Sprinklers _____
Mail: □ No □ Aluminum □ Bronze □ Stainless □ _____ Size _____ □ Bottom Collector
Package: □ No □ Aluminum □ Bronze □ Stainless _____
Rubbish: □ No □ Aluminum □ Aluminized Steel □ Stainless Steel _____ Ga. Diameter _____
□ Bottom Collector □ Sprinklers _____
Compartments & Cubicles: □ No □ Hospital _____ □ Office _____
□ Shower _____ □ Toilet _____ □ _____
Control Boards: □ No □ Yes _____
Decorative Grilles and Screens: □ No □ Yes _____
Directory Boards: □ No □ Exterior □ Interior □ Aluminum □ Bronze □ Stainless □ Lighted _____
Describe _____
Disappearing Stairs: □ No □ Stock □ Custom □ Manual □ Electric _____ Ceiling Height _____
Display Cases: □ No □ Economy □ Deluxe _____
Fire Extinguishers: □ No □ CO_2 □ Dry Chemical □ Foam □ Pressure Water □ Soda Acid □ _____
□ Aluminum □ Copper □ Painted Steel □ Stainless Steel □ _____ Size _____
Cabinets: □ No □ Aluminum □ Painted Steel □ Stainless Steel □ _____
Hose Equipment: □ No □ Blanket □ Cabinets □ Hose _____ Size _____
Protection System: □ No □ Yes _____
Fireplace, Prefabricated: □ No □ Economy □ Deluxe □ Wall Hung □ Free Standing _____
_____

Flagpoles: □ No □ Aluminum □ Bronze □ Fiberglass □ Stainless □ Steel □ Wood □ Tapered □ Sectional _____
□ Ground Set □ Wall Set □ Counterbalanced □ Outriggers _____ Height _____
Bases: □ No □ Economy □ Deluxe _____
Foundation: □ No □ Yes _____
Folding Gates: □ No □ Scissors Type □ Vertical Members □ Stock □ Custom _____ Opening _____
Lockers: □ No □ No Locks □ Keyed □ Combination _____ Tier Size _____ Height _____
Athletic: □ No □ Basket □ Ventilating □ Overhead _____ Size _____
Benches: □ No □ Yes _____
Special Lockers: _____
Mail Specialties Boxes: □ No □ Front Loading □ Rear Loading □ Aluminum □ Stainless □ _____
Size _____
Letter Slot: □ No □ Yes _____ Counter Window: □ No □ Yes _____
Directory: □ No □ Yes _____ Key Keeper _____
Other: _____

SPEC-AID
DIVISION 10: SPECIALTIES

Accordion Folding Partitions: ☐ No ☐ Acoustical ☐ Non Acoustical _____ Weight _____ psf.
Ceiling Height _____ Describe _____

Folding Leaf Partitions: ☐ No ☐ Acoustical ☐ Non Acoustical _____ Weight _____ psf.
Ceiling Height _____ Describe _____

Hospital Partitions: ☐ No ☐ Metal ☐ Curtain Track _____

Movable Office Partitions: ☐ No ☐ Acoustical ☐ Non Acoustical ☐ Asbestos Cement ☐ Hardboard
☐ Laminated Gypsum ☐ Plywood ☐ _____
☐ With Glass ☐ No Glass Describe _____ Partition Height _____
Special Finish: _____
Doors: ☐ No ☐ Yes Type _____ Finish _____ Size _____

Operable Partitions: ☐ No ☐ Yes Type _____

Portable Partitions: ☐ No ☐ Acoustical ☐ Non Acoustical _____ Weight _____ psf.
Partition Height _____ Describe _____

Shower Partitions: ☐ No ☐ Fiberglass ☐ Glass ☐ Marble ☐ Metal ☐ _____ Finish _____
☐ Stock ☐ Custom ☐ Economy ☐ Deluxe Size _____
Doors: ☐ No ☐ Glass ☐ Tempered Glass ☐ Plastic ☐ Curtain Only _____ Size _____
Receptors: ☐ No ☐ Concrete ☐ Metal ☐ Plastic ☐ Terrazzo _____ Size _____
Tub Enclosure: ☐ No ☐ Stock ☐ Custom ☐ Economy ☐ Deluxe _____ Size _____

Toilet Partitions: ☐ No ☐ Fiberglass ☐ Marble ☐ Metal ☐ Slate ☐ Wood ☐ _____
☐ Floor Mounted ☐ Wall Hung ☐ Ceiling Hung _____
Special Finish _____
Doors: ☐ No ☐ Yes _____
Screens: ☐ No ☐ Full Height ☐ Urinal ☐ Floor Mounted ☐ Wall Hung ☐ Ceiling Hung _____

Woven Wire Partitions: ☐ No ☐ Walls ☐ Ceilings ☐ Panel Width _____ Height _____
Doors: ☐ No ☐ Sliding ☐ Swing _____ Windows: ☐ No ☐ Yes _____
☐ Painted ☐ Galvanized _____

Other Partitions: _____

Parts Bins: ☐ No ☐ Yes _____
Scales: ☐ No ☐ Built in ☐ Portable ☐ Beam Type ☐ Dial Type _____ Capacity _____
Platform Size _____ Material _____ Foundations _____
Accessory Items _____
Shelving, Storage: ☐ No ☐ Metal ☐ Wood _____
Signs: Individual Letters: ☐ No ☐ Aluminum ☐ Bronze ☐ Plastic ☐ Stainless ☐ Steel ☐ _____
☐ Cast ☐ Fabricated Describe _____
Plaques: ☐ No ☐ Aluminum ☐ Bronze _____
Signs: ☐ No ☐ Metal ☐ Plastic ☐ Lighted _____

Sun Control Devices: ☐ No ☐ Yes _____
Telephone Enclosures: ☐ No ☐ Indoor ☐ Outdoor _____
Turnstiles: ☐ No ☐ Yes _____
Vending Machines: ☐ No ☐ Yes _____
Wardrobe Specialties: ☐ No ☐ Yes _____
Other Specialties: _____
_____
_____
_____

Page 20 of 30

SPEC-AID

DATE _____

DIVISION 11: ARCHITECTURAL EQUIPMENT

PROJECT _____ LOCATION _____

Appliances, Residential: ☐ No ☐ Yes Allowance _____ ☐ Separate Contract
☐ Cook Tops _____ ☐ Compactors _____ ☐ Dehumidifier _____ ☐ Dishwasher _____
☐ Dryer _____ ☐ Garbage Disposer _____ ☐ Heaters, Electric _____
☐ Hood _____ ☐ Humidifier _____ ☐ Ice Maker _____ ☐ Oven _____
☐ Refrigerator _____ ☐ Sump Pump _____ ☐ Washing Machine _____ ☐ Water Heater _____
☐ Water Softener _____ ☐ _____

Automotive Equipment: ☐ No ☐ Yes Allowance _____ ☐ Separate Contract
☐ Hoists _____ ☐ Lube _____ ☐ Pumps _____ ☐ _____

Bank Equipment: ☐ No ☐ Yes Allowance _____
☐ Counters _____ ☐ Safes _____ ☐ Vaults _____ ☐ Windows _____
☐ _____

Check Room Equipment: ☐ No ☐ Yes Allowance _____ ☐ Separate Contract
Describe _____

Church Equipment: ☐ No ☐ Yes Allowance _____ ☐ Separate Contract
☐ Altar _____ ☐ Baptistries _____ ☐ Bells & Carillons _____ ☐ Confessionals _____
☐ Organ _____ ☐ Pews _____ ☐ Pulpit _____ ☐ Spires _____
☐ Wall Cross _____ ☐ _____

Commercial Equipment: ☐ No ☐ Yes Allowance _____ ☐ Separate Contract
Describe _____

Darkroom Equipment: ☐ No ☐ Yes Allowance _____ ☐ Separate Contract
Describe _____

Data Processing Equipment: ☐ No ☐ Yes Allowance _____ ☐ Separate Contract
Describe _____

Dental Equipment: ☐ No ☐ Yes Allowance _____ ☐ Separate Contract
☐ Chair _____ ☐ Drill _____ ☐ Lights _____ ☐ X-Ray _____
☐ _____

Dock Equipment: ☐ No ☐ Yes Allowance _____ ☐ Separate Contract
☐ Bumpers _____ ☐ Boards _____ ☐ Door Seal _____ ☐ Levelers _____
☐ Lights _____ ☐ Shelters _____ ☐ _____

Food Service Equipment: ☐ No ☐ Yes Allowance _____ ☐ Separate Contract
☐ Bar Units _____ ☐ Cooking Equip. _____ ☐ Dishwashing Equip. _____ ☐ Food Prep. _____
☐ Food Serving _____ ☐ Refrigerated Cases _____ ☐ Tables _____ ☐ _____

Gymnasium Equipment: ☐ No ☐ Yes Allowance _____ ☐ Separate Contract
☐ Basketball Backstops _____ ☐ Benches _____ ☐ Bleachers _____
☐ Divider Curtain _____ ☐ Gymnastic Equip. _____ ☐ Mats _____ ☐ Scoreboards _____
☐ _____

Industrial Equipment: ☐ No ☐ Yes Allowance _____ ☐ Separate Contract
Describe _____

Laboratory Equipment: ☐ No ☐ Yes Allowance _____ ☐ Separate Contract
☐ Casework _____ ☐ Counter Tops _____ ☐ Hoods _____ ☐ Sinks _____
☐ Tables _____ ☐ _____

Laundry Equipment: ☐ No ☐ Yes Allowance _____ ☐ Separate Contract
☐ Dryers _____ ☐ Washers _____ ☐ _____

Library Equipment: ☐ No ☐ Yes Allowance _____ ☐ Separate Contract
☐ Book Shelves _____ ☐ Book Stacks _____ ☐ Card Files _____ ☐ Carrels _____
☐ Charging Desks _____ ☐ Racks _____ ☐ _____

Medical Equipment: ☐ No ☐ Yes Allowance _____ ☐ Separate Contract
☐ Casework _____ ☐ Exam Room _____ ☐ Incubators _____ ☐ Patient Care _____
☐ Radiology _____ ☐ Sterilizers _____ ☐ Surgery Equip. _____ ☐ Therapy Equip. _____
☐ _____

SPEC-AID

DIVISION 11: ARCHITECTURAL EQUIPMENT

Mortuary Equipment: ☐ No ☐ Yes Allowance _____ ☐ Separate Contract
Describe _____

Musical Equipment: ☐ No ☐ Yes Allowance _____ ☐ Separate Contract
Describe _____

Observatory Equipment: ☐ No ☐ Yes Allowance _____ ☐ Separate Contract
Describe _____

Parking Equipment: ☐ No ☐ Yes Allowance _____ ☐ Separate Contract
☐ Automatic Gates _____ ☐ Booths _____ ☐ Control Station _____
☐ Ticket Dispenser _____ ☐ Traffic Detectors _____
☐ _____

Playground Equipment: In Division 2 _____

Prison Equipment: ☐ No ☐ Yes Allowance _____ ☐ Separate Contract
☐ Ceiling Lining _____ ☐ Wall Lining _____ ☐ Bar Walls _____ ☐ Doors _____
☐ Bunks _____ ☐ Lavatory _____ ☐ Water Closet _____ ☐ _____

Residential Equipment: ☐ No ☐ Yes Allowance _____ ☐ Separate Contract
☐ Kitchen Cabinets (Also Div. 6) _____ ☐ Lavatory Cabinets _____ ☐ Kitchen Equipment _____
☐ Laundry Equip. _____ ☐ Unit Kitchens _____ ☐ Vacuum Cleaning _____ ☐ _____
☐ _____

Safes: ☐ No ☐ Yes Allowance _____ ☐ Separate Contract
☐ Office _____ ☐ Money _____ ☐ _____ ☐ Rating _____
Describe _____

Saunas: ☐ No ☐ Yes Allowance _____ ☐ Separate Contract
☐ Built in Place ☐ Prefabricated Size _____ Describe _____
☐ Heater _____ ☐ Seats _____ ☐ Timer _____ ☐ _____

School Equipment: ☐ No ☐ Yes Allowance _____ ☐ Separate Contract
☐ Art & Crafts _____ ☐ Audio-Visual _____ ☐ Language Labs _____ ☐ Vocational _____
☐ Wall Benches _____ ☐ Wall Tables _____ ☐ _____
☐ _____

Shop Equipment: ☐ No ☐ Yes Allowance _____ ☐ Separate Contract
Describe _____

Stage Equipment: ☐ No ☐ Yes Allowance _____ ☐ Separate Contract
Describe _____

Steam Baths: ☐ No ☐ Yes Allowance _____ ☐ Separate Contract
Describe _____

Swimming Pool Equipment: ☐ No ☐ Yes Allowance _____ ☐ Separate Contract
☐ Diving Board _____ ☐ Diving Stand _____ ☐ Life Guard Chair _____ ☐ Ladders _____
☐ Heater _____ ☐ Lights _____ ☐ Pool Cover _____ ☐ Slides _____
☐ _____

Unit Kitchens: ☐ No ☐ Yes Allowance _____ ☐ Separate Contract
Describe _____

Vacuum Cleaning, Central: ☐ No ☐ Yes Allowance _____ ☐ Separate Contract
☐ _____ Valves Describe _____

Waste Disposal Compactors: ☐ No ☐ Yes
Incinerators: ☐ No ☐ Electric ☐ Gas Type Waste _____ Capacity _____

Special Equipment: _____
_____
_____
_____
_____

SPEC-AID   DATE _____

DIVISION 12: FURNISHINGS

PROJECT _____ LOCATION _____

Artwork: ☐ No ☐ Yes Allowance _____ ☐ Separate Contract
☐ Murals _____ ☐ Paintings _____ ☐ Photomurals _____ ☐ Sculptures _____
☐ Stained Glass _____ ☐ _____

Interior Landscaping: ☐ No ☐ Yes Allowance _____ **Blinds, Exterior:** ☐ No ☐ Yes Allowance _____ ☐ Separate Contract
☐ Solid ☐ Louvered ☐ Aluminum ☐ Nylon ☐ Vinyl ☐ Wood ☐ _____
Describe _____

Blinds, Interior: ☐ No ☐ Yes Allowance _____ ☐ Separate Contract
Folding: ☐ No ☐ Stock ☐ Custom ☐ Wood ☐ _____
Describe _____
Venetian: ☐ No ☐ Stock ☐ Custom ☐ Aluminum ☐ Plastic ☐ Steel ☐ Wood ☐ _____
Describe _____
Vertical: ☐ No ☐ Aluminum ☐ Cloth ☐ Vinyl ☐ _____
Describe _____
Other: _____

Cabinets: ☐ No ☐ Yes Allowance _____ ☐ Separate Contract
☐ Classroom _____
☐ Dormitory _____
☐ Hospital _____
☐ _____

Carpets: In Division 9 _____

Dormitory Units: ☐ No ☐ Yes Allowance _____ ☐ Separate Contract
☐ Beds _____ ☐ Desks _____ ☐ Wardrobes _____ ☐ _____
☐ _____

Drapery & Curtains: ☐ No ☐ Yes Allowance _____ ☐ Separate Contract
Describe _____

Floor Mats: ☐ No ☐ Yes Allowance _____ ☐ Separate Contract
☐ Recessed ☐ Non Recessed _____
☐ Link ☐ Solid _____

Furniture: ☐ No ☐ Yes Allowance _____ ☐ Separate Contract
☐ Beds _____ ☐ Chairs _____ ☐ Chests _____ ☐ Desks _____
Sofas _____ ☐ Tables _____ ☐ _____
☐ _____

Seating Auditorium: ☐ No ☐ Yes Allowance _____ ☐ Separate Contract
Describe _____
Classroom: ☐ No ☐ Yes Allowance _____ ☐ Separate Contract
Describe _____
Stadiuim: ☐ No ☐ Yes Allowance _____ ☐ Separate Contract
Describe _____

Shades: ☐ No ☐ Yes Allowance _____ ☐ Separate Contract
☐ Stock ☐ Custom ☐ Lightproof ☐ Fireproof _____
☐ Cotton ☐ Fiberglass ☐ Vinyl ☐ Woven Aluminum ☐ _____
Describe _____

Wardrobes: ☐ No ☐ Yes Allowance _____ ☐ Separate Contract
☐ Classroom _____ ☐ Dormitory _____ ☐ Hospital _____ ☐ _____
Describe _____

Other Furnishings: _____
_____
_____

Page 23 of 30

SPEC-AID

DATE _____

DIVISION 13: SPECIAL CONSTRUCTION

PROJECT _____ LOCATION _____

Acoustical Echo Chamber: ☐ No ☐ Yes Allowance _____ ☐ Separate Contract
    Describe _____

    Enclosures: ☐ No ☐ Yes Allowance _____ ☐ Separate Contract
    Describe _____

    Panels: ☐ No ☐ Yes Allowance _____ ☐ Separate Contract
    Describe _____

Air Curtains: ☐ No ☐ Yes Allowance _____ ☐ Separate Contract
    ☐ Heated Air ☐ Unheated Air ☐ Recirculating ☐ Non Recirculating ☐ _____
    Describe _____

Air Inflated Buildings: ☐ No ☐ Yes Describe _____ ☐ Separate Contract

Anechoic Chambers: ☐ No ☐ Yes Allowance _____ ☐ Separate Contract
    Describe _____

Audiometric Rooms: ☐ No ☐ Yes Allowance _____ ☐ Separate Contract
    Describe _____

Bowling Alleys: ☐ No ☐ Yes Allowance _____ ☐ Separate Contract
    Describe _____

Broadcasting Studio: ☐ No ☐ Yes Allowance _____ ☐ Separate Contract
    Describe _____

Chimneys: ☐ No ☐ Yes Allowance _____ ☐ Separate Contract
    Concrete: ☐ No ☐ Unlined ☐ Lined ☐ _____ Diameter_____ Height_____
    Metal: ☐ No ☐ Insulated ☐ Not Insulated ☐ U.L. Listed ☐ Not U.L. Listed ☐ _____
    Describe _____ Diameter_____ Height_____
    Radial Brick: ☐ No ☐ Unlined ☐ Lined ☐ _____ Diameter_____ Height_____
    Foundation: ☐ No ☐ Yes _____

Clean Rooms: ☐ No ☐ Yes Allowance _____ ☐ Separate Contract
    Describe _____

Comfort Stations: ☐ No ☐ Yes Describe _____

Dark Rooms: ☐ No ☐ Yes Allowance _____ ☐ Separate Contract
    Describe _____

Domes, Observation: ☐ No ☐ Yes Allowance _____ ☐ Separate Contract
    Describe _____

Garage: ☐ No ☐ Yes Describe _____ Cars_____

Garden House: ☐ No ☐ Yes Allowance _____ ☐ Separate Contract
    Describe _____

Grandstand: ☐ No ☐ Yes Describe _____ Seats_____

Greenhouse: ☐ No ☐ Yes Allowance _____ ☐ Separate Contract
    Describe _____

Hangars: ☐ No ☐ Yes Describe _____ Planes_____

Hyperbaric Rooms: ☐ No ☐ Yes Allowance _____ ☐ Separate Contract
    Describe _____

Incinerators (See also Division 10): ☐ No ☐ Yes Allowance _____ ☐ Separate Contract
    Describe _____ Capacity_____

Insulated Rooms: ☐ No ☐ Yes Allowance _____ ☐ Separate Contract
    Doors: ☐ No ☐ Cooler ☐ Freezer ☐ Manual ☐ Electric _____
    ☐ Galvanized ☐ Stainless Describe _____
    Coolers: ☐ No ☐ Yes Describe _____
    Freezers: ☐ No ☐ Yes Describe _____
    Partitions: ☐ No ☐ Yes ☐ Stock ☐ Custom Describe _____
    Other: _____

Page 24 of 30

SPEC-AID
DIVISION 13: SPECIAL CONSTRUCTION

Integrated Ceilings: ☐ No ☐ Yes Module _____ Ceiling Height _____
  Lighting: ☐ No ☐ Yes Describe _____ Foot Candles _____
  Heating: ☐ No ☐ Yes Describe _____
  Ventilating: ☐ No ☐ Yes Describe _____
  Air Conditioning: ☐ No ☐ Yes Describe _____
Music Practice Rooms: ☐ No ☐ Yes Allowance _____ ☐ Separate Contract
  Describe _____
Pedestal Floors: ☐ No ☐ Yes Allowance _____ ☐ Separate Contract
  ☐ Aluminum ☐ Plywood ☐ Steel ☐ _____ Panel Size _____ Height _____
  ☐ High Density Plastic ☐ Vinyl Tile ☐ V.A. Tile ☐ _____
  Describe _____
Portable Booths: ☐ No ☐ Yes Allowance _____ ☐ Separate Contract
  Describe _____
Prefabricated Structures: ☐ No ☐ Yes Allowance _____ ☐ Separate Contract
  Describe _____
Radiation Protection, Fluoroscopy Room: ☐ No ☐ Yes _____
  Nuclear Reactor: ☐ No ☐ Yes _____
  Radiological Room: ☐ No ☐ Yes _____
  X-Ray Room: ☐ No ☐ Yes _____
  Other: _____
Radio Frequency Shielding: ☐ No ☐ Yes Allowance _____ ☐ Separate Contract
  Describe _____
Radio Tower: ☐ No ☐ Yes Allowance _____ ☐ Separate Contract
  ☐ Guyed ☐ Self Supporting Wind Load _____ psf. _____ Height _____
  Foundations _____
Saunas and Steam Rooms: ☐ No ☐ Yes Allowance _____ ☐ Separate Contract
  Describe _____
Silos: ☐ No ☐ Yes Allowance _____ ☐ Separate Contract
  ☐ Concrete ☐ Steel ☐ Wood ☐ _____ Diameter _____ Height _____
  Foundations _____
Squash & Hand Ball Courts: ☐ No ☐ Yes Allowance _____ ☐ Separate Contract
  Describe _____
Storage Vaults: ☐ No ☐ Yes Allowance _____ ☐ Separate Contract
  Describe _____
Swimming Pool Enclosure: ☐ No ☐ Yes Allowance _____ ☐ Separate Contract
  Describe _____
Swimming Pool Equipment: In Division 11 _____
Swimming Pools: ☐ No ☐ Yes Allowance _____ ☐ Separate Contract
  ☐ Aluminum ☐ Concrete ☐ Gunite ☐ Plywood ☐ Steel ☐ _____
  ☐ Lined ☐ Unlined _____
  Deck: ☐ No ☐ Concrete ☐ Stone _____ Size _____
  Bath Houses: ☐ No ☐ Yes _____ Fixtures _____
Tanks: ☐ No ☐ Yes Allowance _____ ☐ Separate Contract
  ☐ Concrete ☐ Fiberglass ☐ Steel ☐ Wood ☐ _____ Capacity _____
  ☐ Fixed Roof ☐ Floating Roof ☐ _____ Height _____
  Foundations: _____
Therapeutic Pools: ☐ No ☐ Yes Describe _____
Vault Front: ☐ No ☐ Yes Allowance _____ ☐ Separate Contract
  Describe _____ Hour Test _____
Zoo Structures: ☐ No ☐ Yes Describe _____
Other Special Construction: _____

SPEC-AID

DATE _____

DIVISION 14: CONVEYING SYSTEMS

PROJECT _____ LOCATION _____

Ash Hoist: ☐ No ☐ Yes Allowance _____ ☐ Separate Contract
Describe _____

Conveyers: ☐ No ☐ Yes Allowance _____ ☐ Separate Contract
Describe _____

Correspondence Lift: ☐ No ☐ Yes Allowance _____ ☐ Separate Contract
Describe _____

Dumbwaiters: ☐ No ☐ Yes Allowance _____ ☐ Separate Contract
Capacity _____ Size _____ Number _____ Floors _____
Stops _____ Speed _____ Finish _____
Describe _____

Elevators, Freight: ☐ No ☐ Yes Allowance _____ ☐ Separate Contract
☐ Hydraulic ☐ Electric ☐ Geared ☐ Gearless _____
Capacity _____ Size _____ Number _____ Floors _____
Stops _____ Speed _____ Finish _____
Machinery Location _____ Door Type _____
Signals _____ Special Requirements _____

Elevators, Passenger: ☐ No ☐ Yes Allowance _____ ☐ Separate Contract
☐ Hydraulic ☐ Electric ☐ Geared ☐ Gearless _____
Capacity _____ Size _____ Number _____ Floors _____
Stops _____ Speed _____ Finish _____
Machinery Location _____ Door Type _____
Signals _____ Special Requirements _____

Escalators: ☐ No ☐ Yes Allowance _____ ☐ Separate Contract
Capacity _____ Size _____ Number _____ Floors _____
Story Height _____ Speed _____ Finish _____
Machinery Location _____ Incline Angle _____
Special Requirements _____

Hoists & Cranes: ☐ No ☐ Yes Allowance _____ ☐ Separate Contract
Describe _____

Lists: ☐ No ☐ Yes Allowance _____ ☐ Separate Contract
Describe _____

Material Handling Systems: ☐ No ☐ Yes Allowance _____ ☐ Separate Contract
☐ Automated ☐ Non Automated ☐ _____
Describe _____

Moving Stairs & Sidewalks ☐ No ☐ Yes Allowance _____ ☐ Separate Contract
Capacity _____ Size _____ Number _____ Floors _____
Story Height _____ Speed _____ Finish _____
Machinery Location _____ Incline Angle _____
Special Requirements _____

Pneumatic Tube System: ☐ No ☐ Yes Allowance _____ ☐ Separate Contract
☐ Automatic ☐ Manual ☐ _____ Size _____ Stations _____
Length _____ Special Requirements _____

Vertical Conveyer: ☐ No ☐ Yes Allowance _____ ☐ Separate Contract
☐ Automatic ☐ Non Automatic ☐ _____
Describe _____

Other Conveying: _____

Page 26 of 30

SPEC-AID   DATE _____

DIVISION 15: MECHANICAL

PROJECT _____   LOCATION _____

Building Drainage: Design Rainfall _____ ☐ Roof Drains _____ ☐ Court Drains _____
   ☐ Floor Drains _____ ☐ Yard Drains _____ ☐ Lawn Drains _____ ☐ Balcony Drains _____
   ☐ Area Drains _____ ☐ Sump Drains _____ Shower Drains _____ ☐ _____
   ☐ Drain Piping: Size _____ Describe _____
   ☐ Drain Gates _____ ☐ Clean Outs _____ ☐ Grease Traps _____

Sanitary System: ☐ No ☐ Yes ☐ Site Main _____ ☐ Manholes _____
   ☐ Sump Pumps _____ ☐ Bilge Pumps _____ ☐ Ejectors _____
   ☐ Soils, Stacks _____ ☐ Wastes, Vents _____ ☐ _____

Domestic Cold Water: ☐ No ☐ Water Meters _____ ☐ Law Sprinkler Connection _____
   ☐ Water Softening _____ ☐ Water Filtering _____
   ☐ Boiler Feed Water _____ ☐ Conditioning Apparatus _____
   ☐ Standpipe System _____ ☐ Hose Bibbs _____
   ☐ Pressure Tank _____ ☐ Booster Pumps _____
   ☐ Reducing Valves _____ ☐ _____

Domestic Hot Water: ☐ No ☐ Electric ☐ Gas ☐ Oil ☐ Solar _____
   ☐ Boiler _____ ☐ Conditioner _____ ☐ Fixture Connections _____
   ☐ Storage Tanks _____ Capacity _____
   ☐ Pumps _____

Piping: ☐ No ☐ Yes Material _____
   ☐ Air Chambers _____ ☐ Escutcheons _____ ☐ Expansion Joints _____
   ☐ Shock Absorbers _____ ☐ Hangers _____
   ☐ Valves _____ ☐ Paint _____

Special Piping: ☐ No ☐ Compressed Air _____ ☐ Vacuum _____
   ☐ Oxygen _____ ☐ Nitrous Oxygen _____
   ☐ Carbon Dioxide _____ ☐ Process Piping _____

Insulation Cold: ☐ No ☐ Yes Material _____ Jacket _____
   Hot: ☐ No ☐ Yes Material _____ Jacket _____

Fixtures Bathtub: ☐ No ☐ C.I. ☐ Steel ☐ Fiberglass ☐ _____ Color _____
   ☐ Curtain ☐ Rod ☐ Enclosure ☐ Wall Shower _____
   Drinking Fountain: ☐ No ☐ Yes ☐ Wall Hung ☐ Pedestal _____
   Hose Bibb: ☐ No ☐ Yes Describe _____
   Lavatory: ☐ No ☐ China ☐ C.I. ☐ Steel ☐ _____ Color _____
   ☐ Wall Hung ☐ Legs ☐ Acid Resisting _____
   Shower: ☐ No ☐ Individual ☐ Group ☐ Heads ☐ _____ Size _____
   Compartment: ☐ No ☐ Metal ☐ Stone ☐ Fiberglass ☐ _____ ☐ Door ☐ Curtain
   Receptor: ☐ No ☐ Plastic ☐ Metal ☐ Terrazzo ☐ _____
   Sinks: ☐ No ☐ Kitchen _____ ☐ Janitor _____
   ☐ Laundry _____ ☐ Pantry _____
   ☐ _____
   Urinals: ☐ No ☐ Floor Mounted ☐ Wall Hung _____
   Screens: ☐ No ☐ Floor Mounted ☐ Wall Hung _____
   Wash Centers: ☐ No ☐ Yes Describe _____
   Wash Fountains: ☐ No ☐ Floor Mounted ☐ Wall Hung _____ Size _____
   Describe _____
   Water Closets: ☐ No ☐ Floor Mounted ☐ Wall Hung Color _____
   Describe _____
   Water Coolers: ☐ No ☐ Floor Mounted ☐ Wall Hung _____ Capacity _____ gph.
   ☐ Water Supply ☐ Bottle ☐ Hot ☐ Compartment _____
   Other Fixtures: _____

SPEC-AID

DIVISION 15: MECHANICAL

Fire Protection: ☐ Carbon Dioxide System _____ ☐ Standpipe _____

☐ Sprinkler System ☐ Wet ☐ Dry _____ Spacing _____

☐ Fire Department Connection _____ ☐ Building Alarm _____

☐ Hose Cabinets _____ ☐ Hose Racks _____

☐ Roof Manifold _____ ☐ Compressed Air Supply _____

☐ Hydrants _____ ☐ _____

Special Plumbing _____

_____

Gas Supply System: ☐ No ☐ Natural Gas ☐ Manufactured Gas _____

Pipe: Schedule _____ Fittings _____

Shutoffs: _____ Master Control Valve: _____

Insulation: _____ Paint: _____

Oil Supply System: ☐ No ☐ Tanks ☐ Above Ground ☐ Below Ground _____

☐ Steel ☐ Plastic ☐ _____ Capacity _____

Heating Plant: ☐ No ☐ Electric ☐ Gas ☐ Oil ☐ Solar _____

☐ Boilers _____ ☐ Pumps _____

☐ PRV Stations _____ ☐ Piping _____

☐ Heat Pumps _____

Cooling Plant: ☐ No ☐ Yes _____ Tons _____

Chillers: ☐ Steam ☐ Water ☐ Air _____

Condenser—Compressor ☐ Air ☐ Water _____

Pumps _____ Cooling Towers _____

System Type: _____

☐ Single Zone _____ ☐ Multi-Zone _____

☐ All Air _____ ☐ Terminal Reheat _____

☐ Double Duct _____ ☐ Radiant Panels _____

☐ Fan Coil _____ ☐ Unit Ventilators _____

☐ Perimeter Radiation _____ ☐ _____

Air Handling Units: Area Served _____ Number _____

Total CFM _____ % Outside Air _____

Cooling, Tons _____ Heating, MBH _____

Filtration _____ Supply Fans _____

Economizer _____

Fans: ☐ No ☐ Return ☐ Exhaust ☐ _____

Describe _____

Distribution: Ductwork _____ Material _____

Terminals: ☐ Diffusers _____ ☐ Registers _____

☐ Grilles _____ ☐ Hoods _____

Volume Dampers: _____

Terminal Boxes: ☐ High Velocity _____ ☐ With Coil _____

☐ Double Duct _____ ☐ _____

Coils: _____

☐ Preheat _____ ☐ Reheat _____

☐ Cooling _____ ☐ _____

Piping: See Previous Page _____

Insulation: Cold: ☐ No ☐ Yes Material _____ Jacket _____

☐ Hot: ☐ No ☐ Yes Material _____ Jacket _____

Automatic Temperature Controls: _____

Air & Hydronic Balancing: _____

Special HVAC: _____

_____

SPEC-AID

DATE _____

DIVISION 16: ELECTRICAL

PROJECT _____ LOCATION _____

Incoming Service: □ Overhead   □ Underground

	Primary	Secondary
Voltage		
Unit Sub-station & Size		
Number of Manholes		
Feeder Size		
Length		
Conduit		
Duct		
Concrete: □ No □ Yes		
Other		

Building Service: Size _____ Amps   Switchboard _____

Panels: □ Distribution _____ Lighting _____ Power _____

Describe _____

Motor Control Center: Furnished by _____

Describe _____

Bus Duct: □ No □ Yes Size _____ Amps Application _____

Describe _____

Cable Tray: □ No □ Yes Describe _____

Emergency System: □ No □ Yes Allowance _____ □ Separate Contract

Generator: □ No □ Diesel □ Gas □ Gasoline _____ Size _____ KW

Transfer Switch: □ No □ Yes Number _____ Size _____ Amps

Area Protection Relay Panels: □ No □ Yes _____

Other _____

Conduit: □ No □ Yes □ Aluminum _____

□ Electric Metallic Tubing _____

□ Galvanized Steel _____

□ Plastic _____

Wire: □ No □ Yes □ Type Installation _____

□ Armored Cable _____

□ Building Wire _____

□ Metallic Sheath Cable _____

□ _____

Underfloor Duct: □ No □ Yes Describe _____

Header Duct: □ No □ Yes Describe _____

Trench Duct: □ No □ Yes Describe _____

Underground Duct: □ No □ Yes Describe _____

Explosion Proof Areas: □ No □ Yes Describe _____

Motors: □ No □ Yes Total H.P. _____ No. of Fractional H.P. _____ Voltage _____

□ 1/2 to 5 H.P. _____ □ 7-1/2 to 25 H.P. _____ □ Over 25 H.P. _____

Describe _____

Starters: Type _____

Supplied by: _____

SPEC-AID

DIVISION 16: ELECTRICAL

Telephone System: ☐ No ☐ Yes   Service Size _____ Length _____

  Manhole: ☐ No   ☐ Yes   Number _____ Termination _____

  Concrete Encased: ☐ No   ☐ Yes   ☐ Rigid Galv.   ☐ Duct   ☐ _____

Fire Alarm System: ☐ No   ☐ Yes   Service Size _____ Length _____ Wire Type _____

  Concrete Encased: ☐ No   ☐ Yes   ☐ Rigid Galv.   ☐ Duct   ☐ _____

  ☐ Stations _____ ☐ Horns _____ ☐ Lights _____ ☐ Combination _____

  Detectors: ☐ Rate of Rise _____ ☐ Fixed _____ ☐ Smoke _____

    Describe _____ Insulation _____ Wire Size _____

  ☐ Zones _____ ☐ Conduit _____ ☐ E.M.T. _____ ☐ Empty _____

    Describe _____

Watchmans Tour: ☐ No   ☐ Yes   ☐ Stations _____ ☐ Door Switches _____

  ☐ Alarm Bells _____ ☐ Key Re-sets _____ ☐ _____

  ☐ Conduit _____ ☐ E.M.T. _____ ☐ Wire _____ ☐ Empty _____

    Describe _____

Clock System: ☐ No   ☐ Yes   ☐ Electronic   ☐ Wired   ☐ _____

  ☐ Single Dial _____ ☐ Double Dial _____ ☐ Program Bell _____

  ☐ Conduit _____ ☐ E.M.T. _____ ☐ Empty _____

    Describe _____

Sound System: ☐ No   ☐ Yes   Type _____ Speakers _____

  ☐ Conduit _____ ☐ Cable _____ ☐ E.M.T. _____ ☐ Empty _____

    Describe _____

Television System: ☐ No   ☐ Yes   Describe _____

  ☐ Antenna _____ ☐ Closed Circuit _____ ☐ Teaching _____ ☐ Security _____

  ☐ Learning Laboratory _____ ☐ _____

  ☐ Conduit _____ ☐ E.M.T. _____ ☐ Wire _____ ☐ Empty _____

Lightning Protection: ☐ No   ☐ Yes   Describe _____

Low Voltage Switching: ☐ No   ☐ Yes   Describe _____

Scoreboards: ☐ No   ☐ Yes   Describe _____ Number _____

Comfort Systems: ☐ No   ☐ Electric Heat   ☐ Snow Melting   ☐ _____

  Describe _____

Other Systems: _____

_____

_____

Lighting Fixtures: ☐ No   ☐ Yes   ☐ Allowance _____ ☐ Separate Contract

  ☐ Economy   ☐ Commercial   ☐ Deluxe   ☐ Explosion Proof   ☐ _____

  ☐ Incandescent _____

  _____ Foot Candles _____

  ☐ Fluorescent _____

  _____ Foot Candles _____

  ☐ Mercury Vapor _____

  _____ Foot Candles _____

  ☐ _____ Foot Candles _____

  ☐ Step Lighting _____ ☐ Planter Lighting _____ ☐ Fountain Lighting _____

  ☐ Site Lighting _____ ☐ Poles _____ ☐ Area Lighting _____ ☐ Flood Lighting _____

  Dimming System: ☐ No   ☐ Yes   ☐ Incandescent   ☐ Fluorescent _____

  Ceilings: ☐ T Bar   ☐ Concealed Spline   ☐ _____

  Emergency Battery Units: ☐ No   ☐ Lead Acid   ☐ Nickel Cadmium   ☐ 6 Volt _____ 12 Volt _____

    Describe _____

Special Considerations: _____

_____

_____

	NO.	DESCRIPTION			UNIT	UNIT COST	NEW SF COST	MODEL SF COST	+/- CHANGE
A SUBSTRUCTURE									
A	1010	Standard Foundations		Bay size:	S.F. Gnd.				
A	1030	Slab on Grade	Material:	Thickness:	S.F. Slab				
A	2010	Basement Excavation	Depth:	Area:	S.F. Gnd.				
A	2020	Basement Walls			L.F. Walls				
B SHELL									
B10 Superstructure									
B	1010	Floor Construction	Elevated floors:		S.F. Floor				
					S.F. Floor				
B	1020	Roof Construction			S.F. Roof				
B20 Exterior Enclosure									
B	2010	Exterior walls	Material:	Thickness: % of wall	S.F. Walls				
			Material:	Thickness: % of wall	S.F. Walls				
B	2020	Exterior Windows	Type:	% of wall	S.F. Wind.				
			Type:	% of wall Each	S.F. Wind.				
B	2030	Exterior Doors	Type:	Number:	Each				
			Type:	Number:	Each				
B30 Roofing									
B	3010	Roof Coverings	Material:		S.F. Roof				
			Material:		S.F. Roof				
B	3020	Roof Openings			S.F. Opng.				
C INTERIORS									
C	1010	Partitions:	Material:	Density:	S.F. Part.				
			Material:	Density:					
C	1020	Interior Doors	Type:	Number:	Each				
C	1030	Fittings			Each				
C	2010	Stair Construction			Flight				
C	3010	Wall Finishes	Material:	% of Wall	S.F. Walls				
			Material:		S.F. Walls				
C	3020	Floor Finishes	Material:		S.F. Floor				
			Material:		S.F. Floor				
			Material:		S.F. Floor				
C	3030	Ceiling Finishes	Material:		S.F. Ceil.				

351

	NO.	SYSTEM/COMPONENT	DESCRIPTION	UNIT	UNIT COST	NEW S.F. COST	MODEL S.F. COST	+/- CHANGE
D SERVICES								
	D10 Conveying							
D	1010	Elevators & Lifts	Type: Capacity: Stops:	Each				
D	1020	Escalators & Moving Walks	Type:	Each				
	D20 Plumbing							
D	2010	Plumbing		Each				
D	2020	Domestic Water Distribution		S.F. Floor				
D	2040	Rain Water Drainage		S.F. Roof				
	D30 HVAC							
D	3010	Energy Supply		Each				
D	3020	Heat Generating Systems	Type:	S.F. Floor				
D	3030	Cooling Generating Systems	Type:	S.F. Floor				
D	3090	Other HVAC Sys. & Equipment		Each				
	D40 Fire Protection							
D	4010	Sprinklers		S.F. Floor				
D	4020	Standpipes		S.F. Floor				
	D50 Electrical							
D	5010	Electrical Service/Distribution		S.F. Floor				
D	5020	Lighting & Branch Wiring		S.F. Floor				
D	5030	Communications & Security		S.F. Floor				
D	5090	Other Electrical Systems		S.F. Floor				
E EQUIPMENT & FURNISHINGS								
E	1010	Commercial Equipment		Each				
E	1020	Institutional Equipment		Each				
E	1030	Vehicular Equipment		Each				
E	1090	Other Equipment		Each				
F SPECIAL CONSTRUCTION								
F	1020	Integrated Construction		S.F.				
F	1040	Special Facilities		S.F.				
G BUILDING SITEWORK								

ITEM		Total Change
17	Total _____ _____ $ _____	
18	Adjusted S.F. cost _____ $ _____ item 17 +/- changes	
19	Building area - from item 11 _____ S.F. x adjusted S.F. cost	$.
20	Basement area - from item 15 _____ S.F. x S.F. cost $	$.
21	Base building sub-total - item 20 + item 19	$
22	Miscellaneous addition (quality, etc.)	$
23	Sub-total - item 22 + 21	$
24	General conditions -25 % of item 23	$
25	Sub-total - item 24 + item 23	$
26	Architects fees _____ % of item 25	$
27	Sub-total - item 26 + item 27	$
28	Location modifier	x
29	Local replacement cost - item 28 x item 27	$
30	Depreciation _____ % of item 29	$
31	Depreciated local replacement cost - item 29 less item 30	$
32	Exclusions	$
33	Net depreciated replacement cost - item 31 less item 32	$

Appendix B

Appendix B

The following tables have been included to assist in the estimating process. They provide labor-hours, technical data, and shortcuts for material takeoffs. These figures have been numbered by MasterFormat Division and are a sampling from *Means Estimating Handbook*. The tables should be used with discretion, taking into account any unique or unusual project requirements. Factors for determining quantities related to carpentry include waste allowance. However, items such as plates, double joists or studs, and corners must be taken off and added separately.

Figure 2.7   Installation Time in Man-Hours for Dewatering

Description	Man-Hours	Unit
Excavate Drainage Trench, 2' Wide		
2' Deep	.178	C.Y.
3' Deep	.160	C.Y.
Sump Pits, by Hand		
Light Soil	1.130	C.Y.
Heavy Soil	2.290	C.Y.
Pumping 8 Hours, Diaphragm or Centrifugal Pump		
Attended 2 hours per day	3.000	Day
Attended 8 hours per day	12.000	Day
Pumping 24 Hours, Attended 24 Hours, 4 Men at		
6 Hour Shifts, 1 Week Minimum	25.140	Day
Relay Corrugated Metal Pipe, Including		
Excavation, 3' Deep		
12" Diameter	.209	L.F.
18" Diameter	.240	L.F.
Sump Hole Construction, Including Excavation,		
with 12" Gravel Collar		
Corrugated Pipe		
12" Diameter	.343	L.F.
18" Diameter	.480	L.F.
Wood Lining, Up to 4'x4'	.080	SFCA
Wellpoint System, Single Stage, Install and		
Remove, per Length of Header		
Minimum	.750	L.F.
Maximum	2.000	L.F.
Wells, 10' to 20' Deep with Steel Casing		
2' Diameter		
Minimum	.145	V.L.F.
Average	.245	V.L.F.
Maximum	.490	V.L.F.

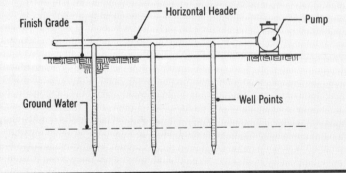

Figure 2.9   Excavating

The selection of equipment used for structural excavation and bulk excavation or for grading is determined by the following factors.

1. Quantity of material
2. Type of material
3. Depth or height of cut
4. Length of haul
5. Condition of haul road
6. Accessibility of site
7. Moisture content and dewatering requirements
8. Availability of excavating and hauling equipment

Some additional costs must be allowed for hand trimming the sides and bottom of concrete pours and other excavation below the general excavation.

When planning excavation and fill, the following should also be considered.

1. Swell factor
2. Compaction factor
3. Moisture content
4. Density requirements

A typical example for scheduling and estimating the cost of excavation of a 15' deep basement on a dry site when the material must be hauled off the site, is outlined below.

Assumptions:

1. Swell factor, 18%
2. No mobilization or demobilization
3. Allowance included for idle time and moving on job
4. No dewatering, sheeting, or bracing
5. No truck spotter or hand trimming

Number of B.C.Y. per truck   = 1.5 C.Y. bucket x 8 passes = 12 loose C.Y.

$$= 12 \times \frac{100}{118} = 10.2 \text{ B.C.Y. per truck}$$

Truck haul cycle:

Load truck 8 passes	= 4 minutes
Haul distance 1 mile	= 9 minutes
Dump time	= 2 minutes
Return 1 mile	= 7 minutes
Spot under machine	= 1 minute
	23 minute cycle

Fleet Haul Production per Day in B.C.Y.

$$4 \text{ trucks} \times \frac{50 \text{ min. hr.}}{23 \text{ min. haul cycle}} \times 8 \text{ hrs.} \times 10.2 \text{ B.C.Y.} = 4 \times 2.2 \times 8 \times 10.2 = 718 \text{ B.C.Y./day}$$

Note: B.C.Y. = Bank Measure Cubic Yards

Figure 3.3 Formwork Labor Hours

Man-hours for formwork vary greatly from one component to another. This table is convenient for planning the number of forms to be fabricated or provided, depending on the number of reuses anticipated.

Item	Unit	Fabricate	Erect & Strip	Clean & Move	Total Hours 1 Use	2 Use	3 Use	4 Use
Beam and Girder, interior beams, 12" wide	100 S.F.	6.4	8.3	1.3	16.0	13.3	12.4	12.0
Hung from steel beams		5.8	7.7	1.3	14.8	12.4	11.6	11.2
Beam sides only, 36" high		5.8	7.2	1.3	14.3	11.9	11.1	10.7
Beam bottoms only, 24" wide		6.6	13.0	1.3	20.9	18.1	17.2	16.7
Box out for openings		9.9	10.0	1.1	21.0	16.6	15.1	14.3
Buttress forms, to 8' high		6.0	6.5	1.2	13.7	11.2	10.4	10.0
Centering, steel, 3/4" rib lath			1.0		1.0			
3/8" rib lath or slab form			0.9		0.9			
Chamfer strip or keyway	100 L.F.		1.5		1.5	1.5	1.5	1.5
Columns, fiber tube 8" diameter			20.6		20.6			
12"			21.3		21.3			
16"			22.9		22.9			
20"			23.7		23.7			
24"			24.6		24.6			
30"			25.6		25.6			
Round Steel, 12" diameter			22.0		22.0	22.0	22.0	22.0
16"			25.6		25.6	25.6	25.6	25.6
20"			30.5		30.5	30.5	30.5	30.5
24"			37.7		37.7	37.7	37.7	37.7
Plywood 8" x 8"	100 S.F.	7.0	11.0	1.2	19.2	16.2	15.2	14.7
12" x 12"		6.0	10.5	1.2	17.7	15.2	14.4	14.0
16" x 16"		5.9	10.0	1.2	17.1	14.7	13.8	13.4
24" x 24"		5.8	9.8	1.2	16.8	14.4	13.6	13.2
Steel framed plywood 8" x 8"			10.0	1.0	11.0	11.0	11.0	11.0
12" x 12"			9.3	1.0	10.3	10.3	10.3	10.3
16" x 16"			8.5	1.0	9.5	9.5	9.5	9.5
24" x 24"			7.8	1.0	8.8	8.8	8.8	8.8
Drop head forms, plywood		9.0	12.5	1.5	23.0	19.0	17.7	17.0
Coping forms		8.5	15.0	1.5	25.0	21.3	20.0	19.4
Culvert, box			14.5	4.3	18.8	18.8	18.8	18.8
Curb forms, 6" to 12" high, on grade		5.0	8.5	1.2	14.7	12.7	12.1	11.7
On elevated slabs		6.0	10.8	1.2	18.0	15.5	14.7	14.3
Edge forms to 6" high, on grade	100 L.F.	2.0	3.5	0.6	6.1	5.6	5.4	5.3
7" to 12" high	100 S.F.	2.5	5.0	1.0	8.5	7.8	7.5	7.4
Equipment foundations	"	10.0	18.0	2.0	30.0	25.5	24.0	23.3
Flat slabs, including drops		3.5	6.0	1.2	10.7	9.5	9.0	8.8
Hung from steel		3.0	5.5	1.2	9.7	8.7	8.4	8.2
Closed deck for domes		3.0	5.8	1.2	10.0	9.0	8.7	8.5
Open deck for pans		2.2	5.3	1.0	8.5	7.9	7.7	7.6
Footings, continuous, 12" high		3.5	3.5	1.5	8.5	7.3	6.8	6.6
Spread, 12" high		4.7	4.2	1.6	10.5	8.7	8.0	7.7
Pile caps, square or rectangular		4.5	5.0	1.5	11.0	9.3	8.7	8.4
Grade beams, 24" deep		2.5	5.3	1.2	9.0	8.3	8.0	7.9
Lintel or Sill forms		8.0	17.0	2.0	27.0	23.5	22.3	21.8
Spandrel beams, 12" wide		9.0	11.2	1.3	21.5	17.5	16.2	15.5
Stairs			25.0	4.0	29.0	29.0	29.0	29.0
Trench forms in floor		4.5	14.0	1.5	20.0	18.3	17.7	17.4
Walls, Plywood, at grade, to 8' high		5.0	6.5	1.5	13.0	11.0	9.7	9.5
8' to 16'		7.5	8.0	1.5	17.0	13.8	12.7	12.1
16' to 20'		9.0	10.0	1.5	20.5	16.5	15.2	14.5
Foundation walls, to 8' high		4.5	6.5	1.0	12.0	10.3	9.7	9.4
8' to 16' high		5.5	7.5	1.0	14.0	11.8	11.0	10.6
Retaining wall to 12' high, battered		6.0	8.5	1.5	16.0	13.5	12.7	12.3
Radial walls to 12' high, smooth		8.0	9.5	2.0	19.5	16.0	14.8	14.3
But in 2' chords		7.0	8.0	1.5	16.5	13.5	12.5	12.0
Prefabricated modular, to 8' high		—	4.3	1.0	5.3	5.3	5.3	5.3
Steel, to 8' high		—	6.8	1.2	8.0	8.0	8.0	8.0
8' to 16' high		—	9.1	1.5	10.6	10.3	10.2	10.2
Steel framed plywood to 8' high		—	6.8	1.2	8.0	7.5	7.3	7.2
8' to 16' high		—	9.3	1.2	10.5	9.5	9.2	9.0

Figure 3.16 Data on Handling and Transporting Concrete

On most construction jobs, the method of concrete placement is an important consideration for the estimator and the project manager. This table is a guide for planning time and equipment for concrete placement.

Quantity	Rate
Unloading 6 cubic yard mixer truck, 1 operator	Minimum time 2 minutes, average time 7 minutes
Wheeling, using 4-1/2 cubic feet wheelbarrows, 1 laborer	
up to 100'	Average 1-1/2 cubic yards per hour
up to 200'	Average 1 cubic yard per hour
Wheeling, using 8 cubic feet hand buggies, 1 laborer	
up to 100'	Average 5 cubic yards per hour
up to 200'	Average 3 cubic yards per hour
Transporting, using 28 cubic feet power buggies, 1 laborer	
up to 500'	Average 20 cubic feet per hour
up to 1000'	Average 15 cubic feet per hour
Portable conveyor, 16" belt, 30° elevation, 1 laborer	
100 FPM belt speed	30 C.Y./Hr. max., 15 C.Y./Hr. average
200 FPM belt speed	60 C.Y./Hr. max., 30 C.Y./Hr. average
300 FPM belt speed	90 C.Y./Hr. max., 45 C.Y./Hr. average
400 FPM belt speed	120 C.Y./Hr. max., 60 C.Y./Hr. average
500 FPM belt speed	150 C.Y./Hr. max., 75 C.Y./Hr. average
600 FPM belt speed	180 C.Y./Hr. max., 90 C.Y./Hr. average
Feeder conveyors, up to 600', 5 laborers	
500 FPM belt speed	150 C.Y./Hr. max., 50 C.Y./Hr. average
600 FPM belt speed	180 C.Y./Hr. max., 60 C.Y./Hr. average
Side discharge conveyor, 1 laborer	
fed by portable conveyor	40–60 C.Y./Hr. average
fed by feeder conveyor	80–100 C.Y./Hr. average
fed by crane & bucket	35–60 C.Y./Hr. average
Mobile crane, 1 operator* 2 laborers	
1 cubic yard bucket	40 C.Y./Hr. average
2 cubic yard bucket	60 C.Y./Hr. average
Tower crane 1 operator* 2 laborers	
1 cubic yard bucket	35–40 C.Y./Hr. average
Small line pumping systems, 1 foot vertical = 6 feet horizontally 1 90 degree bend = 40 feet horizontally 1 45 degree bend = 20 feet horizontally 1 30 degree bend = 13 feet horizontally 1 foot rubber hose = 1-1/2 feet of steel tubing 1 operator*	
average output, 1000' horizontally	40–50 C.Y./Hr. Average

*In some regions an oiler or helper may be needed, according to union rules.

(*courtesy Concrete Estimating Handbook, Michael F. Kenny, Van Nostrand Reinhold Company*).

Figure 4.16 Man-Hours Required for the Installation of Brick Masonry

Description	Man-Hours	Unit
Brick Wall		
Veneer		
4" Thick		
Running Bond		
Standard Brick (6.75/S.F.)	.182	S.F.
Engineer Brick (5.63/S.F.)	.154	S.F.
Economy Brick (4.50/S.F.)	.129	S.F.
Roman Brick (6.00/S.F.)	.160	S.F.
Norman Brick (4.50/S.F.)	.125	S.F.
Norwegian Brick (3.75/S.F.)	.107	S.F.
Utility Brick (3.00/S.F.)	.089	S.F.
Common Bond, Standard Brick (7.88/S.F.)	.216	S.F.
Flemish Bond, Standard Brick (9.00/S.F.)	.267	S.F.
English Bond, Standard Brick (10.13/S.F.)	.286	S.F.
Stack Bond, Standard Brick (6.75/S.F.)	.200	S.F.
6" Thick		
Running Bond		
S.C.R. Brick (4.50/S.F.)	.129	S.F.
Jumbo Brick (3.00/S.F.)	.092	S.F.
Backup		
4" Thick		
Running Bond		
Standard Brick (6.75/S.F.)	.167	S.F.
Solid, Unreinforced		
8" Thick Running Bond (13.50/S.F.)	.296	S.F.
12" Thick, Running Bond (20.25/S.F.)	.421	S.F.
Solid, Rod Reinforced		
8" Thick, Running Bond	.308	S.F.
12" Thick, Running Bond	.444	S.F.
Cavity		
4" Thick		
4" Backup	.242	S.F.
6" Backup	.276	S.F.
Brick Chimney		
16" x 16", Standard Brick w/8" x 8" Flue	.889	V.L.F.
16" x 16", Standard Brick w/8" x 12" Flue	1.000	V.L.F.
20" x 20", Standard Brick w/12" x 12" Flue	1.140	V.L.F.
Brick Column		
8" x 8", Standard Brick 9.0 V.L.F.	.286	V.L.F.
12" x 12", Standard Brick 20.3 V.L.F.	.640	V.L.F.
20" x 20", Standard Brick 56.3 V.L.F.	1.780	V.L.F.
Brick Coping		
Precast, 10" Wide, or Limestone, 4" Wide	.178	L.F.
Precast 14" Wide, or Limestone, 6" Wide	.200	L.F.

(continued on next page)

Figure 4.16 Man-Hours Required for the
Installation of Brick Masonry (continued)

Description	Man-Hours	Unit
Brick Fireplace		
30" x 24" Opening, Plain Brickwork	40.000	Ea.
Firebox Only, Fire Brick (110/Ea.)	8.000	Ea.
Brick Prefabricated Wall Panels, 4" Thick		
Minimum	.093	S.F.
Maximum	.144	S.F.
Brick Steps	53.330	M
Window Sill		
Brick on Edge	.200	L.F.
Precast, 6" Wide	.229	L.F.
Needle Brick and Shore, Solid Brick		
8" Thick	6.450	Ea.
12" Thick	8.160	Ea.
Repoint Brick		
Hard Mortar		
Running Bond	.100	S.F.
English Bond	.123	S.F.
Soft Mortar		
Running Bond	.080	S.F.
English Bond	.098	S.F.
Toothing Brick		
Hard Mortar	.267	V.L.F.
Soft Mortar	.200	V.L.F.
Sandblast Brick		
Wet System		
Minimum	.024	S.F.
Maximum	.057	S.F.
Dry System		
Minimum	.013	S.F.
Maximum	.027	S.F.
Sawing Brick, Per Inch of Depth	.027	L.F.
Steam Clean, Face Brick	.033	S.F.
Wash Brick, Smooth	.014	S.F.

Figure 4.24 Installation Time in Man-Hours for Block Walls, Partitions, and Accessories

Description	Man-Hours	Unit
Foundation Walls, Trowel Cut Joints, Parged 1/2" Thick, 1 Side, 8" x 16" Face		
Hollow		
8" Thick	.093	S.F.
12" Thick	.122	S.F.
Solid		
8" Thick	.096	S.F.
12" Thick	.126	S.F.
Backup Walls, Tooled Joint 1 Side, 8" x 16" Face		
4" Thick	.091	S.F.
8" Thick	.100	S.F.
Partition Walls, Tooled Joint 2 Sides 8" x 16" Face		
Hollow		
4" Thick	.093	S.F.
8" Thick	.107	S.F.
12" Thick	.141	S.F.
Solid		
4" Thick	.096	S.F.
8" Thick	.111	S.F.
12" Thick	.148	S.F.
Stud Block Walls, Tooled Joints 2 Sides 8" x 16" Face		
6" Thick and 2", Plain	.098	S.F.
Embossed	.103	S.F.
10" Thick and 2", Plain	.108	S.F.
Embossed	.114	S.F.
6" Thick and 2" Each Side, Plain	.114	S.F.
Acoustical Slotted Block Walls Tooled 2 Sides		
4" Thick	.127	S.F.
8" Thick	.151	S.F.
Glazed Block Walls, Tooled Joint 2 Sides 8" x 16", Glazed 1 Face		
4" Thick	.116	S.F.
8" Thick	.129	S.F.
12" Thick	.171	S.F.
8" x 16", Glazed 2 Faces		
4" Thick	.129	S.F.
8" Thick	.148	S.F.
8" x 16", Corner		
4" Thick	.140	Ea.

(continued on next page)

Figure 4.24 Installation Time in Man-Hours for Block Walls, Partitions, and Accessories (continued)

Description	Man-Hours	Unit
Structural Facing Tile, Tooled 2 Sides		
5" x 12", Glazed 1 Face		
4" Thick	.182	S.F.
8" Thick	.222	S.F.
5" x 12", Glazed 2 Faces		
4" Thick	.205	S.F.
8" Thick	.246	S.F.
8" x 16", Glazed 1 Face		
4" Thick	.116	S.F.
8" Thick	.129	S.F.
8" x 16", Glazed 2 Faces		
4" Thick	.123	S.F.
8" Thick	.137	S.F.
Exterior Walls, Tooled Joint 2 Sides, Insulated		
8" x 16" Face, Regular Weight		
8" Thick	.110	S.F.
12" Thick	.145	S.F.
Lightweight		
8" Thick	.104	S.F.
12" Thick	.137	S.F.
Architectural Block Walls, Tooled Joint 2 Sides		
8" x 16" Face		
4" Thick	.116	S.F.
8" Thick	.138	S.F.
12" Thick	.181	S.F.
Interlocking Block Walls, Fully Grouted		
Vertical Reinforcing		
8" Thick	.131	S.F.
12" Thick	.145	S.F.
16" Thick	.173	S.F.
Bond Beam, Grouted, 2 Horizontal Rebars		
8" x 16" Face, Regular Weight		
8" Thick	.133	L.F.
12" Thick	.192	L.F.
Lightweight		
8" Thick	.131	L.F.
12" Thick	.188	L.F.
Lintels, Grouted, 2 Horizontal Rebars		
8" x 16" Face, 8" Thick	.119	L.F.
16" x 16" Face, 8" Thick	.131	L.F.
Control Joint 4" Wall	.013	L.F.
8" Wall	.020	L.F.
Grouting Bond Beams and Lintels		
8" Deep Pumped, 8" Thick	.018	L.F.
12" Thick	.025	L.F.
Concrete Block Cores Solid		
4" Thick By Hand	.035	S.F.
8" Thick Pumped	.038	S.F.
Cavity Walls 2" Space Pumped	.016	S.F.
6" Space	.034	S.F.

(continued on next page)

Figure 4.24 Installation Time in Man-Hours for Block
Walls, Partitions, and Accessories (continued)

Description	Man-Hours	Unit
Joint Reinforcing		
Wire Strips Regular Truss to 6" Wide	.267	C.L.F.
12" Wide	.400	C.L.F.
Cavity Wall with Drip Section to 6" Wide	.267	C.L.F.
12" Wide	.400	C.L.F.
Lintels Steel Angles Minimum	.008	lb.
Maximum	.016	lb.
Wall Ties	.762	C
Coping For 12" Wall Stock Units, Aluminum	.200	L.F.
Precast Concrete	.188	L.F.
Structural Reinforcing, Placed Horizontal,		
#3 and #4 Bars	.018	lb.
#5 and #6 Bars	.010	lb.
Placed Vertical, #3 and #4 Bars	.023	lb.
#5 and #6 Bars	.012	lb.
Acoustical Slotted Block		
4" Thick	.127	S.F.
6" Thick	.138	S.F.
8" Thick	.151	S.F.
12" Thick	.163	S.F.
Lightweight Block		
4" Thick	.090	S.F.
6" Thick	.095	S.F.
8" Thick	.100	S.F.
10" Thick	.103	S.F.
12" Thick	.130	S.F.
Regular Block		
Hollow		
4" Thick	.093	S.F.
6" Thick	.100	S.F.
8" Thick	.107	S.F.
10" Thick	.111	S.F.
12" Thick	.141	S.F.
Solid		
4" Thick	.095	S.F.
6" Thick	.105	S.F.
8" Thick	.113	S.F.
12" Thick	.150	S.F.
Glazed Concrete Block		
Single Face 8" x 16"		
2" Thick	.111	S.F.
4" Thick	.116	S.F.
6" Thick	.121	S.F.
8" Thick	.129	S.F.
12" Thick	.171	S.F.
Double Face		
4" Thick	.129	S.F.
6" Thick	.138	S.F.
8" Thick	.148	S.F.

(continued on next page)

Figure 4.24 Installation Time in Man-Hours for Block Walls, Partitions, and Accessories (continued)

Description	Man-Hours	Unit
Joint Reinforcing Wire Strips		
4" and 6" Wall	.267	C.L.F.
8" Wall	.320	C.L.F.
10" and 12" Wall	.400	C.L.F.
Steel Bars Horizontal		
#3 and #4	.018	lb.
#5 and #6	.010	lb.
Vertical		
#3 and #4	.023	lb.
#5 and #6	.012	lb.
Grout Cores Solid		
By Hand 6" Thick	.035	S.F.
Pumped 8" Thick	.038	S.F.
10" Thick	.039	S.F.
12" Thick	.040	S.F.

Figure 5.22 Installation Time in Man-Hours for the Erection of Steel Superstructure Systems

Description	Man-Hours	Unit
Steel Columns Concrete Filled		
4" Diameter	.072	L.F.
5" Diameter	.055	L.F.
6-5/8" Diameter	.047	L.F.
Steel Pipe		
6" Diameter	6.000	ton
12" Diameter	2.000	ton
Structural Tubing		
6" x 6"	6.000	ton
10" x 10"	2.000	ton
Wide Flange		
W8 x 31	3.355	ton
W10 x 45	2.412	ton
W12 x 50	2.171	ton
W14 x 74	1.538	ton
Beams WF Average	4.000	ton
Steel Joists H Series Horizontal Bridging		
To 30' Span	6.667	ton
30' to 50' Span	6.353	ton
(Includes One Row of Bolted Cross Bridging for Spans Over 40' Where Required)		
LH Series Bolted Cross Bridging		
Spans to 96'	6.154	ton
DLH Series Bolted Cross Bridging		
Spans to 144' Shipped in 2 Pieces	6.154	ton
Joist Girders	6.154	ton
Trusses, Factory Fabricated, with Chords	7.273	ton
Metal Decking Open Type		
1-1/2" Deep		
22 Gauge	.007	S.F.
18 and 20 Gauge	.008	S.F.
3" Deep		
20 and 22 Gauge	.009	S.F.
18 Gauge	.010	S.F.
16 Gauge	.011	S.F.
4-1/2" Deep		
20 Gauge	.012	S.F.
18 Gauge	.013	S.F.
16 Gauge	.014	S.F.
7-1/2" Deep		
18 Gauge	.019	S.F.
16 Gauge	.020	S.F.

Figure 5.22 Installation Time in Man-Hours for the Erection of Steel Superstructure Systems (continued)

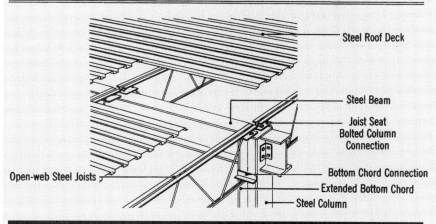

- Steel Roof Deck
- Steel Beam
- Joist Seat
- Bolted Column Connection
- Bottom Chord Connection
- Extended Bottom Chord
- Steel Column
- Open-web Steel Joists

Figure 5.24 Installation Time in Man-Hours for Fireproofing Structural Steel

Description	Man-Hours	Unit
Fireproofing - 10" Column Encasements		
Perlite Plaster	.273	V.L.F.
1" Perlite on 3/8" Gypsum Lath	.345	V.L.F.
Sprayed Fiber	.131	V.L.F.
Concrete 1-1/2" Thick	.716	V.L.F.
Gypsum Board 1/2" Fire Resistant,		
1 Layer	.364	V.L.F.
2 Layer	.428	V.L.F.
3 Layer	.530	V.L.F.
Fireproofing – 16" x 7" Beam Encasements		
Perlite Plaster on Metal Lath	.453	L.F.
Gypsum Plaster on Metal Lath	.408	L.F.
Sprayed Fiber	.079	L.F.
Concrete 1-1/2" Thick	.554	L.F.
Gypsum Board 5/8" Fire Resistant	.488	L.F.

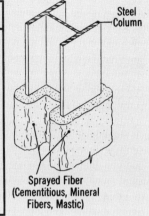

Sprayed Fiber on Columns

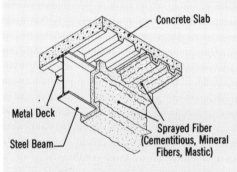

Sprayed Fiber on Beams and Girders

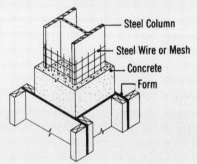

Concrete Encasement on Columns

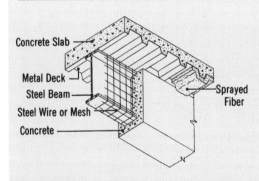

Concrete Encasement on Beams and Girders

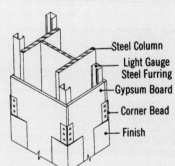

Gypsum Board on Columns

Figure 5.24 Installation Time in Man-Hours for
Fireproofing Structural Steel (continued)

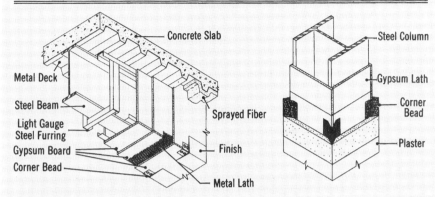

Gypsum Board on Beams and Girders   Plaster on Gypsum Lath — Columns

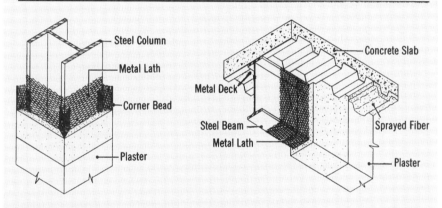

Plaster on Metal Lath — Columns   Plaster on Metal Lath — Beams and Girders

Figure 6.16   Flat Roof Framing

This table can be used to compute the amount of lumber in board feet required to frame a flat roof, based on the size lumber required.

Flat Roof Framing			
Joist Size	Inches On Center	Board Feet per Square Foot of Ceiling Area	Nails Lbs. per MBM
2" x 6"	12"	1.17	10
	16"	.91	10
	20"	.76	10
	24"	.65	10
2" x 8"	12"	1.56	8
	16"	1.21	8
	20"	1.01	8
	24"	.86	8
2" x 10"	12"	1.96	6
	16"	1.51	6
	20"	1.27	6
	24"	1.08	6
2" x 12"	12"	2.35	5
	16"	1.82	5
	20"	1.52	5
	24"	1.30	5
3" x 8"	12"	2.35	5
	16"	1.82	5
	20"	1.52	5
	24"	1.30	5
3" x 10"	12"	2.94	4
	16"	2.27	4
	20"	1.90	4
	24"	1.62	4

MBM; MFBM = Thousand Feet Board Measure

Figure 6.20 Pitched Roof Framing

This table is used to compute the board feet of lumber required per square foot of roof area, based on the required spacing and the lumber size to be used.

	Rafters Including Collar Ties, Hip and Valley Rafters, Ridge Poles							
	Spacing Center to Center							
Rafter Size	12"		16"		20"		24"	
	Board Feet per Square Foot of Roof Area	Nails Lbs. per MBM	Board Feet per Square Foot of Roof Area	Nails Lbs. per MBM	Board Feet per Square Foot of Roof Area	Nails Lbs. per MBM	Board Feet per Square Foot of Roof Area	Nails Lbs. per MBM
2" x 4"	.89	17	.71	17	.59	17	.53	17
2" x 6"	1.29	12	1.02	12	.85	12	.75	12
2" x 8"	1.71	9	1.34	9	1.12	9	.98	9
2" x 10"	2.12	7	1.66	7	1.38	7	1.21	7
2" x 12"	2.52	6	1.97	6	1.64	6	1.43	6
3" x 8"	2.52	6	1.97	6	1.64	6	1.43	6
3" x 10"	3.13	5	2.45	5	2.02	5	1.78	5

Figure 6.25   Board Feet Required for On-the-Job Cut Bridging

Based on the size of the lumber used for joists, the total lengths of lumber for various sized bridging can be obtained from this chart.

Cross Bridging—Board Feet per Square Foot of Floors, Ceiling or Flat Roof Area Nails — Pounds Per MBM of Bridging							
		1" x 3"		1" x 4"		2" x 3"	
Joist Size	Spacing	B.F.	Nails	B.F.	Nails	B.F.	Nails
2" x 8"	12"	.04	147	.05	112	.08	77
	16"	.04	120	.05	91	.08	61
	20"	.04	102	.05	77	.08	52
	24"	.04	83	.05	63	.08	42
2" x 10"	12"	.04	136	.05	103	.08	71
	16"	.04	114	.05	87	.08	58
	20"	.04	98	.05	74	.08	50
	24"	.04	80	.05	61	.08	41
2" x 12"	12"	.04	127	.05	96	.08	67
	16"	.04	108	.05	82	.08	55
	20"	.04	94	.05	71	.08	48
	24"	.04	78	.05	59	.08	39
3" x 8"	12"	.04	160	.05	122	.08	84
	16"	.04	127	.05	96	.08	66
	20"	.04	107	.05	81	.08	54
	24"	.04	86	.05	65	.08	44
3" x 10"	12"	.04	146	.05	111	.08	77
	16"	.04	120	.05	91	.08	62
	20"	.04	102	.05	78	.08	52
	24"	.04	83	.05	63	.08	42

Figure 6.33 Wood Siding Factors

This table shows the factor by which area to be covered is multiplied to determine exact amount of surface material needed.

Item	Nominal Size	Width Overall	Face	Area Factor
Shiplap	1" x 6" 1 x 8 1 x 10 1 x 12	5-1/2" 7-1/4 9-1/4 11-1/4	5-1/8" 6-7/8 8-7/8 10-7/8	1.17 1.16 1.13 1.10
Tongue and Grooved	1 x 4 1 x 6 1 x 8 1 x 10 1 x 12	3-3/8 5-3/8 7-1/8 9-1/8 11-1/8	3-1/8 5-1/8 6-7/8 8-7/8 10-7/8	1.28 1.17 1.16 1.13 1.10
S4S	1 x 4 1 x 6 1 x 8 1 x 10 1 x 12	3-1/2 5-1/2 7-1/4 9-1/4 11-1/4	3-1/2 5-1/2 7-1/4 9-1/4 11-1/4	1.14 1.09 1.10 1.08 1.07
Solid Paneling	1 x 6 1 x 8 1 x 10 1 x 12	5-7/16 7-1/8 9-1/8 11-1/8	5-7/16 6-3/4 8-3/4 10-3/4	1.19 1.19 1.14 1.12
Bevel Siding*	1 x 4 1 x 6 1 x 8 1 x 10 1 x 12	3-1/2 5-1/2 7-1/4 9-1/4 11-1/4	3-1/2 5-1/2 7-1/4 9-1/4 11-1/4	1.60 1.33 1.28 1.21 1.17

Note: This area factor is strictly so-called milling waste. The cutting and fitting waste must be added.

*1" lap

(*from Western Wood Products Association*)

Figure 6.10   Exterior Wall Stud Framing

This table allows you to compute the board feet of lumber required for each square foot of exterior wall to be framed based on the lumber size and spacing design.

Stud Size	Inches On Center	Studs Including Corner Bracing		Horizontal Bracing Midway Between Plates	
		Board Feet per Square Foot of Ext. Wall Area	Lbs. of Nails per MBM of Stud Framing	Board Feet per Square Foot of Ext. Wall Area	Lbs. of Nails per MBM of Bracing
2" x 3"	16"	.78	30	.03	117
	20"	.74	30	.03	97
	24"	.71	30	.03	85
2" x 4"	16"	1.05	22	.04	87
	20"	.98	22	.04	72
	24"	.94	22	.04	64
2" x 6"	16"	1.51	15	.06	59
	20"	1.44	15	.06	48
	24"	1.38	15	.06	43

Figure 6.12 Floor Framing

This table can be used to compute the board feet required for each square foot of floor area, based on the size of lumber to be used and the spacing design.

Joist Size	Inches On Center	Floor Joists			Block Over Main Bearing	
		Board Feet per Square Foot of Floor Area	Nails Lbs. per MBM	Board Feet per Square Foot of Floor Area	Nails Lbs. per MBM Blocking
2" x 6"	12"	1.28	10	.16	133
	16"	1.02	10	.03	95
	20"	.88	10	.03	77
	24"	.78	10	.03	57
2" x 8"	12"	1.71	8	.04	100
	16"	1.36	8	.04	72
	20"	1.17	8	.04	57
	24"	1.03	8	.05	43
2" x 10"	12"	2.14	6	.05	79
	16"	1.71	6	.05	57
	20"	1.48	6	.06	46
	24"	1.30	6	.06	34
2" x 12"	12"	2.56	5	.06	66
	16"	2.05	5	.06	47
	20"	1.77	5	.07	39
	24"	1.56	5	.07	29
3" x 8"	12"	2.56	5	.04	39
	16"	2.05	5	.05	57
	20"	1.77	5	.06	45
	24"	1.56	5	.06	33
3" x 10"	12"	3.20	4	.05	72
	16"	2.56	4	.07	46
	20"	2.21	4	.07	36
	24"	1.95	4	.08	26

Figure 6.26 Board Feet of Sheathing and Subflooring

This table is used to compute the board feet of sheathing or subflooring required per square foot of roof, ceiling or flooring.

Type	Size	Board Feet per Square Foot of Area	Diagonal Lbs. Nails per MBM Lumber Joist, Stud or Rafter Spacing			
			12"	16"	20"	24"
Surface 4 Sides (S4S)	1" x 4"	1.22	58	46	39	32
	1" x 6"	1.18	39	31	25	21
	1" x 8"	1.18	30	23	19	16
	1" x 10"	1.17	35	27	23	19
Tongue and Groove (T&G)	1" x 4"	1.36	65	51	43	36
	1" x 6"	1.26	42	33	27	23
	1" x 8"	1.22	31	24	20	17
	1" x 10"	1.20	36	28	24	19
Shiplap	1" x 4"	1.41	67	53	45	37
	1" x 6"	1.29	43	33	28	23
	1" x 8"	1.24	31	24	20	17
	1" x 10"	1.21	36	28	24	19

Figure 6.5 Factors for the Board Foot Measure of Floor or Ceiling Joists

This table allows you to compute the amount of lumber (in board feet) for the floor or ceiling joists of a known area, based on the size of lumber and the spacing. To use this chart, simply multiply the area of the room or building in question by the factor across from the appropriate board size and spacing.

Ceiling Joists			
Joist Size	Inches On Center	Board Feet per Square Foot of Ceiling Area	Nails Lbs. per MBM
2" x 4"	12"	.78	17
	16"	.59	19
	20"	.48	19
2" x 6"	12"	1.15	11
	16"	.88	13
	20"	.72	13
	24"	.63	13
2" x 8"	12"	1.53	9
	16"	1.17	9
	20"	.96	9
	24"	.84	9
2" x 10"	12"	1.94	7
	16"	1.47	7
	20"	1.21	7
	24"	1.04	7
3" x 8"	12"	2.32	6
	16"	1.76	6
	20"	1.44	6
	24"	1.25	6

Figure 6.11 Furring Quantities

This table provides multiplication factors for converting square feet of wall area requiring furring into board feet of lumber. These figures are based on lumber size and spacing.

Size	Board Feet per Square Feet of Wall Area				Lbs. Nails per MBM of Furring
	Spacing Center to Center				
	12"	16"	20"	24"	
1" x 2"	.18	.14	.11	.10	55
1" x 3"	.28	.21	.17	.14	37

Figure 6.9 Partition Framing

This table can be used to compute the board feet of lumber required for each square foot of wall area to be framed, based on the lumber size and spacing design.

Stud Size	Inches on Center	Studs Including Sole and Cap Plates		Horizontal Bracing in All Partitions		Horizontal Bracing in Bearing Partitions Only	
		Board Feet per Square Foot of Partition Area	Lbs. Nails per MBM of Stud Framing	Board Feet per Square Foot of Partition Area	Lbs. Nails per MBM of Bracing	Board Feet per Square Foot of Partition Area	Lbs. Nails per MBM of Bracing
2" x 3"	12"	.91	25	.04	145	.01	145
	16"	.83	25	.04	111	.01	111
	20"	.78	25	.04	90	.01	90
	24"	.76	25	.04	79	.01	79
2" x 4"	12"	1.22	19	.05	108	.02	108
	16"	1.12	19	.05	87	.02	87
	20"	1.05	19	.05	72	.02	72
	24"	1.02	19	.05	64	.02	64
2" x 6"	16"	1.38	19			.04	59
	20"	1.29	16			.04	48
	24"	1.22	16			.04	43
2" x 4" Staggered	8"	1.69	22				
3" x 4"	16"	1.35	17				
2" x 4" 2" Way	16"	1.08	19				

Figure 7.1 Common Insulating Materials Used in Construction

This table lists the common types of materials used in construction, their forms and their major uses.

Physical Form	Type of Material	Major Uses
Powders	Diatomaceous earth; sawdust; silica aerogel	Filler
Loose fibrous materials	Cork granules; glass wool; mineral wool; shredded bark; vermiculite; perlite; mica	Flat areas such as air spaces above ceilings, adjacent to roof, to reduce conduction and convection
Batt or blanket insulation[a]	Glass wool; mineral wool; wood fibers, etc., enclosed by paper, cloth, wire mesh, or aluminum	Air spaces, particularly in vertical walls and flat surfaces, to reduce conduction and convection
Board, sheet, and integrant insulation[a]	Cork; fiber; paper pulp; cellular glass; glass fiber	Sheathing for walls, to increase strength as well as reduce heat loss; rigid insulation on roofs; perimeter insulation along edges of slab floors
Reflective insulation	Aluminum foil, often combined in layers with one or more adjoining air spaces or combined with sheets of paper	Used principally for all types of refrigerated or controlled environmental spaces
Special types of block and brick insulation and refractories	Insulation block made of cork, expanded glass, 85% magnesia, or vermiculite; insulating refractory block or brick made of diatomaceous earth or kaolin (clay); heavy refractories made of fire clay, magnesite, or silica	Special controlled temperature and high-temperature insulation problems; for example, refrigerator rooms, pipes, ducts, boilers, fire chambers of boilers, and chimneys
Foam-type insulation	Rigid boards; 2-component on-job application with special applicators; polystyrene and polyurethane	Interior-applied insulation for walls above or below grade; roof insulation; air space and perimeter insulation

[a]These are available with vapor barrier as part of insulation.
(*courtesy Construction Materials, Caleb Hornbostel, John Wiley & Sons*)

Figure 7.27 Installation Time in Man-Hours for Shingle and Tile Roofing

Description	Man-Hours	Unit
Shingles		
Aluminum	3.200	Sq.
Ridge Cap or Valley	.047	L.F.
Shakes	3.478	Sq.
Fiberglass		
325 lb. per Sq.	2.000	Sq.
165 lb. per Sq.	2.286	Sq.
Hip and Ridge	8.000	C.L.F.
Asphalt Standard Strip		
Class A 210 to 235 lb. per Sq.	1.455	Sq.
Class C 235 to 240 lb. per Sq.	1.600	Sq.
Standard Laminated		
Class A 240 to 260 lb. per Sq.	1.778	Sq.
Class C 260 to 300 lb. per Sq.	2.000	Sq.
Premium Laminated		
Class A 260 to 300 lb. per Sq.	2.286	Sq.
Class C 300 to 385 lb. per Sq.	2.667	Sq.
Hip and Ridge Roll	.020	L.F.
Slate Including Felt Underlay	4.571	Sq.
Steel	3.636	Sq.
Wood		
5" Exposure	3.325	Sq.
5-1/2" Exposure	3.034	Sq.
Panelized 8' Strips 7" Exposure	2.667	Sq.
Laps Rakes or Valleys	.040	L.F.
Tiles		
Aluminum		
Mission	3.200	Sq.
Spanish	2.667	Sq.
Clay 8-1/4" x 11"	4.848	Sq.
Spanish	4.444	Sq.
Mission	6.957	Sq.
French	5.926	Sq.
Norman	8.000	Sq.
Williamsburg	5.926	Sq.
Concrete 13" x 16-1/2"	5.926	Sq.
Steel	3.200	Sq.

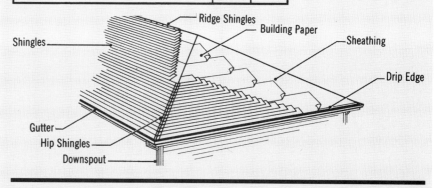

Figure 8.7 Installation Time in Man-Hours for Exterior Doors and Entry Systems

Description	Man-Hours	Unit
Glass Entrance Door, Including Frame and Hardware		
Balanced, Including Glass		
3' x 7'		
Economy	17.780	Ea.
Premium	22.860	Ea.
Hinged, Aluminum		
3' x 7'	8.000	Ea.
3' x 7', 3' Transom	8.890	Ea.
6' x 7'	12.310	Ea.
6' x 10', 3' Transom	14.550	Ea.
Stainless Steel, Including Glass		
3' x 7'		
Minimum	10.000	Ea.
Average	11.430	Ea.
Maximum	13.330	Ea.
Tempered Glass		
3' x 7'	8.000	
6' x 7'	11.430	Ea.
Hinged, Automatic, Aluminum		
6' x 7'	22.860	Ea.
Revolving, Aluminum, 7'-0" Diameter x 7' High		
Minimum	42.667	Ea.
Average	53.333	Ea.
Maximum	71.111	Ea.
Stainless Steel, 7'-0" Diameter x 7' High	106.667	Ea.
Bronze, 7'-0" Diameter x 7' High	213.333	Ea.
Glass Storefront System, Including Frame and Hardware		
Hinged, Aluminum, Including Glass, 400 S.F.		
w/3' x 7' Door		
Commercial Grade	.107	S.F.
Institutional Grade	.123	S.F.
Monumental Grade	.139	S.F.
w/6' x 7' Door		
Commercial Grade	.119	S.F.
Institutional Grade	.139	S.F.
Monumental Grade	.160	S.F.
Sliding, Automatic, 12' x 7'-6" w/5' x 7' door	22.860	Ea.
Mall Front, Manual, Aluminum		
15' x 9'	12.310	Ea.
24' x 9'	22.860	Ea.
48' x 9' w/Fixed Panels	17.780	Ea.

Figure 8.7 Installation Time in Man-Hours for
Exterior Doors and Entry Systems (continued)

Description	Man-Hours	Unit
Tempered All-Glass w/Glass Mullions,		
up to 10' High	.185	S.F.
up to 20' High, Minimum	.218	S.F.
Average	.240	S.F.
Maximum	.300	S.F.
Entrance Frames, Aluminum, 3' x 7'	2.290	Ea.
3' x 7', 3' Transom	2.460	Ea.
6' x 7'	2.670	Ea.
6' x 7', 3' Transom	2.910	Ea.
Glass, Tempered, 1/4" Thick	.133	S.F.
1/2" Thick	.291	S.F.
3/4" Thick	.457	S.F.
Insulating, 1" Thick	.213	S.F.
Overhead Commercial Doors		
Frames Not Included		
Stock Sectional Heavy Duty Wood		
1-3/4" Thick		
8' x 8'	8.000	Ea.
10' x 10'	8.889	Ea.
12' x 12'	10.667	Ea.
Fiberglass and Aluminum Heavy Duty Sectional		
12' x 12'	10.667	Ea.
20' x 20', Chain Hoist	32.000	Ea.
Steel 24' Gauge Sectional Manual		
8' x 8' High	8.000	Ea.
10' x 10' High	8.889	Ea.
12' x 12' High	10.667	Ea.
20' x 14' High, Chain Hoist	22.857	Ea.
For Electric Trolley Operator to 14' x 14'	4.000	Ea.
Over 14' x 14'	8.000	Ea.

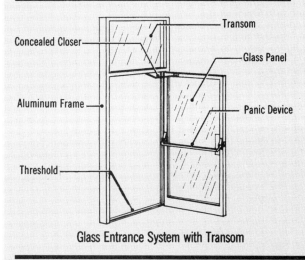

Glass Entrance System with Transom

Figure 8.8 Installation Time in Man-Hours
for Interior Doors and Frames

Description	Man-Hours	Unit
Architectural, Flush, Interior, Hollow Core, Veneer Face		
Up to 3'-0" x 7' x 0"	1.020	Ea.
4'-0" x 7'-0"	1.080	Ea.
High Pressured Plastic Laminate Face		
Up to 2'-6" x 6'-8"	1.000	Ea.
3'-0" x 7'-0"	1.153	Ea.
4'-0" x 7'-0"	1.234	Ea.
Particle Core, Veneer Face		
2'-6" x 6'-8"	1.067	Ea.
3'-0" x 6'-8"	1.143	Ea.
3'-0" x 7'-0"	1.231	Ea.
4'-0" x 7'-0"	1.333	Ea.
M.D.O. on Hardboard Face		
3'-0" x 7'-0"	1.333	Ea.
4'-0" x 7'-0"	1.600	Ea.
High Pressure Plastic Laminate Face		
3'-0" x 7'-0"	1.455	Ea.
4'-0" x 7'-0"	2.000	Ea.
Flush, Exterior, Solid Core, Veneer Face		
2'-6" x 7'-0"	1.067	Ea.
3'-0" x 7'-0"	1.143	Ea.
Decorator, Hand Carved		
Solid Wood		
Up to 3'-0" x 7'-0"	1.143	Ea.
3'-6" x 7'-0"	1.231	Ea.
Fire Door, Flush, Mineral Core		
B Label, 1 Hour, Veneer Face		
2'-6" x 6'-8"	1.143	Ea.
3'-0" x 7'-0"	1.333	Ea.
4'-0" x 7'-0"	1.333	Ea.
High Pressure Plastic Laminate Face		
3'-0" x 7'-0"	1.455	Ea.
4'-0" x 7'-0"	1.600	Ea.
Residential, Interior		
Hollow Core or Panel		
Up to 2'-8" x 6'-8"	.889	Ea.
3'-0" x 6'-8"	.941	Ea.
Bi-Folding Closet		
3'-0" x 6'-8"	1.231	Ea.
5'-0" x 6'-8"	1.455	Ea.
Interior Prehung, Hollow Core or Panel		
Up to 2'-8" x 6'-8"	.800	Ea.
3'-0" x 6'-8"	.842	Ea.
Exterior, Entrance, Solid Core or Panel		
Up to 2'-8" x 6'-8"	1.000	Ea.
3'-0" x 6'-8"	1.067	Ea.
Exterior Prehung, Entrance		
Up to 3'-0" x 7'-0"	1.000	Ea.

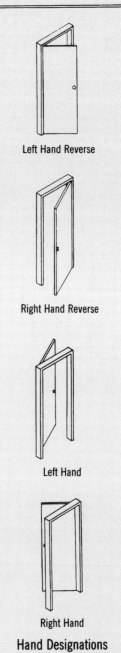

Left Hand Reverse

Right Hand Reverse

Left Hand

Right Hand

Hand Designations

Figure 8.8 Installation Time in Man-Hours
for Interior Doors and Frames (continued)

Description	Man-Hours	Unit
Hollow Metal Doors Flush		
Full Panel, Commercial		
20 Gauge		
2'-0" x 6'-8"	.800	Ea.
2'-6" x 6'-8"	.888	Ea.
3'-0" x 6'-8" or 3'-0" x 7'-0"	.941	Ea.
4'-0" x 7'-0"	1.066	Ea.
18 Gauge		
2'-6" x 6'-8" or 2'-6" x 7'-0"	.941	Ea.
3'-0" x 6'-8" or 3'-0" x 7'-0"	1.000	Ea.
4'-0" x 7'-0"	1.066	Ea.
Residential		
24 Gauge		
2'-8" x 6'-8"	1.000	Ea.
3'-0" x 7'-0"	1.066	Ea.
Bifolding		
3'-0" x 6'- 8"	1.000	Ea.
5'-0" x 6'-8"	1.143	Ea.
Steel Frames		
18 Gauge		
3'-0" Wide	1.000	Ea.
6'-0" Wide	1.142	Ea.
16 Gauge		
4'-0" Wide	1.066	Ea.
8'-0" Wide	1.333	Ea.
Transom Lite Frames		
Fixed Add	.103	S.F.
Movable Add	.123	S.F.

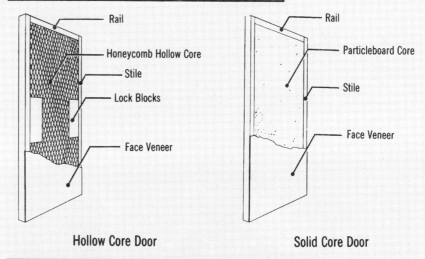

Hollow Core Door   Solid Core Door

Figure 9.1   Partition Density Guide

This table allows you to estimate the average quantities of partitions found in various types of buildings when no interior plans are available, such as in a conceptual estimating situation. To use this chart, pick out the type of structure to be estimated from the left-most column. Moving across, choose from the second column the number of stories of the building. Read across to the "Partition Density" column. This figure represents the number of square feet of floor area for every linear foot of partition.

Example: The project to be estimated is a three-story office building of 10,000 square feet per floor (total of 30,000 square feet). To estimate the approximate quantity of interior partitions in the entire building, first look in column 1 for "Office." Reading across to the "Stories" column, find the "3-5 Story" line. The "Partition Density" column for that line indicates 20 S.F./L.F. To find the number of partitions, divide the building area by the density factor: 30,000 S.F./20 = 1,500 L.F. of partition. Note that the right-most column gives a breakdown of average partition types. For our office building, the average mix is 30% concrete block (generally found at stairwells and service area partitions) and 70% drywall. Our drywall total would then be 1,050 L.F. To determine the total square footage of partitions, multiply this linear footage by the specified (or assumed) partition height. Note that the remainder of the partitions (450 L.F.) must be added to the masonry portion of the estimate, or at least accounted for in the estimate.

Building Type	Stories	Partition Density	Description of Partition
Apartments	1 story	9 S.F./L.F.	Plaster, wood doors & trim
	2 story	8 S.F./L.F.	Drywall, wood studs, wood doors & trim
	3 story	9 S.F./L.F.	Plaster, wood studs, and wood doors & trim
	5 story	9 S.F./L.F.	Plaster, wood studs, wood doors & trim
	6–15 story	8 S.F./L.F.	Drywall, wood studs, wood doors & trim
Bakery	1 story	50 S.F./L.F.	Conc. block, paint, door & drywall, wood studs
	2 story	50 S.F./L.F.	Conc. block, paint, door & drywall, wood studs
Bank	1 story	20 S.F./L.F.	Plaster, wood studs, wood doors & trim
	2-4 story	15 S.F./L.F.	Plaster, wood studs, wood doors & trim
Bottling Plant	1 story	50 S.F./L.F.	Conc. block, drywall, wood studs, wood trim
Bowling Alley	1 story	50 S.F./L.F.	Conc. block, wood & metal doors, wood trim
Bus Terminal	1 story	15 S.F./L.F.	Conc. block, ceramic tile, wood trim
Cannery	1 story	100 S.F./L.F.	Drywall on metal studs
Car Wash	1 story	18 S.F./L.F.	Concrete block, painted & hollow metal door
Dairy Plant	1 story	30 S.F./L.F.	Concrete block, glazed tile, insulated cooler doors
Department Store	1 story	60 S.F./L.F.	Drywall, wood studs, wood doors & trim
	2-5 story	60 S.F./L.F.	30% concrete block, 70% drywall, wood studs
Dormitory	2 story	9 S.F./L.F.	Plaster, concrete block, wood doors & trim
	3-5 story	9 S.F./L.F.	Plaster, concrete block, wood doors & trim
	6–15 story	9 S.F./L.F.	Plaster, concrete block, wood doors & trim
Funeral Home	1 story	15 S.F./L.F.	Plaster on concrete block & wood studs, paneling
	2 story	14 S.F./L.F.	Plaster, wood studs, paneling & wood doors
Garage Sales & Service	1 story	30 S.F./L.F.	50% conc. block, 50% drywall, wood studs
Hotel	3-8 story	9 S.F./L.F.	Plaster, conc. block, wood doors & trim
	9-15 story	9 S.F./L.F.	Plaster, conc. block, wood doors & trim
Laundromat	1 story	25 S.F./L.F.	Drywall, wood studs, wood doors & trim
Medical Clinic	1 story	6 S.F./L.F.	Drywall, wood studs, wood doors & trim
	2-4 story	6 S.F./L.F.	Drywall, wood studs, wood doors & trim
Motel	1 story	7 S.F./L.F.	Drywall, wood studs, wood doors & trim
	2–3 story	7 S.F./L.F.	Concrete block, drywall on wood studs, wood paneling

(continued on next page)

Figure 9.1 Partition Density Guide (continued)

Building Type	Stories	Partition Density	Description of Partition
Movie 200–600 seats Theater 601–1400 seats 1401–2200 seats	1 story	18 S.F./L.F. 20 S.F./L.F. 25 S.F./L.F.	Concrete block, wood, metal, vinyl trim Concrete block, wood, metal, vinyl trim Concrete block, wood, metal, vinyl trim
Nursing Home	1 story 2–4 story	8 S.F./L.F. 8 S.F./L.F.	Drywall, wood studs, wood doors & trim Drywall, wood studs, wood doors & trim
Office	1 story 2 story 3–5 story 6–10 story 11–20 story	20 S.F./L.F. 20 S.F./L.F. 20 S.F./L.F. 20 S.F./L.F. 20 S.F./L.F.	30% concrete block, 70% drywall on wood studs 30% concrete block, 70% drywall on wood studs 30% concrete block, 70% movable partitions 30% concrete block, 70% movable partitions 30% concrete block, 70% movable partitions
Parking Ramp (Open) Parking Garage	2–8 story 2–8 story	60 S.F./L.F. 60 S.F./L.F.	Stair and elevator enclosures only Stair and elevator enclosures only

Pre-Engineered   Store
                    Office
                    Shop
	1 story
	1 story
	1 story
		60 S.F./L.F.
		15 S.F./L.F.
		15 S.F./L.F.
			Drywall on wood studs, wood doors & trim
			Concrete block, movable wood partitions
			Movable wood partitions

Radio & TV Broadcasting & TV Transmitter	1 story 1 story	25 S.F./L.F. 40 S.F./L.F.	Concrete block, metal and wood doors Concrete block, metal and wood doors
Self Service Restaurant Cafe & Drive-in Restaurant Restaurant with seating Supper Club Bar or Lounge	1 story 1 story 1 story 1 story 1 story	15 S.F./L.F. 18 S.F./L.F. 25 S.F./L.F. 25 S.F./L.F. 24 S.F./L.F.	Concrete block, wood and aluminum trim Drywall, wood studs, ceramic & plastic trim Concrete block, paneling, wood studs & trim Concrete block, paneling, wood studs & trim Plaster or gypsum lath, wood studs
Retail Store or Shop	1 story	60 S.F./L.F.	Drywall wood studs, wood doors & trim

Service Station   Masonry
                  Metal panel
                    Frame
	1 story
	1 story
	1 story
		15 S.F./L.F.
		15 S.F./L.F.
		15 S.F./L.F.
			Concrete block, paint, door & drywall, wood studs
			Concrete block paint door & drywall, wood studs
			Drywall, wood studs, wood doors & trim

Shopping Center (strip) (group)	1 story 1 story 2 story	30 S.F./L.F. 40 S.F./L.F. 40 S.F./L.F.	Drywall, wood studs, wood doors & trim 50% concrete block, 50% drywall, wood studs 50% concrete block, 50% drywall, wood studs
Small Food Store	1 story	30 S.F./L.F.	Concrete block drywall, wood studs, wood trim
Store/Apt. Masonry above Frame Frame	2 story 2 story 3 story	10 S.F./L.F. 10 S.F./L.F. 10 S.F./L.F.	Plaster, wood studs, wood doors & trim Plaster wood studs, wood doors & trim Plaster, wood studs and wood doors & trim
Supermarkets	1 story	40 S.F./L.F.	Concrete block, paint, drywall & porcelain panel
Truck Terminal	1 story	0	
Warehouse	1 story	0	

Figure 9.12 Installation Time in Man-Hours for Wood Stud Partition Systems

Description	Man-Hours	Unit
Wood Partitions Studs with Single Bottom Plate and Double Top Plate		
2" x 3" or 2" x 4" Studs		
12" On Center	.020	S.F.
16" On Center	.016	S.F.
24" On Center	.013	S.F.
2" x 6" Studs		
12" On Center	.023	S.F.
16" On Center	.018	S.F.
24" On Center	.014	S.F.
Plates		
2" x 3"	.019	L.F.
2" x 4"	.020	L.F.
2" x 6"	.021	L.F.
Studs		
2" x 3"	.013	L.F.
2" x 4"	.012	L.F.
2" x 6"	.016	L.F.
Blocking	.032	L.F.
Grounds 1" x 2"		
For Casework	.024	L.F.
For Plaster	.018	L.F.
Insulation Fiberglass Batts	.005	S.F.
Metal Lath Diamond Expanded		
2.5 lb. per S.Y.	.094	S.Y.
3.4 lb. per S.Y.	.100	S.Y.
Gypsum Lath		
3/8" Thick	.094	S.Y.
1/2" Thick	.100	S.Y.
Gypsum Plaster		
2 Coats	.381	S.Y.
3 Coats	.460	S.Y.
Perlite or Vermiculite Plaster		
2 Coats	.435	S.Y.
3 Coats	.541	S.Y.
Wood Fiber Plaster		
2 Coats	.556	S.Y.
3 Coats	.702	S.Y.

Figure 9.12 Installation Time in Man-Hours
for Wood Stud Partition Systems (continued)

Description	Man-Hours	Unit
Drywall Gypsum Plasterboard Including Taping		
3/8" Thick	.015	S.F.
1/2" or 5/8" Thick	.017	S.F.
For Thin Coat Plaster Instead of Taping Add	.013	S.F.
Prefinished Vinyl Faced Drywall	.015	S.F.
Sound-deadening Board	.009	S.F.
Walls in Place		
2" x 4" Studs with 5/8"		
Gypsum Drywall Both Sides Taped	.053	S.F.
2" x 4" Studs with 2 Layers Gypsum Drywall		
Both Sides Taped	.078	S.F.

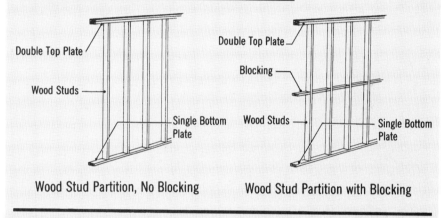

Wood Stud Partition, No Blocking   Wood Stud Partition with Blocking

Figure 10.16   Installation Time for Metal Lockers

Metal Lockers	Crew Makeup	Daily Output	Man-Hours	Unit
LOCKERS Steel, baked enamel, 60" or 72", single tier				
Minimum	1 Sheet Metal Worker	14	.571	Opng.
Maximum		12	.667	Opng.
2 tier, 60" or 72" total height, minimum		26	.308	Opng.
Maximum		20	.400	Opng.
5 tier box lockers, minimum		30	.267	Opng.
Maximum		24	.333	Opng.
6 tier box lockers, minimum		36	.222	Opng.
Maximum		30	.267	Opng.
Basket rack with 32 baskets, 9" x 13" x 8" basket		50	.160	Basket
24 baskets, 12" x 13" x 8" basket		50	.160	Basket
Athletic, wire mesh, no lock, 18" x 18" x 72" basket		12	.667	Ea.
Overhead locker baskets on chains, 14" x 14" baskets	3 Sheet Metal Workers	96	.250	Basket
Overhead locker framing system, add		600	.040	Basket
Locking rail and bench units, add		120	.200	Basket
Locker bench, laminated maple, top only	1 Sheet Metal Worker	100	.080	L.F.
Pedestals, steel pipe	"	25	.320	Ea.
Teacher and pupil wardrobes, enameled				
22" x 15" x 61" high, minimum	1 Sheet Metal Worker	10	.800	Ea.
Average		9	.889	Ea.
Maximum		8	1.000	Ea.
Duplex lockers with 2 doors, 72" high, 15" x 15"		10	.800	Ea.
15" x 21"		10	.800	Ea.

Figure 10.25 Installation Time for Panel Partitions

Panel Partitions	Crew Makeup	Daily Output	Man-Hours	Unit
PARTITIONS, FOLDING LEAF Acoustic, wood				
Vinyl faced, to 18' high, 6 psf, minimum	2 Carpenters	60	.267	S.F.
Average		45	.356	S.F.
Maximum		30	.533	S.F.
Formica or hardwood finish, minimum		60	.267	S.F.
Maximum		30	.533	S.F.
Wood, low acoustical type, 4.5 psf, to 14' high		50	.320	S.F.
Steel, acoustical, 9 to 12 lb. per S.F., vinyl faced				
Minimum		60	.267	S.F.
Maximum		30	.533	S.F.
Aluminum framed, acoutical, to 12' high, 5.5 psf				
Minimum		60	.267	S.F.
Maximum		30	.533	S.F.
6.5 lb. per S.F., minimum		60	.267	S.F.
Maximum		30	.533	S.F.
PARTITIONS, OPERABLE Acoustic air wall, 1-5/8" thick				
Minimum		375	.043	S.F.
Maximum		365	.044	S.F.
2-1/4" thick, minimum		360	.044	S.F.
Maximum		330	.048	S.F.
Overhead track type, acoustical, 3" thick, 11 psf				
Minimum		350	.046	S.F.
Maximum		300	.053	S.F.

Figure 11.5 Installation Time for Theater/Stage Equipment

Theater/ Stage Equipment	Crew Makeup	Daily Output	Man-Hours	Unit
STAGE EQUIPMENT Control boards				
with dimmers & breakers, minimum	1 Electrician	1	8.000	Ea.
Average		.50	16.000	Ea.
Maximum	↓	.20	40.000	Ea.
Curtain track, straight, light duty	2 Carpenters	20	.800	L.F.
Heavy duty		18	.889	L.F.
Curved sections		12	1.330	L.F.
Curtains, velour, medium weight		600	.027	S.F.
Asbestos		50	.320	S.F.
Silica based yarn, fireproof	↓	50	.320	S.F.
Lights, border, quartz, reflector, vented,				
colored or white	1 Electrician	20	.400	L.F.
Spotlight, follow spot, with transformer, 2100 watt		4	2.000	Ea.
Stationary spot, fresnel quartz, 6″ lens		4	2.000	Ea.
8″ lens		4	2.000	Ea.
Ellipsoidal quartz, 1000W, 6″ lens		4	2.000	Ea.
12″ lens		4	2.000	Ea.
Strobe light, 1 to 15 flashes per second, quartz		3	2.670	Ea.
Color wheel, portable, five hole, motorized	↓	4	2.000	Ea.
Telescoping platforms, extruded alum., straight				
Minimum	4 Carpenters	157	.204	S.F. Stg.
Maximum		77	.416	S.F. Stg.
Pie-shaped, minimum		150	.213	S.F. Stg.
Maximum		70	.457	S.F. Stg.
Band risers, steel frame, plywood deck, minimum		275	.116	S.F. Stg.
Maximum	↓	138	.232	S.F. Stg.
Chairs for above, self-storing, minimum	2 Carpenters	43	.372	Ea.
Maximum	″	40	.400	Ea.
Rule of thumb: total stage equipment, minimum	4 Carpenters	100	.320	S.F. Stg.
Maximum	″	25	1.280	S.F. Stg.
MOVIE EQUIPMENT				
Lamphouses, incl. rectifiers, xenon, 1000 watt	1 Electrician	2	4.000	Ea.
1600 watt		2	4.000	Ea.
2000 watt		1.50	5.330	Ea.
4000 watt	↓	1.50	5.330	Ea.
Projection screens, rigid, in wall, acrylic, 1/4″ thick	2 Glaziers	195	.082	S.F.
1/2″ thick	″	130	.123	S.F.
Electric operated, heavy duty, 400 S.F.	2 Carpenters	1	16.000	Ea.
Sound systems, incl. amplifier, single system				
Minimum	1 Electrician	.90	8.890	Ea.
Dolby/Super Sound, maximum		.40	20.000	Ea.
Dual system, minimum		.70	11.430	Ea.
Dolby/Super Sound, maximum		.40	20.000	Ea.
Speakers, recessed behind screen, minimum		2	4.000	Ea.
Maximum	↓	1	8.000	Ea.
Seating, painted steel, upholstered, minimum	2 Carpenters	35	.457	Ea.
Maximum	″	28	.571	Ea.

Figure 11.13 Installation Time for
Loading Dock Equipment

Loading Dock Equipment	Crew Makeup	Daily Output	Man-Hours	Unit
LOADING DOCK Bumpers, rubber blocks 4-1/2" thick				
10" high, 14" long	1 Carpenter	26	.308	Ea.
24" long		22	.364	Ea.
36" long		17	.471	Ea.
12" high, 14" long		25	.320	Ea.
24" long		20	.400	Ea.
36" long		15	.533	Ea.
Rubber blocks 6" thick, 10" high, 14" long		22	.364	Ea.
24" long		18	.444	Ea.
36" long		13	.615	Ea.
20" high, 11" long		13	.615	Ea.
Extruded rubber bumpers, T section				
22" x 22" x 3" thick		41	.195	Ea.
Molded rubber bumpers, 24" x 12" x 3" thick		20	.400	Ea.
Welded installation of above bumpers	1 Welder Foreman 1 Gas Welding Machine	8	1.000	Ea.
For drilled anchors, add per anchor	1 Carpenter	36	.222	Ea.
Door seal for door perimeter				
12" x 12", vinyl covered	"	26	.308	L.F.
Levelers, hinged for trucks, 10 ton capacity, 6' x 8'	2 Skilled Workers 1 Helper	1.90	12.630	Ea.
7' x 8'		1.90	12.630	Ea.
Hydraulic, 10 ton capacity, 6' x 8'		1.90	12.630	Ea.
7' x 8'		1.90	12.630	Ea.
Lights for loading docks, single arm, 24" long	1 Electrician	3.80	2.110	Ea.
Double arm, 60" long	"	3.80	2.110	Ea.
Shelters, fabric, for truck or train, scissor arms				
Minimum	1 Carpenter	1	8.000	Ea.
Maximum	"	.50	16.000	Ea.
DOCK BUMPERS Bolts not incl. 2" x 6" to 4" x 8"				
Average	1 Carpenter Power Tools	.30	26.670	M.B.F.

Figure 12.3 Installation Time for Hospital Casework

Hospital Casework	Crew Makeup	Daily Output	Man-Hours	Unit
CABINETS				
Hospital, base cabinets, laminated plastic	2 Carpenters	10	1.600	L.F.
Enameled steel		10	1.600	L.F.
Stainless steel		10	1.600	L.F.
Cabinet base trim, 4" high, enameled steel		200	.080	L.F.
Stainless steel		200	.080	L.F.
Counter top, laminated plastic, no backsplash		40	.400	L.F.
With backsplash		40	.400	L.F.
For sink cutout, add		12.20	1.310	Ea.
Stainless steel counter top		40	.400	L.F.
Nurses station, door type, laminated plastic		10	1.600	L.F.
Enameled steel		10	1.600	L.F.
Stainless steel		10	1.600	L.F.
Wall cabinets, laminated plastic		15	1.070	L.F.
Enameled steel		15	1.070	L.F.
Stainless steel		15	1.070	L.F.
Kitchen, base cabinets, metal, minimum		30	.533	L.F.
Maximum		25	.640	L.F.
Wall cabinets, metal, minimum		30	.533	L.F.
Maximum		25	.640	L.F.
School, 24" deep		15	1.070	L.F.
Counter height units		20	.800	L.F.
Wood, custom fabricated, 32" high counter		20	.800	L.F.
Add for counter top		56	.286	L.F.
84" high wall units		15	1.070	L.F.

Figure 12.4 Installation Time for Display Casework

Display Casework	Crew Makeup	Daily Output	Man-Hours	Unit
DISPLAY CASES Free standing, all glass				
Aluminum frame, 42″ high x 36″ x 12″ deep	2 Carpenters	8	2.000	Ea.
70″ high x 48″ x 18″ deep	"	6	2.670	Ea.
Wall mounted, glass front, aluminum frame				
Non-illuminated, one section 3′ x 4′ x 1′-4″	2 Carpenters	5	3.200	Ea.
5′ x 4′ x 1′-4″		5	3.200	Ea.
6′ x 4′ x 1′-4″		4	4.000	Ea.
Two sections, 8′ x 4′ x 1′-4″		2	8.000	Ea.
10′ x 4′ x 1′-4″		2	8.000	Ea.
Three sections, 16′ x 4′ x 1′-4″		1.50	10.670	Ea.
Table exhibit cases, 2′ wide, 3′ high, 4′ long, flat top		5	3.200	Ea.
3′ wide, 3′ high, 4′ long, sloping top		3	5.330	Ea.

Figure 13.12 Installation Time for
Pre-Engineered Buildings

Pre-Engineered Buildings	Crew Makeup	Daily Output	Man-Hours	Unit
HANGARS Prefabricated steel T hangars Galv. steel roof & walls, incl. electric bi-folding doors, 4 or more units, not including floors or foundations, minimum	1 Struc. Steel Foreman 4 Struc. Steel Workers 1 Equip. Oper. (crane) 1 Equip. Oper. Oiler 1 Crane, 90 Ton	1,275	.044	S.F. Flr.
Maximum		1,063	.053	S.F. Flr.
With bottom rolling doors, minimum		1,386	.040	S.F. Flr.
Maximum	↓	966	.058	S.F. Flr.
Alternate pricing method: Galv. roof and walls, electric bi-folding doors, minimum	1 Struc. Steel Foreman 4 Struc. Steel Workers 1 Equip. Oper. (crane) 1 Equip. Oper. Oiler 1 Crane, 90 Ton	1.06	52.830	Plane
Maximum		.91	61.540	Plane
With bottom rolling doors, minimum		1.25	44.800	Plane
Maximum	↓	.97	57.730	Plane
Circular type, prefab., steel frame, plastic skin, electric door, including foundations, 80′ diameter, for up to 5 light planes Minimum	1 Struc. Steel Foreman 4 Struc. Steel Workers 1 Equip. Oper. (crane) 1 Equip. Oper. Oiler 1 Crane, 90 Ton	.50	112.000	Total
Maximum	"	.25	224.000	Total

Figure 13.18 Installation Time for Ground Storage Tanks

Ground Storage Tanks	Crew Makeup	Daily Output	Man-Hours	Unit
TANKS Not incl. pipe or pumps				
Wood tanks, ground level, 2" cypress, 3000 gallons	3 Carpenters 1 Building Laborer Power Tools	.19	168.000	Ea.
2-1/2" cypress, 10,000 gallons		.12	267.000	Ea.
3" redwood or 3" fir, 20,000 gallons		.10	320.000	Ea.
30,000 gallons		.08	400.000	Ea.
45,000 gallons	↓	.07	457.000	Ea.
Vinyl coated fabric pillow tanks, freestanding				
5000 gallons	4 Building Laborers	4	8.000	Ea.
Supporting embankment not included				
25,000 gallons	6 Building Laborers	2	24.000	Ea.
50,000 gallons	8 Building Laborers	1.50	42.670	Ea.
100,000 gallons	9 Building Laborers	.90	80.000	Ea.
150,000 gallons		.50	144.000	Ea.
200,000 gallons		.40	180.000	Ea.
250,000 gallons	↓	.30	240.000	Ea.

Figure 14.6 Installation Time for Elevators

Elevators	Crew Makeup	Daily Output	Man-Hours	Units
ELEVATORS				
2 story, hydraulic, 4000 lb. capacity, minimum	3 Elevator Constructors 1 Elevator Apprentice Hand Tools	.09	356.000	Ea.
Maximum		.09	356.000	Ea.
10,000 lb. capacity, minimum		.07	457.000	Ea.
Maximum		.07	457.000	Ea.
6 story, hydraulic, 4000 lb. capacity		.04	800.000	Ea.
10,000 lb. capacity		.04	800.000	Ea.
6 story geared electric 4000 lb. capacity		.04	800.000	Ea.
10,000 lb. capacity		.04	800.000	Ea.
12 story gearless electric 4000 lb. capacity		.03	1066.000	Ea.
10,000 lb. capacity		.03	1066.000	Ea.
20 story gearless electric 4000 lb. capacity		.02	1600.000	Ea.
10,000 lb. capacity		.02	1600.000	Ea.
Passenger, 2 story hydraulic, 2000 lb. capacity		.07	457.000	Ea.
5000 lb. capacity		.07	457.000	Ea.
6 story hydraulic, 2000 lb. capacity		.04	800.000	Ea.
5000 lb. capacity		.04	800.000	Ea.
6 story geared electric, 2000 lb. capacity		.04	800.000	Ea.
5000 lb. capacity		.04	800.000	Ea.
12 story gearless electric, 2000 lb. capacity		.03	1066.000	Ea.
5000 lb. capacity		.03	1066.000	Ea.
20 story gearless electric, 2000 lb. capacity		.02	1600.000	Ea.
5000 lb. capacity		.02	1600.000	Ea.
Passenger, pre-engineered, 5 story, hydraulic				
2500 lb. capacity		.04	800.000	Ea.
For less than 5 stops deduct		.29	110.000	Stop
10 story, geared traction, 200 FPM				
2500 lb. capacity		.02	1600.000	Ea.
For less than 10 stops, deduct		.34	94.120	Stop
For 4500 lb. capacity, general purpose		.02	1600.000	Ea.
For hospital		.02	1600.000	Ea.
Residential, cab type, 1 floor, 2 stop, minimum	2 Elevator Constructors	.20	80.000	Ea.
Maximum		.10	160.000	Ea.
2 floor, 3 stop, minimum		.12	133.000	Ea.
Maximum		.06	267.000	Ea.
Stair climber (chair lift) single seat, minimum		1	16.000	Ea.
Maximum		.20	80.000	Ea.
Wheelchair, porch lift, minimum		1	16.000	Ea.
Maximum		.50	32.000	Ea.
Stair lift, minimum		1	16.000	Ea.
Maximum		.20	80.000	Ea.

Figure 14.8 Installation Time for Electric Dumbwaiters

Electric Dumbwaiters	Crew Makeup	Daily Output	Man-Hours	Unit
DUMBWAITERS 2 stop, electric, minimum	2 Elevator Constructors	.13	123.000	Ea.
Maximum		.11	145.000	Ea.
For each additional stop, add	↓	.54	29.630	Stop

Figure 15.11 Pipe Sizing for Heating

Heating Load, BTU/HR.	GPM Circulated (20°T.D.)	Recommended Connecting Tubing Size (Type M) for Various Heating Loads and Connecting Tubing Lengths. (Figures based on 10,000 BTU per GPM, or on temperature drop of 20° thru the circuit.)				
		Total Length Ft. Connecting Tubing				
		0–50	50–100	100–150	150–200	200–300
		Tubing, Nominal O.D., Type M				
5,000	0.5	3/8	3/8	1/2	1/2	1/2
10,000	1.0	3/8	3/8	1/2	1/2	1/2
15,000	1.5	1/2	1/2	1/2	1/2	3/4
20,000	2.0	1/2	1/2	1/2	3/4	3/4
30,000	3.0	1/2	1/2	3/4	3/4	3/4
40,000	4.0	1/2	3/4	3/4	3/4	3/4
50,000	5.0	3/4	3/4	3/4	1	1
60,000	6.0	3/4	3/4	1	1	1
75,000	7.5	3/4	1	1	1	1
100,000	10.0	1	1	1	1-1/4	1-1/4
125,000	12.5	1	1	1-1/4	1-1/4	1-1/4
150,000	15.0	1	1-1/4	1-1/4	1-1/4	1-1/2
200,000	20.0	1-1/4	1-1/4	1-1/2	1-1/2	1-1/2
250,000	25.0	1-1/4	1-1/2	1-1/2	2	2
300,000	30.0	1-1/2	1-1/2	2	2	2
400,000	40.0	2	2	2	2	2
500,000	50.0	2	2	2	2-1/2	2-1/2
600,000	60.0	2	2	2-1/2	2-1/2	2-1/2
800,000	80.0	2-1/2	2-1/2	2-1/2	2-1/2	3
1,000,000	100.0	2-1/2	2-1/2	3	3	3
1,250,000	125.0	2-1/2	3	3	3	3-1/2
1,500,000	150.0	3	3	3	3-1/2	3-1/2
2,000,000	200.0	3	3-1/2	3-1/2	4	4
2,500,000	250.0	3-1/2	3-1/2	4	4	5
3,000,000	300.0	3-1/2	4	4	4	5
4,000,000	400.0	4	4	5	5	5
5,000,000	500.0	5	5	5	6	6
6,000,000	600.0	5	6	6	6	8
8,000,000	800.0	8	6	8	8	8
10,000,000	1,000.0	8	8	8	8	10

Figure 15.37 Installation Time in Man-Hours for Plumbing Fixtures

Description	Man-Hours	Unit
For Setting Fixture and Trim		
Bath Tub	3.636	Ea.
Bidet	3.200	Ea.
Dental Fountain	2.000	Ea.
Drinking Fountain	2.500	Ea.
Lavatory		
Vanity Top	2.500	Ea.
Wall Hung	2.000	Ea.
Laundry Sinks	2.667	Ea.
Prison/Institution Fixtures		
Lavatory	2.000	Ea.
Service Sink	5.333	Ea.
Urinal	4.000	Ea.
Water Closet	2.759	Ea.
Combination Water Closet and Lavatory	3.200	Ea.
Shower Stall	8.000	Ea.
Sinks		
Corrosion Resistant	5.333	Ea.
Kitchen, Countertop	3.330	Ea.
Kitchen, Raised Deck	7.270	Ea.
Service, Floor	3.640	Ea.
Service, Wall	4.000	Ea.
Urinals		
Wall Hung	5.333	Ea.
Stall Type	6.400	Ea.
Wash Fountain, Group	9.600	Ea.
Water Closets		
Tank Type, Wall Hung	3.019	Ea.
Floor Mount, One Piece	3.019	Ea.
Bowl Only, Wall Hung	2.759	Ea.
Bowl Only, Floor Mount	2.759	Ea.
Gang, Side by Side, First	2.759	Ea.
Each Additional	2.759	Ea.
Gang, Back to Back, First Pair	5.520	Pair
Each Additional Pair	5.520	Pair
Water Conserving Type	2.963	Ea.
Water Cooler	4.000	Ea.

Figure 16.14   Installation Time in Man-Hours for Conduit

Description	Man-Hours	Unit
Rigid Galvanized Steel 1/2" Diameter	.089	L.F.
1-1/2" Diameter	.145	L.F.
3" Diameter	.320	L.F.
6" Diameter	.800	L.F.
Aluminum 1/2" Diameter	.080	L.F.
1-1/2" Diameter	.123	L.F.
3" Diameter	.178	L.F.
6" Diameter	.400	L.F.
IMC 1/2" Diameter	.080	L.F.
1-1/2" Diameter	.133	L.F.
3" Diameter	.267	L.F.
4" Diameter	.320	L.F.
Plastic Coated Rigid Steel 1/2" Diameter	.100	L.F.
1-1/2" Diameter	.178	L.F.
3" Diameter	.364	L.F.
6" Diameter	.800	L.F.
EMT 1/2" Diameter	.047	L.F.
1-1/2" Diameter	.089	L.F.
3" Diameter	.160	L.F.
4" Diameter	.200	L.F.
PVC Nonmetallic 1/2" Diameter	.042	L.F.
1-1/2" Diameter	.080	L.F.
3" Diameter	.145	L.F.
6" Diameter	.267	L.F.

Rigid Steel, Plastic Coated Coupling

PVC Conduit

PVC Elbow

Aluminum Conduit

EMT Set Screw Connector

Aluminum Elbow

EMT Connector

Rigid Steel, Plastic Coated Conduit

EMT to Conduit Adapter

Rigid Steel, Plastic Coated Elbow

EMT to Greenfield Adapter

Figure 16.30 Installation Time in Man-Hours for Wiring Devices

Description	Man-Hours	Unit
Receptacle 20A 250V	.290	Ea.
Receptacle 30A 250V	.530	Ea.
Receptacle 50A 250V	.720	Ea.
Receptacle 60A 250V	1.000	Ea.
Box, 4" Square	.400	Ea.
Box, Single Gang	.290	Ea.
Box, Cast Single Gang	.660	Ea.
Cover, Weatherproof	.120	Ea.
Cover, Raised Device	.150	Ea.
Cover, Brushed Brass	.100	Ea.

30 Amp, 125 Volt, Nema 5

50 Amp, 125 Volt, Nema 5

20 Amp, 250 Volt, Nema 6

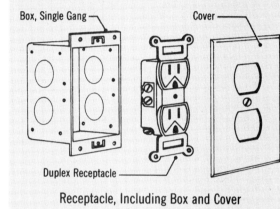

Box, Single Gang

Cover

Duplex Receptacle

Receptacle, Including Box and Cover

Receptacles

Index

Index

Notes

Notes

Notes